Protein–Ligand Interactions:
hydrodynamics and calorimetry

The Practical Approach Series

Related **Practical Approach** Series Titles

Protein Purification Techniques I*
Protein Purification Applications II*
DNA Protein Interactions
Spectrophotometry & Spectrofluorimetry
Protein Localization by Fluorescence Microscopy
Crystallization of Nucleic Acids and Proteins 2/e
High Resolution Chromatography
Gel Electrophoresis of Proteins 3/e
RNA–Protein Interactions
Enzyme Assays

* indicates a forthcoming title

Please see the **Practical Approach** series website at
http://www.oup.co.uk/pas
for full contents lists of all Practical Approach titles.

Protein–Ligand Interactions: hydrodynamics and calorimetry

A Practical Approach

Edited by

Stephen E. Harding
NCMH Physical Biochemistry Laboratory,
University of Nottingham, School of Biosciences,
Sutton Bonington, Leicestershire

and

Babur Z. Chowdhry
School of Chemical and Life Sciences,
University of Greenwich, Wellington Street,
Woolwich, London

UNIVERSITY PRESS

Great Clarendon Street, Oxford OX2 6DP

Oxford University Press is a department of the University of Oxford.
It furthers the University's objective of excellence in research,
scholarship, and education by publishing worldwide in

Oxford New York

Athens Auckland Bangkok Bogotá Buenos Aires Calcutta Cape Town
Chennai Dar es Salaam Delhi Florence Hong Kong Istanbul Karachi
Kuala Lumpur Madrid Melbourne Mexico City Mumbai Nairobi Paris
São Paulo Singapore Taipei Tokyo Toronto Warsaw

with associated companies in Berlin Ibadan

Oxford is a registered trade mark of Oxford University Press in the UK
and in certain other countries

Published in the United States by Oxford University Press Inc., New York

© Oxford University Press, 2001

The moral rights of the author have been asserted

Database right Oxford University Press (maker)

First published 2001

All rights reserved. No part of this publication may be reproduced,
stored in a retrieval system, or transmitted, in any form or by any
means, without the prior permission in writing of Oxford University
Press, or as expressly permitted by law, or under terms agreed with the
appropriate reprographics rights organization. Enquiries concerning
reproduction outside the scope of the above should be sent to the Rights
Department, Oxford University Press, at the address above

You must not circulate this book in any other binding or cover and you
must impose this same condition on any acquirer

British Library Cataloguing in Publication Data
Data available

Library of Congress Cataloguing in Publication Data

1 3 5 7 9 10 8 6 4 2

ISBN 0-19-963749-0 (Hbk.)
ISBN 0-19-963746-6 (Pbk.)

Typeset in Swift by Footnote Graphics, Warminster, Wilts
Printed in Great Britain on acid-free paper
by The Bath Press, Bath, Avon

Preface

Interactions involving proteins with other molecules (termed as 'ligands'), whether they be nucleic acid, carbohydrate, vitamin, hormone, steroid, cofactor, metal ion, peptide, or even other proteins, underpin the whole of Biological Science. Nature, when it manufactures or engineers a protein does not generally do so with a single polypeptide chain as the intended end-point but more often than not with a composite structure involving other molecules of life in view. A thorough practical grounding in the experimental and computational methodologies used to investigate such interactions is therefore highly appropriate to the modern day protein scientist, and these two volumes in the Internationally renowned 'Practical Approach' series have been constructed with this intention. The volumes consider the range of principal techniques for studying interactions of proteins with ligands and from a practical standpoint: 22 chapters, produced by experts in the various areas cover: equilibrium dialysis, chromatography, analytical ultracentrifugation, surface plasmon resonance, capillary electrophoresis, electro-optics, X-ray and neutron scattering, isothermal titration calorimetry, differential scanning microcalorimetry, X-ray crystallography, molecular modelling, circular dichroism, fluorescence, stopped flow Fourier transform infrared, Raman and mass spectroscopy, paramagnetic probes, nuclear magnetic resonance, and atomic force microscopy. The volumes are designed for the Academic and Industrialist alike, intelligible by young (Masters, PhD, Postdoctoral) scientists and yet penetrating enough for established scientists.

The whole subject of Protein Science is one of the most intellectually attractive, diverse, and exciting subjects in contemporary science. It is theoretically as well as experimentally demanding and its scope is truly enormous. In fact it encompasses and impacts upon virtually every branch of science, from fundamental theoretical blue skies research at the subatomic level through chemical synthesis, molecular characterization/biology, clinical medicine, biotechnology, and the existing and emerging biophysical technologies (e.g. proteonomics) at the other end of the spectrum.

Proteins are so numerous and diverse in e.g. size, shape, geometry, topology, and other physico-chemical properties that there are very few methodologies/ techniques which have, at some stage or another, not been applied in order to parse their structure–function relationships. Moreover the *in vivo* function of proteins is superbly—and obviously optimally—conditioned by the interrelationship between structure, short- and long-range interactions, together with environmental properties toward accomplishment of evolutionary directed function. In addition their microscopic and macroscopic structural/electronic behaviour is precisely modulated, both spatially and temporally, by interactions with other macromolecules (nucleic acids/proteins/carbohydrates/lipids or lipid assemblies) and small molecular weight intra- and extracellular (metabolite) molecules. The latter are often referred to as ligands, although this term can encompass virtually any kind of molecule and is not well defined. Understanding such interactions, at a detailed level (micro- and macroscopic) is essential for gaining an intrinsic, curiosity driven knowledge base relating to cellular behaviour. In addition the applied clinical, agricultural, pharmaceutical, and biotechnology knowledge base has and continues (be it more slowly than often envisaged) to be extended by the need to design specific ligands which will alter/modify intrinsic protein function.

The many scientific and technical disciplines embodied in the subject of protein science and, more specifically, protein–ligand interactions, are making significant contributions to the goal of elucidating structure–function relationships in proteins. For example spectroscopists define electronic structure, enzymologists determine equilibrium/rate parameters, and molecular/structural biologists perturb structure and observe the effect on reactivity and structure at different levels. The theoretical/physical organic chemist can attempt to synthesize and assemble minimal reactivity site representations and determine and/or model intrinsic geometric, electronic, and reactivity characteristics involved in proteins and protein–ligand interactions. All this, and more, is directed towards what is undoubtedly the ultimate goal of protein research: to define function in terms of the detailed inter-relationship between structure/dynamics/kinetics and thermodynamics of a system under known physico-chemical conditions and systematic variations therein.

The field of protein–ligand interaction studies is at a propitious stage of development as well as being in a very active growth phase. It is hoped that both of these volumes will be used by both professional and inexperienced research workers wishing to characterize protein–ligand systems by using a multidisciplinary technique approach: such an approach is more or less inevitable given that the research ethos is now (rightly) more problem solving orientated rather than technique orientated. Both volumes are meant to complement and enhance the information content contained within the other titles in the Practical Approach series.

The techniques used to probe protein–ligand interactions have and continue to improve dramatically over recent years due to the development of new (or continued evolution of) reagents, protocols, strategies, and instrumental techniques/theoretical concepts. The methods presented here are not a comprehensive compilation of all the numerous techniques, which can be used to probe protein–ligand interactions. That would require an encyclopaedia! They are, instead, a selection of what we believe are the most useful and easily applied methods—many of which are now routinely used in laboratories world-wide for research into protein–ligand interactions. It is hoped that the two volumes will act as an introduction for investigating the basic principles and practical application of the subject before embarking on more detailed and/or other techniques, which are not, as yet, so commonly used. Towards this end the co-operation of all contributing authors together with the quality of their presentations are greatly appreciated.

Nottingham and Greenwich S. E. H. and B. Z. C.
October 2000

Contents

List of protocols *page xvii*
Abbreviations *xxi*

1 Protein–ligand interactions and their analysis 1
Babur Z. Chowdhry and Stephen E. Harding

1 Introduction *1*
2 A brief comparison of the 'high resolution' methods *6*
3 Other spectroscopic techniques *12*
4 Non-spectral methods *13*
5 Computational methods/molecular modelling *14*
6 Information sources *16*
7 Additional remarks: the layout of the volumes *16*
   References *17*

2 Equilibrium dialysis and rate dialysis 19
Bent Honoré

1 Introduction *19*
2 Equilibrium dialysis *21*
   Principle *21*
   Measurement of free ligand by UV or visible light absorption *21*
   Running an equilibrium dialysis experiment *23*
   Running control experiments for equilibrium dialysis *27*
   Measurement of ligand concentration by using a radiolabelled ligand *31*
3 Rate dialysis *34*
   Principle and theory *34*
   Running experiments using rate dialysis *37*
   Rate dialysis in microchambers *40*
4 Rate dialysis for measurement of free ligand concentrations in complex mixtures, e.g. serum *41*
   Acknowledgements *45*
   References *45*

3 Quantitative characterization of ligand binding by chromatography   47
Donald J. Winzor

1   Introduction   47
2   General experimental aspects   48
    Zonal and frontal affinity chromatography techniques   48
    Comparison of advancing and trailing elution profiles: the concept of non-enantiography   48
    Elution volume: an equilibrium parameter   50
    Measurement of binding constants   51
3   Frontal gel chromatographic studies of ligand binding   52
    Evaluation of the binding constant for a 1:1 interaction   52
    Interaction of a univalent ligand with a multivalent acceptor   56
    Allowance for multivalence of the ligand   57
    Evaluation of K_{AS} from the dependence of the constituent elution volume of acceptor upon ligand concentration   59
4   Zonal gel chromatographic studies of ligand binding   61
    The Hummel and Dreyer technique   61
    A limitation of the Hummel and Dreyer technique   62
5   Evaluation of binding constants by quantitative affinity chromatography   63
    General theoretical aspects   64
    Evaluation of ligand binding constants by frontal affinity chromatography   66
    Zonal quantitative affinity chromatography   68
    Other forms of quantitative affinity chromatography   70
6   Concluding remarks   72
    References   72

4 Sedimentation velocity analytical ultracentrifugation   75
Stephen E. Harding and Donald J. Winzor

1   Introduction: sedimentation velocity and sedimentation equilibrium   75
2   Basic principles of sedimentation velocity   76
    Measurement of a sedimentation coefficient   76
    Measurement of molecular mass by sedimentation velocity   78
3   General experimental aspects   82
    Optical systems for sedimentation velocity and sedimentation equilibrium   82
    Sedimentation velocity optical records   83
    Data capture   84
    Two complications   85
    Co-sedimentation diagrams   85
    Concentration dependence of the sedimentation coefficient   85
    Sedimentation coefficient ratios   88
    Sedimentation velocity fingerprinting   89
4   Sedimentation velocity analysis of the shape of a molecular complex   89
    Direct evaluation of molecular shape   90
    Selecting a plausible structure which best agrees with the data   90

5 Sedimentation velocity studies of ligand binding   *91*
   Interactions of a (protein) acceptor with a small ligand   *91*
   Interactions of an acceptor with a macromolecular ligand   *92*
   Sedimentation velocity studies of weak interactions   *94*
   The shape of sedimenting boundaries for acceptor–ligand systems   *95*
6 The study of ligand-mediated conformational changes   *98*
7 Concluding remarks   *101*
   References   *101*

5 Sedimentation equilibrium in the analytical ultracentrifuge   *105*
Donald J. Winzor and Stephen E. Harding

1 Introduction   *105*
2 Experimental aspects of sedimentation equilibrium   *106*
   Procedural details of a sedimentation equilibrium experiment   *107*
   Extraction of the molecular weight of a single solute   *109*
   Extraction of point average molecular weights for interacting systems   *115*
   Direct curve-fitting of concentration distributions   *116*
   Omega and psi analyses of sedimentation equilibrium distributions   *118*
3 Sedimentation equilibrium studies of ligand binding   *121*
   Evaluation of the concentration distributions of individual species   *121*
   Direct modelling of sedimentation equilibrium distributions   *124*
4 Ligand perturbation of acceptor self-association   *126*
   Characterization of acceptor self-association by sedimentation equilibrium   *126*
   Displacement of an acceptor self-association equilibrium by ligand binding   *130*
   Illustrative studies of preferential ligand binding by sedimentation equilibrium   *130*
5 Concluding remarks   *132*
   References   *133*

6 Surface plasmon resonance   *137*
P. Anton Van Der Merwe

1 Introduction   *137*
2 Principles and applications of surface plasmon resonance   *137*
   Principles   *137*
   Applications   *139*
3 General principles of BIAcore experiments   *141*
   A typical experiment   *141*
   Preparation of materials and buffers   *141*
   Monitoring the dips   *141*
4 Ligand   *143*
   Direct versus indirect immobilization   *143*
   Covalent immobilization   *143*
   Non-covalent immobilization (ligand capture)   *149*
   Activity of immobilized ligand   *152*
   Control surfaces   *152*
   Reusing sensor chips   *153*

5 Analyte *153*
   Purity, activity, and concentration *153*
   Valency *153*
   Refractive index effect and control analytes *154*
   Low molecular weight analytes *155*
6 Qualitative analysis: do they interact? *155*
   Positive and negative controls *155*
   Qualitative comparisons using a multivalent analyte *155*
7 Quantitative measurements *155*
   Affinity *156*
   Kinetics *160*
   Stoichiometry *165*
   Thermodynamics *165*
   Activation energy *166*
8 Conclusion *166*
9 Appendix *167*
   Physical basis of SPR *167*
   Samples and buffers *168*
   Additional information *168*

Acknowledgements *169*

References *169*

7 Capillary electrophoresis *171*
Niels H. H. Heegaard

1 Introduction *171*
2 The capillary electrophoresis technique *171*
3 Why use electrophoresis for the study of reversibly interacting molecules? *173*
4 Instrumentation and experimental variables *175*
   Apparatus *175*
   Experimental variables *177*
5 Capillary electrophoretic binding experiments *181*
   Experimental approaches for electrophoretic binding studies *181*
   Demonstration of ligand binding *182*
   Determination of binding constants *184*
6 Conclusions *192*

Acknowledgements *193*

References *193*

8 Molecular electro-optics *197*
Dietmar Porschke

1 Introduction *197*
2 Equations for representing the field-induced orientation of molecules *200*
   Dichroism amplitude *200*
   Dichroism decay *201*

3 Instruments *203*
 General *203*
 Pulse generators *204*
 Measuring cells *205*
 Spectrophotometric detection *205*
 Transient data storage and data processing *207*
 Automatic data collection *207*

4 Experimental procedures *207*
 Sample preparation *207*
 Measurements *210*

5 Data evaluation *211*
 Amplitudes: the stationary dichroism *211*
 Interpretation *214*
 Simulations of electro-optical data from macromolecular structures *215*

6 Various spectroscopic techniques *216*
 Birefringence *216*
 Fluorescence *218*
 Light scattering *219*

7 Kinetics *219*
 References *220*

9 High-flux X-ray and neutron solution scattering *223*
Stephen J. Perkins

1 Introduction *223*

2 Applications of scattering *226*
 Complementary structural approaches *226*
 Properties of X-ray scattering *227*
 Properties of neutron scattering *227*

3 X-ray and neutron facilities *229*
 High flux sources *229*
 X-ray instrumentation *232*
 Neutron reactor instrumentation *234*
 Spallation neutron instrumentation *236*

4 Experimental approach *238*
 Applications for X-ray and neutron beamtime *238*
 Preparation of protein–ligand complexes for scattering *239*
 Data collection strategies *242*
 Guinier analyses of the reduced scattering curve *248*
 Distance distribution function analyses *251*

5 Structural modelling of scattering data *252*
 Sphere modelling of scattering curves: use of atomic structures *252*
 Spherical harmonics modelling *255*
 Other modelling using spheres, shells, or ellipsoids *255*
 Comparison with sedimentation and diffusion coefficients *256*

6 Conclusions and worked examples *256*
 Small ligand binding to a protein (AmiC) *257*
 Ligand binding to a homodimeric complex (serum amyloid P component) *258*

A heterodimeric complex (factor VIIa–tissue factor complex) 259
A protein–DNA complex (RuvA–Holliday junction complex) 259

Acknowledgements 260

References 260

10 Isothermal titration calorimetry of biomolecules 263
Ronan O'Brien, Babur Z. Chowdhry, and John E. Ladbury

1 Introduction 263

2 Principle of operation 264
    Instrumentation 264
    Instrumental calibration 267

3 Full thermodynamic characterization of biological interactions 268
    Derivation of thermodynamic parameters 268
    Heats of ionization and pH dependence 269
    Determination of ΔG and ΔS 272
    Determination of the change in heat capacity 272

4 Preparation for an ITC experiment 273
    Concentration requirements 273
    Sample preparation 273
    Use of mixed solvent systems 274
    Concentration determination 275

5 ITC experimental protocol 276
    Performing a titration 276
    Trouble-shooting 277
    Controls 278

6 ITC data handling 278
    Data fitting 278
    Data analysis 279
    Choosing the appropriate model 279

7 Structure/thermodynamic relationships 281
    Heat capacity versus buried surface area 281
    Thermodynamic insights into the mechanisms of interaction involving DNA 282

8 Extending the range of binding constants determination 282

9 Advantages and disadvantages 283

References 284

11 Differential scanning microcalorimetry 287
Alan Cooper, Margaret A. Nutley, and Abdul Wadood

1 Introduction 287

2 DSC basics 288
    Instrumentation 287
    Quantitative analysis of DSC data—practical considerations 290
    Concentration measurements 293
    Units 293
    Scan rates/reversibility 294
    The baseline problem 294

3 Thermodynamic background *297*
    Heat capacity, enthalpy, and entropy *297*
    Equilibrium and free energy *299*
    Temperature dependence of thermodynamic quantities *299*
    The van't Hoff enthalpy *301*
4 Effects of ligand binding *302*
    Ligand binding and folding equilibrium *303*
    Change in T_m *307*
    Effects when ligand binds to both N and U *308*
    One peak or two? *308*
    Hydrogen ions as ligands: the effect of pH on protein stability *313*
    Non-specific metal ion effects *315*

Acknowledgements *317*

References *318*

List of suppliers *319*

Index *327*

Protocol list

Equilibrium dialysis
    Measurement of drug concentration by UV or visible light absorption spectroscopy *23*
    Setting up a dialysis experiment with spectroscopic measurement of the ligand *24*
    Control experiments for equilibrium dialysis using spectroscopic measurement of the ligand *30*
    Setting up dialysis experiments with a radiolabelled ligand *32*
    Determination of the dialysis rate constant, k, for the ligand *36*
    Setting up a ligand binding experiment using rate dialysis *38*
    Setting up rate dialysis using microchambers *40*

Rate dialysis for measurement of free ligand concentrations in complex mixtures, e.g. serum
    Measurement of free ligand concentrations in serum using rate dialysis *43*

Frontal gel chromatographic studies of ligand binding
    Determination of the binding constant for 1:1 complex formation between acceptor and macromolecular ligand by frontal gel chromatography *54*
    Determination of the stoichiometry and strength of an acceptor–ligand interaction by frontal gel chromatography *56*

Zonal gel chromatographic studies of ligand binding
    Characterization of ligand binding by the Hummel and Dreyer procedure *61*

Evaluation of binding constants by quantitative affinity chromatography
    Determination of binding constants by frontal affinity chromatography *67*
    Evaluation of binding constants by zonal affinity chromatography *69*

Basic principles of sedimentation velocity
    Sedimentation velocity: basic operation and measurement of a sedimentation coefficient *79*

Experimental aspects of sedimentation equilibrium
    Sedimentation equilibrium: basic operation *108*
    Sedimentation equilibrium: software *113*
    Characterization of a macromolecular interaction between dissimilar reactants by sedimentation equilibrium *120*

General principles of BIAcore experiments
    Normal and abnormal 'dips' *142*

PROTOCOL LIST

Ligand
An approach to covalent coupling of a protein   144
Determining the optimum pre-concentration conditions for a protein   147
Amine coupling   148
Establishing regeneration conditions   148
Developing a new antibody-mediated indirect coupling method   151
Quantitating binding levels   152
Reusing sensor chips   153

Quantitative measurements
Affinity measurements   158
Kinetic measurements   161

Capillary electrophoretic binding experiments
Capillary electrophoretic procedure to demonstrate binding   183
Capillary electrophoresis migration shift binding assay   188
Measurement of strong binding by capillary electrophoresis   191

Experimental procedures
Dialysis of macromolecules   208
Degassing of solutions   209

Data evaluation
Measure dichroism amplitude   212
Check dichroism optics   212
Check for reaction effects/field induced conformation   213

Various spectroscopic techniques
Adjustment of the optics for measurements of birefringence   217
Measure/calculate birefringence   218

X-ray and neutron facilities
Contact details for synchrotron X-ray and neutron facilities (September 1999)   229
Details of other synchrotron X-ray and neutron facilities world-wide   230

Experimental approach
Applications for beamtime: experimental reports   238
Pre-experimental organization   239
General experimental procedures   240
X-ray data collection   243
Neutron data collection   243

ITC experimental protocol
Setting up a titration   276

ITC data handling
Checking for additional equilibria   280

DSC basics
DSC of protein unfolding — basic procedure using lysozyme as a model   289

Effects of ligand binding
Ligand binding to folded protein—DSC of RNase with 2'-CMP   304
Ligand binding to unfolded protein—DSC of lysozyme with cyclodextrin   306
DSC of protein–DNA complex   310
Spreadsheet simulation/calculation   312

Abbreviations

a,b,c	semi-axial dimensions of an ellipsoidal protein
a,b,c	unit cell dimensions (Å)
a,b,c	unit cell parameters (Vol. 2, Chap. 1)
a,b,c	unit cell vectors/axes
AC	alternating current
AFM	atomic force microscopy
A_x	absorption (UV/visible) of component x or at wavelength x
CARS	coherent anti-Stokes Raman spectroscopy
CAT	chloramphenicol acetyl transferase
Cc	correlation coefficient
CCD	charge-coupled device
CCD	charge-coupled device area detector
CCDB	Cambridge Crystallographic Data Base
CCDC	Cambridge Crystallographic Data Centre
CD	circular dichroism
CE	capillary electrophoresis
CID	collision-induced dissociation
CMP	cytidine monophosphate
CoA	coenzyme A
Con A	concanavalin A
C_p	excess molar heat capacity (J.K^{-1} mol^{-1}; erg.K^{-1} mol^{-1})
CSLM	confocal laser scanning microscopy
C_x	weight (mass) concentration (g.ml^{-1}) of species x
c_x or [x]	molar concentration (M or mol.l^{-1}) of species x
d(hkl)	crystal plane interplanar spacing (Å)
d^*	reciprocal lattice vector
DB3	antibody DB3 (and its Fab' fragment)
Dc	crystal density
DC	direct current
DM	density modification
D_{max}	maximum dimension of a protein
dmin	resolution of X-ray data (Å)

ABBREVIATIONS

DMS	polydimethylsiloxane
DMSO	dimethyl sulfoxide
DSC	differential scanning calorimeter
DTE	dithioerythritol
DTNB	5,5'-dithiobis (2-nitrobenzoic acid)
DTT	dithiothreitol
D_x	translational diffusion coefficient of component x ($cm^2.s^{-1}$)
E	electric field strength ($V.cm^{-1}$)
EDTA	ethylenediamine tetraacetic acid
EPR	electron paramagnetic resonance
ESI	electrospray ionization
ESRF	European Synchrotron Radiation Facility
EtOH	ethanol
$\mathbf{F}(hkl)$	X-ray structure factor
f/f_o	frictional ratio of macromolecule to that of a spherical particle of the same anhydrous mass and volume
Fab'	antibody Fab' fragment
FAB	fast-atom bombardment
FCS	fetal calf serum
FF	force field
FFT	fast Fourier transform
fL	fluorescent ligand
FMS	polytrifluoropropylmethylsiloxane
FPE	fluoresceinphosphatidylethanolamine
FPLC	fast protein liquid chromatography
FSD	Fourier self-deconvolution
FT	Fourier transform
FTICR	Fourier transform ion cyclotron resonance
G	molar Gibbs free energy ($J.mol^{-1}$, $erg.mol^{-1}$)
GA	genetic algorithm
GdnCl	guanidinium chloride
H	molar enthalpy ($J.mol^{-1}$, $erg.mol^{-1}$)
h	Planck constant
Hb	haemoglobin
HSA	human serum albumin
HSQC	heteronuclear single quantum coherence
HUVECS	human umbilical vein endothelial cells
I	intensity
I	ionic strength ($mol.l^{-1}$)
IL	interleukin
ILL	Institut Laue Langevin (Grenoble, France)
IPDA	intensified photodiode array
IR	isomorphous replacement
IRS	internal reflection spectroscopy
ITC	isothermal titration calorimetery

ABBREVIATIONS

J_x	absolute concentration (fringe units) of component x
j_x	relative concentration to the meniscus (fringe units) of component x
k	dialysis rate constant (s^{-1})
K, K_a	association constant (M^{-1} ($l.mol^{-1}$) or $ml.mol^{-1}$)
k_B	Boltzmann's constant
K_D, K_d	dissociation constant (M ($mol.l^{-1}$) or $mol.ml^{-1}$)
k_{off}	dissociation rate constant (s^{-1})
k_{off}	first-order dissociation rate constant (s^{-1})
k_{on}	association rate constant (s^{-1})
k_{on}	first-order association rate constant (s^{-1})
k_s	Gralén coefficient, $ml.g^{-1}$
K_{xy}	intrinsic binding constant between acceptor x and ligand y
L	ligand
l, L	litre
mAb	monoclonal antibody
MALDI	matrix-assisted laser desorption/ionization
MC	Monte Carlo
MD	molecular dynamics
MetJ	methionine repressor protein
MIR	multiple isomorphous replacement
MLA	mistletoe lectin A-chain
MLB	mistletoe lectin B-chain
MLI	mistletoe lectin
MLVs	multilamellar vesicular
MPD	methylpentanediol
MR	molecular replacement
MS-MS	tandem mass spectrometry
$M_{w,x}$	weight average molecular weight (molar mass) of component x
MWPC	multiwire proportional counter
$M_x\ M_{r,x}$	molecular weight (Da) or molar mass ($g.mol^{-1}$) of component x
N	Newton
N	native folded polypeptide
N, N_A	Avogadro's number (6.02252×10^{23} mol^{-1})
NA	numerical aperture
NAG	N-acetyl glucosamine
NCS	non-crystallographic symmetry
NDSB	non-detergent sulfobetaines
nfL	non-fluorescent ligand
NMR	nuclear magnetic resonance
NOEs	nuclear Overhauser enhancement effects
NOESY	nuclear Overhauser enhancement effects spectroscopy
P	protein
P	Patterson function
p	Perrin shape function

ABBREVIATIONS

PADS	peroxylamine disulfonate or Fremy's salt
PBS	phosphate-buffered saline
PC	phosphatidylcholine
PDB	Protein Data Bank
PEG	polyethylene glycol
δ	phase shift or optical retardation (Chap. 8)
PLVs	phospholipid vesicles
PMT	photomultiplier tube
PROXYL	(2,2,5,5-tetramethyl-1-pyrrolidinyloxy)
PS	phosphatidylserine
PSCARS	polarization-sensitive coherent anti-Stokes Raman spectroscopy
PSI	photosystem I
Q	scattering vector (nm^{-1})
QSAR	quantitative structure–activity relationship
R	crystallographic R-factor
R	aggregate radius (Vol. 2, Chap. 1)
R	combined structure factor index (Vol. 2, Chap. 1)
r	fluorescence anisotropy
R	gas constant (8.3143 $J.K^{-1}\ mol^{-1}$)
R	rotation function
RCA	ricin agglutinin A-chain
REPs	resonance enhancement profiles
Rfree	free R-factor
R_G, R_g	radius of gyration (Å, nm, cm)
RIDS	reaction-induced difference spectroscopy
Rint	internal consistency R factor
Rmerge	merging R factor
σ	RMS density (Patterson or electron density)
RMS	root mean square
rmsd	root mean square deviation
RNase	ribonuclease
RNase A	ribonuclease A
ROA	Raman optical activity
ROESY	rotating frame nuclear Overhauser enhancement effects spectroscopy
RTA	ricin A-chain
RTB	ricin B-chain
RU	response units or resonance units
s	equilibrium solubility
S	supersaturation
S	electron spin
S	entropy ($J.mol^{-1}K^{-1}$, $erg.mol^{-1}K^{-1}$)
s	reciprocal of the Bragg spacing ($Å^{-1}$, nm^{-1}, cm^{-1})
s	solvent fraction of the cell mass (Vol. 2, Chap. 1)
SAP	human serum amyloid P component

ABBREVIATIONS

SCRs	structurally conserved regions
SDS	sodium dodecyl sulfate
SERRS	surface-enhanced resonance Raman spectroscopy
SERS	surface-enhanced Raman spectroscopy
SIR	single isomorphous replacement
SPR	surface plasmon resonance
SR	synchrotron radiation
SRS	synchrotron radiation source (Daresbury, UK)
ST-EPR	saturation-transfer electron paramagnetic resonance
SVD	singular value decomposition
s_x	sedimentation coefficient of component x (sec or Svedberg units, $S = 1 \times 10^{-13}$ sec)
T	translation function (in MR)
T_1	spin–lattice relaxation time
T	temperature
TEMPO	(2,2,6,6-tetramethyl-1-piperidinyloxy)
TF	tissue factor
TFE	trifluoroethanol
TOF	time-of-flight
TR^3	time-resolved resonance Raman
trNOE	transferred nuclear Overhauser enhancement effects
U	unfolded polypeptide
U_i	energy of interaction (induced dipole)
U_p	energy of interaction (permanent dipole)
UV	ultra-violet
UVRR	ultra-violet resonance Raman
Vx	elution volume (ml) of species x
$\bar{v}_x$	partial specific volume (ml.g^{-1}) of component x
Web	World Wide Web
X	interaction constant (equilibrium constant) in weight concentration units ml.g^{-1} or l.g^{-1}
z_x, Z_x	net charge (valency) on a species, x
λ	wavelength (Å, nm, cm)
θ (hkl)	Bragg angle for X-ray reflection from hkl planes
φ (hkl)	phase angle of structure factor **F**(hkl)
ρ (xyz)	electron density at point (x,y,z)
α, β, γ	unit cell inter-axial angles
ψ_x	psi (sedimentation equilibrium) function for component x
$[\Theta]$	molar ellipticity
$|F|$	structure factor amplitude
$1/T_2$	transverse relaxation rate
^2H_2O	deuterium oxide
3D	three-dimensional
$\Delta\delta$	chemical shift separation
$\Delta\varepsilon$	molar differential extinction coefficient (M^{-1}cm^{-1}, l.mol^{-1}cm^{-1}), sometimes abbreviated as ε

ABBREVIATIONS

α	polarizability (Chap. 8)
δ	hydration
ε_λ	extinction coefficient (either ml.g^{-1} cm^{-1}, or M^{-1}.cm^{-1}) at a wavelength λ
ϕ_p	maximum packing fraction
$\lambda/4$ plate	quarter-wave plate
μ_p	dipole moment (permanent dipole)
τ_c	rotational correlation time (s)
ω	angular velocity of centrifuge rotor (radians.s^{-1})

Chapter 1
Protein–ligand interactions and their analysis

Babur Z. Chowdhry
School of Chemical and Life Sciences, University of Greenwich, Wellington Street, Woolwich, London SE18 6PF, UK.

Stephen E. Harding
NCMH Physical Biochemistry Laboratory, University of Nottingham, School of Biosciences, Sutton Bonington, Leicestershire LE12 5RD, UK.

1 Introduction

In each cell of an organism, a myriad of reactions, covalent and non-covalent, occur at any given point in time. These reactions are co-ordinated and regulated, both spatially and temporally. Each reaction has a specific purpose, occurs as a result of finely-tuned inter- and intramolecular recognition mechanisms, and forms part of an intricate network of interdependent multi-component linear/non-linear reactions in interconnected compartments (organelles etc.) of the cell. Moreover the frontiers of viability of such reactions—and the living organisms that depend on them—are marked by extreme conditions: 1–12 for pH, −5–110 °C for temperature, 0.1–120 MPa for hydrostatic pressure, and 0.6–1.0 for water activity. Amongst the many molecules that participate in such reactions in the complex—and, as yet hardly understood, milieu of the cell—are proteins.

Proteins play a pivotal, indeed essential role, in cellular (and in multicellular organisms, extracellular) activity. Their numerous biological functions together with the molecular basis of their biophysical properties/behaviour are therefore of multidisciplinary interest within the framework of what are generally called the biological (biochemistry, pharmacology, physiology, immunology, etc.) and physical (physics, chemistry, mathematics, computing, etc.) sciences.

Both our intrinsic curiosity-driven philosophical knowledge base and our need or desire to modify (or use), from an applied perspective (medical, agricultural, biotechnological), the properties or behaviour of proteins are increasingly dependent on the ability to modulate the physico-chemical, and hence biological behaviour of these molecules. Because *protein–ligand interactions play a key role in cellular metabolism*, detailed knowledge of such interactions, at both a microscopic and macroscopic level, is also required. Just two examples, which immediately stand out from the many, are antigen–antibody interactions and proteins which act as receptors (membrane or non-membrane bound): neurotransmission depends on the ability of small and large molecular weight molecules to recognize and bind to specific sites on large membrane bound receptors. Another example: enzyme–ligand interactions form a very large group of important complexes, which have been investigated for many years. In these systems not only is the substrate(s) of obvious importance, but so are other molecules, which regulate enzyme activity such as coenzymes and positive and negative modulators in both allosteric and non-allosteric proteins. Other important systems, which are now being studied at a molecular level, include ligand binding to structural proteins, protein–DNA binding, protein–saccharide, protein–protein, and protein–peptide interactions. The formation—and maintenance—of the quaternary structure of multisubunit proteins, the self-assembly of large structures such as microtubules or chromatin and transcription factor–promoter interactions are other examples; although the list is actually almost endless.

The term 'ligand' in biological systems can have many different meanings. It is usually, in its broadest sense, used to mean any molecule which interacts with a given molecule (in this case a protein). The term 'ligand' thus includes other macromolecules (peptides/proteins/nucleic acids/lipids/carbohydrates or mixed molecular species thereof) as well as 'small' molecular mass (arbitrarily $< \sim$ 1–2 kDa) molecules. Ligands therefore comprise a very large and structurally diverse group of molecules which, unsurprisingly also display a wide variety of physicochemical characteristics. This makes it difficult to understand, delineate, and draw generalized conclusions concerning their biophysical properties. Perhaps the major criterion, in the context of the discussion that follows, is that ligands can interact in a (potentially) reversible, non-covalent manner with a protein and thereby modulate its biological role in a controllable way (i.e. without the requirement to make or break covalent bonds).

To *fully* understand protein–ligand interactions requires that, at a minimum, the following criteria be met. First that the biophysical properties of both the protein and the ligand under investigation are examined independently and that their biophysical behaviour be fully understood: this requirement applies to both existing and/or newly designed and (bio)synthesized peptides/proteins and ligands. *Currently, and contrary to popular opinion, we are far from achieving such an aim, except at an extremely superficial level.* Knowledge of the structure/conformation, if possible, at the atomic level of the protein and the ligand in the unbound form is a prerequisite for protein–ligand interaction studies. Although it may appear trivial to the reader, the high level resolution structures of most proteins (either by X-ray crystallography or high resolution NMR) have not, to date, been established. It is currently estimated that such data exists for only about 30 peptides/proteins (1) but it is seriously arguable, in the sense of the generally accepted meaning of the term high (atomic) resolution (i.e. < 0.8 Å), that even this target has not been met. Additionally the process of forming a complex between a small ligand molecule and a protein is a complicated equilibrium process. Both the ligand and the protein in the solvated state probably exist as an equilibrium mixture of several conformers. Admittedly for many 'small mol-

Table 1 Parameters which have to be defined for the protein, the ligand, and the protein–ligand complex

---
Chemical composition, concentration/purity (contaminants) /stability/specific activity or assay for, e.g. structural proteins and ligand/state of structural integrity. Ability to freeze or air dry and pre-history of protein and ligand samples may be important.

Solubility (in different solvents: aqueous, organic, and mixed phase); state of aggregation.

Structure: size, shape, geometry, topology; protein sequence, and post-translational modifications must be known; ability to use molecular biology techniques (strategies for over-expression and mutagenesis of recombinant protein are critical); combinatorial methods for over production of ligand.

Dynamics of protein/ligand/ protein–ligand 'complex'.

---

ecular weight ligands' high resolution X-ray and/or NMR structures do exist. The fact that our knowledge base is deficient in this area has significant implications for understanding protein–ligand interactions, which unfortunately will not be touched upon, in this brief overview. In addition a myriad of other properties of the reactants also require to be established (see *Table 1*).

Secondly the protein–ligand complex must also be fully characterized. Normally, at a simplistic level, the non-covalent interaction between a protein (P) and a ligand (L) is often represented as follows:

$$P + L \leftrightarrow PL \quad [1]$$

However as Williams and Westwell have pointed out (2) the interaction should actually be written in the following form:

$$P + L \leftrightarrow P'L' \quad [2]$$

Equation 2 may, at first glance appear identical to *Equation 1*. However the formulism of *Equation 2* recognizes, or is taken to mean, that once P and L have undergone an interaction or association, they no longer exist. They have, instead been replaced by the modified entities P' and L'. (Tight complexes often result when protein–ligand interactions are significantly stronger than ligand–solvent and protein–solvent interactions.)

The well described equilibrium molar association or binding constant, K_a (other popular symbols are K_b or just K) corresponding to *Equation 2* is then described by:

$$K_a = [P'L'] / ([P].[L]) \quad [3]$$

where the square brackets means the molar concentration (M, or mol.l^{-1}), which of course is not an SI unit. For the system described by *Equation 2* (or *Equation 1*) the units[1] of K_a will be M^{-1} (l.mol^{-1}). The more popularly used 'molar dissociation constant' K_d is simply the reciprocal of K_a, so for the system of *Equation 2*:

$$K_d = [P].[L] / [P'L'] \quad [4]$$

---

[1] In strict thermodynamic terms K_a and K_d are both dimensionless. Dimensions are however normally added, but only to indicate the dimensions of the quantities used to calculate them. See Price, N. C., Dwek, R. A., Wormald, M. R. and Ratcliffe, R. G., *Principles and Problems in Physical Chemistry*, 3rd edition, Oxford University Press, 2001.

with the units of K_d, M or mol.l^{-1}. Traditionally, K_ds of < 5 μM are regarded as strong interactions, and > 50 μM as 'weak'. Any technique chosen to measure K_d (or K_a) has to cope with a concentration/concentration range where all species (P, L, and P'L') are present: this means to probe the K_d for strong interactions, low concentrations are required and for weak interactions, high concentrations. Of course there are other reactions more complicated than that described by *Equations 1* or *2*. A more generalized form of *Equation 2* for a binary system is:

$$nP + mL \leftrightarrow P'_n L'_m \qquad [5]$$

and the corresponding K_a will be:

$$K_a = [P'_n L'_m] / ([P]^n \cdot [L]^m) \qquad [6]$$

Other variants of Equations 3-6 include the use of weight concentrations C (g/l or g/ml: again, not SI units) rather than molar concentrations: (the K_a, K_d notation is then replaced by, e.g. X_a, X_d).

Establishing the stoichiometry and association/dissociation equilibrium constants (or the corresponding rate constants) is only one step: the goal of research into protein-ligand interactions is to understand, again it must be emphasized, in minute molecular detail the relationship between function and molecular recognition, structure, kinetics, energetics, and dynamics of as many defined systems as possible. Use can then be made of this plethora of knowledge such that either the behaviour of previously unknown systems can be hypothesized in advance of experimental information being gained or that completely new systems can be designed. The pursuit of this goal is a time-consuming and difficult task! Especially when it is realized that the inter-relationship between all of the foregoing parameters (intermolecular recognition-structure-dynamics-kinetics-thermodynamics) need to be assessed and need to be further ascertained over a wide variety of environmental solution conditions. *Table 2* cites the most commonly used experimental variables (but of course, there are countless others) and *Table 3* provides an overall summary of the factors that need to be taken into account when applying a technique (or, preferably, a collection of

Table 2 Experimental variables in the analysis of the protein, the ligand, and the protein-ligand complex

(a) Reagent variables
Protein/ligand concentration
Ligand structural variants (analogues/homologues)
Protein structural variants (site/group specific, conservative/non-conservative mutants; denatured form/partially unfolded forms, presence or absence of cofactors)
(b) Environmental solution condition variables
Temperature, pH, buffers (ionic strength and nature of buffers)
Osmolytes, solvents including co-solvents, salt /ion concentration
Denaturants (urea /GdnCl (and other lyotropic /chaotropic agents)/detergents, surfactants), stabilizers (azide, glycerol etc.); mercaptoethanol, DTT; gaseous phase used for experiments, etc.
(c) Combinations of the above

Table 3 Factors to be taken into account in technique(s) used for investigating protein/ligand /protein–ligand complexes

Design of experimental protocol
Information content
Availability, cost
Complementary/alternative techniques[a]
Analytical parameters associated with the technique(s): *concentration of material required, compared to other techniques,* sensitivity, detection limit, accuracy, precision, reproducibility/repeatability, use or necessity for isotopic or non-isotopic labelling, ease of use, computing resources, methods/complexity and different methods of data analysis, errors/error propagation/statistical analysis of results (if possible); time scales of experimental (spectroscopic) techniques used.
Ability to use the technique in the presence of buffer additives and/or environmental solution conditions given in *Table 2*.
The experimentalist should insure that at every stage of the experiment and (particularly) in the data analysis, the correct (appropriate) theoretical models are used. In using instrumental techniques the 'black box data analysis' syndrome should be avoided. Critically compare and contrast with work on other related systems. Each method will have its own advantages and drawbacks for a given system. The ability to interpret experimental data both intrinsically and in terms of a theoretical basis is very important. The systematic/accumulated instrumental errors are often important, but usually ignored.

[a] Try, if possible to use more than one technique (method) to verify the overall results or the values for particular parameters.

techniques) to the analysis of protein–ligand interactions. *It is extremely important to cross-reference parameters in Tables 1, 2, and 3.*

Within the requirement that the changes in the conversion of P to P' and L to L' (i.e. changes in the reacting species in the 'complex' compared to their individual states) need to be ascertained as a step towards a full understanding of protein–ligand interactions, the following factors include some of the important parameters which require consideration. The changes in structure, in either species, may range from extremely subtle (i.e. an 'insignificant' structural reorganization; 'simply' a tightening or loosening of the internal structure) to large scale changes in both local and global conformational properties. The list of changes which occur not only include changes in structure (secondary/tertiary/quaternary), conformation, size, shape geometry, and topology but also include changes in the charge distribution, the state of hydration and protonation, and the partial molar volume as well as changes in the surface accessible surface area, polarity (hydrophobicity), and intra- and intermolecular entropy factors (e.g. rotational and translational motional properties of molecules: and the dynamics of water molecules). The molecular nature of the binding interface, the identification and quantification of the (individually) 'weak' cooperative molecular forces holding the protein-ligand complex together (such as hydrogen bonding, dipole–dipole interactions, van der Waals forces—induction and dispersion forces alone or combined—hydrophobic, electrostatic interactions, etc.) and the number/role, if any, of water molecules at the binding interface are other important factors which have to be considered. Association of other proteins or other ligand species subsequent to the interaction of the initial species may also be involved; thus making the problem even more complicated! Furthermore if multisite ligand binding to either a monomeric or multimeric (homomeric/heteromeric subunit) protein is involved then cooperativity effects and linkage phenomena (allosteric effects) have also to be taken into account and examined by the appropriate experimental/theoretical techniques. Williams and Westwell (2) have also emphasized the important, but often overlooked, point that the binding energy between two reacting species is not only a property of the interface between them, but also depends on the modifications (as alluded to above) of the internal structures of one or both of the reacting partners.

Obviously in order to achieve the objectives of investigating protein–ligand interactions a multidisciplinary technique/methods approach has to be adopted as we stress in *Table 3*. Some of the methods which are used to examine protein and/or protein–ligand interactions have been compared and contrasted in *Table 4* which has been based to a certain extent on a recent review by Philo (ref. 3), and the contributions to this book (other useful sources can be found in refs. 4 and 5).

2 A brief comparison of the 'high resolution' methods

Since structural elucidation is so important in protein–ligand interactions it makes sense to compare and contrast the application of the two most powerful and commonly used methods for attaining this goal, namely X-ray crystallography and high resolution NMR. Both methods complement each other. There is no doubt, however that, currently at least, X-ray techniques can give a better-defined structure in less time compared to NMR techniques. However the two methods should be considered convergent (they do after all use similar software tools) in that for molecules (that are amenable to NMR analysis) the X-ray structure, if available, can be a substantial aid in the elucidation of the NMR structure. In addition both methods require, as a general rule of thumb, the same amount of material for analysis; approximately 10 mg per 10 kDa of protein. However the fact that NMR can be used for obtaining information on the dynamics of a system, the association constant for a protein–ligand interaction and, for small molecular weight proteins and/or protein–ligand complexes, the structure of different conformations in equilibrium should also be taken into account.

The advantages of X-ray crystallography, for structure determination of proteins, ligands, and their complexes are as follows:

(a) It is a well-established technique.

(b) More mathematically direct image construction is required, compared to NMR.

(c) Objective interpretation of data (usually) easier than NMR.

(d) Raw data processing highly automated.

(e) Quality indicators available (resolution, *R*-factor).

Table 4 Techniques for characterizing protein–ligand interactions (adapted and extended from ref. 3; see also refs 4, 5)[a]

Technique	Basis for detection	Interaction information	Minimum protein required[b]	Analysis time	Comments
Hydrodynamic and calorimetric (Volume 1)					
Equilibrium dialysis and rate dialysis; affinity (zonal and frontal) gel chromatography	Partitioning of free and bound ligand	Stoichiometry, equilibrium binding strength (K_a or K_d). Range of $K_d \sim 10^{-1}$ to 10^{-13} M	10 µl	15 min to 24 h	Chapters 2 and 3. Relatively straightforward compared to other techniques, but still very useful and adaptable.
Analytical size exclusion chromatography	Change in hydrodynamic size	Stoichiometry, strength ($K_d \sim 10^{-4}$ to 10^{-13} M)	1000 µl	30–60 min	Chapter 3. Wide range of potential on-line or off-line detection methods (including multi-angle laser light scattering for absolute M_r); may be labour-intensive depending on degree of automation.
Sedimentation velocity (in analytical ultracentrifuge)	Change in sedimentation coefficient	Stoichiometry, strength ($K_d \sim 10^{-3}$ to 10^{-9} M), conformation changes	10 µg; 400 µl	2–5 h (3–7 samples)	Chapter 4. Newer software makes data analysis fast and user-friendly; generally not as good as sedimentation equilibrium for K_d.
Sedimentation equilibrium (in analytical ultracentrifuge)	Change in solution mass	Stoichiometry, strength ($K_d \sim 10^{-3}$ to 10^{-9} M)	10 µg; 100 µl	12–24 h per rotor speed (9–21 samples)	Chapter 5. Powerful method when there is a significant change in mass; much more accessible with modern hardware and software.
Surface plasmon resonance	Change in mass bound to surface	Stoichiometry, strength ($K_d \sim 10^{-3}$ to 10^{-13} M), kinetics	0.01 µg; 100 µl	15–30 min (1 analyte, up to 4 surfaces)	Chapter 6. High throughput; fairly insensitive to contaminants; potential problems from surface coupling.
Capillary electrophoresis	Change in size, shape, or charge	Equilibrium binding strength ($K_d \sim 10^{-3}$ to 10^{-7} M), kinetics	µM; nl	Minutes	Chapter 7. Particularly useful if only tiny volumes are available, or material available at low purity.

Table 4 Continued

Technique	Basis for detection	Interaction information	Minimum protein required[b]	Analysis time	Comments
Hydrodynamic and calorimetric (Volume 1) (continued)					
Electro-optics	Change in rotational diffusion and/or electrical properties	Conformation change on binding. Kinetics of slow (~ minutes) reactions	~ mg/ml, 100–200 μl	Minutes (per concentration)	Chapter 8. Electric dichroism (and decay) or electric birefringence (and decay). Risks of sample damage through field effects now minimized by better shielding.
X-ray and neutron scattering	Change in solution mass and conformation	Conformation change on binding	~ mg/ml	Minutes to hours	Chapter 9. Some risk of radiation damage to sample during measurement.
Isothermal titration microcalorimetry, and Differential scanning calorimetry	Heat release or uptake; affect on thermal transitions of ligand binding	Stoichiometry, strength (K_d ~ 10^{-3} to 10^{-9} M), ΔH; ΔC_p	0.1–1 mg/ml; 0.5–2 ml	3 h	Chapters 10 and 11. Universal signal; works well for small molecules; useful additional thermodynamic information.
Structural and spectroscopic (Volume 2)					
X-ray crystallography	Change in structure	Conformation change on binding	~ 100 mg for crystals	Weeks to months	Chapter 1. Protein and protein–ligand complex must be crystallizable
Molecular modelling	Change in structure	Conformation change on binding	Nothing	Days to weeks	Chapter 2. X-ray or NMR structure needed to begin with; in cases where there is no structure available, one can resort to the techniques of protein structure prediction.
Circular dichroism	Change in optical activity	Secondary structure change. Interaction constants (K_d ~ 10^{-3} to 10^{-7} M)	< 1 mg/ml (~ 10 μM)	~ 2 h	Chapter 3. Optimum A_{max} ~ 0.8.
Fluorescence methods	Change in fluorescence property. Rotational diffusion (for polarization anisotropy decay measurements)	Change in surface electrical potential or solution conformation. Kinetics from quenching studies	~ 1 mg/ml (~ 10 μM) Depends on amount of chromophore	Hours	Chapter 4. Applicability depends on nature of protein and/or ligand; chromophore label may be required.

Technique	Principle	Information	Concentration/amount	Time	Notes
Stopped flow	Rapid mixing of reactants	Association and dissociation rate constants, k_{on} and k_{off} respectively	~ 1 mg/ml (~ 10 µM) Depends on amount of chromophore	Hours	Chapter 5. Alternative to rapid-reaction methods are relaxation methods in which an equilibrium mixture is perturbed by rapid changes in temperature (T-jump) or pressure (P-jump).
Fourier transform infra-red spectroscopy	Changes in the amide I absorption band	Secondary structure change	10 µl at 5–10 mg/ml	12–24 h	Chapter 6. Especially suited for membrane as well as water soluble proteins. Suitable for small and large proteins (and complexes thereof).
Raman spectroscopy	Change in inelastic light scattering or 'Raman spectrum'	Secondary structure, k (rate constants) and equilibrium constants ($K_d > 10^{-3}$ M)	Higher concentrations needed (several mg/ml) 30 µl	Minutes	Chapter 7. Various forms, such as difference, resonance, time resolved, surface enhanced, coherent anti-Stokes.
Mass spectrometry	Change in molecular mass	Stoichiometry	< 0.001 µg; <1 µl	Minutes to hours	Chapter 8. May not detect weaker interactions. Electrospray ionization (ESI) and matrix-assisted laser desorption/ionization (MALDI). Maximum M_r of complex ~ 1–2 × 10^5 Da.
Electron paramagnetic spin resonance	Changes in electronic environment, rotational diffusion	Changes in structure. Dynamics (rate of change of spectral line shape)	50–100 µg aliquots (1–2 mg/ml)	~ 2 h	Chapter 9. Particularly relevant to membrane proteins. Chemical labelling often necessary.
Nuclear magnetic resonance	Change in chemical shift, diffusion rates (rotational and translational T_2 relaxation)	Changes in structure. Dynamics (k_{on}, k_{off}), equilibria ($K_d > 10^{-3}$ M– ~ 10^{-9} M)	50–100 µM for 1D; 3 mM for 2D; 0.5 ml	Hours (weeks/months for high resolution structure)	Chapter 10. Sample requirements depend on resolution required. 2D high resolution work much more demanding and restricted to small proteins.
Atomic force microscopy	Surface imaging of the force between molecules and the changes on ligand binding	Change in structure	~ 1 mg/ml	Minutes	Chapter 11. Can visualize single molecule: particularly useful for membrane proteins inaccessible by X-ray crystallography.

[a] Abbreviations: ΔH enthalpy change; ΔC_p heat capacity increment; K_a association constant; K_d dissociation constant; k_{on} rate of association; k_{off} rate of dissociation.
[b] Concentration depends on the strength of the interaction being analysed: weaker interactions require higher concentrations.

(f) Mutant proteins, different ligands, and homologues structures (as low as 25% sequence identity) may be extracted by using molecular replacement and subsequent use of electron density/electron density difference methods as an aid for comparison.

(g) Large molecules and assemblies can be determined, e.g. virus particles (however ordering between such assemblies is not usually as good as—i.e. they often diffract less well than—small molecular weight molecules).

(h) Surface water molecules relatively well defined.

(i) Often produces a single structural model that is easy to visualize and interpret.

(j) Use of synchrotron radiation combined with cryo-conditions usually speeds up data acquisition, improves resolution and stability of crystals.

The disadvantages of X-ray crystallographic methods include the following:

(a) Protein/protein–ligand complex has to form stable crystals that diffract well and X-ray crystallography does not directly yield hydrogen atom positions. (Neutron diffraction techniques make this possible but usually only for small molecules and large size crystals are required.)

(b) Need heavy atom derivatives that form isomorphous crystals, unless molecular replacement methods are used.

(c) Crystal production can be difficult and time-consuming and often impossible.

(d) Unnatural, non-physiological environment, i.e. may not wholly represent structure as it exists in solution.

(e) Difficulty in apportioning uncertainty between static and dynamic disorder.

(f) Surface residues may be influenced by crystal packing.

(g) Large molecular weight flexible modular proteins can be problematic and the final structural model represents a time-averaged structure, where details of molecular mobility remain unresolved.

The advantages of NMR for structural elucidation are as follows:

(a) Experiments are conducted under solution conditions which are 'closer to biological conditions' than X-ray methods (i.e. free from artefacts resulting from crystallization).

(b) Can provide information on dynamics (relaxation, rotational, and translational diffusion measurements, proton exchange rates, and the vibrational motion of atoms) and identify individual side chain motions.

(c) Secondary structure can often be derived from limited experimental data.

(d) Increasingly used to monitor conformational change on ligand binding to protein or to fragments thereof.

(e) Good for comparing and checking the correct fold of mutants (useful for protein folding/time resolved studies).

(f) Solution conditions can be explicitly chosen and readily changed, e.g. pH, temperature, etc. (but high concentrations of buffers may be a problem).

(g) 'Internal' water molecules can be detected (as in X-ray), but surface water molecules can be problematic because of fast exchange (lifetimes can still be ascertained).

(h) H-bonds can now be detected (rather than inferred from distance constraints as in X-ray).

(i) The increasing availability of high resolution instruments (e.g. 700 MHz and above) combined with new data acquisition/analysis techniques and greater computing power is and will continue to allow significant improvements in the application of NMR techniques.

The disadvantages of NMR for structural elucidation include the following:

(a) Usually require concentrated solutions, which may present a danger of aggregation: this problem is manifested in e.g. many antibody systems.

(b) Currently limited to determination of relatively small proteins (< 20 kDa), but there are exceptions.

(c) Surface residues *generally* less well defined than in X-ray crystallography because of mobility of surface residues and experimentally fewer interactions (NOEs).

(d) Often produces an ensemble of possible structures rather than one model (this can be advantageous) but conformational variability can make data interpretation difficult and complete structure determination required if protein sequence homology is less than $\sim 60\%$.

(e) Labelling often used, e.g. ^{2}H, ^{13}C, and ^{15}N and may cause problems in terms of protein yield and expense.

It has to be emphasized that true high resolution structures of the reacting species and the complex are sorely needed in order for example to:

(a) Obtain statistically unbiased data on both protein stereochemistry/precise geometry and the validity of the parameters used to obtain the data (i.e. their refinement).

(b) Check the application of 'normal mode' calculations.

(c) Calculate charge density distributions.

(d) Analyse hydration shells around protein molecules and also for a more accurate view of hydrogen bonds.

(e) Obtain precise information on the direction of amino acid side chain/domain movement.

(f) Model alternative conformations/multi-conformational species.

(g) Help to quantitate the physico-chemical parameters involved in the interaction, especially weak non-covalent interactions.

(h) Obtain precise surface areas.

Not only are these parameters required for intrinsic reasons but also for helping to analyse data obtained from other techniques and thereby improve the theoretical underpinning of experimental observations.

3 Other spectroscopic techniques

One of the virtues of spectroscopic techniques used in examining proteins, ligands, and protein–ligand interactions is that they are usually non-destructive and sometimes non-invasive, however some procedures do require the attachment of labels at specific sites on one or both of the interacting species.

Relatively soft ionization methods introduced in *mass spectrometry* (MS) allow a gentle phase transfer from solution into the gas phase of the mass spectrometer such that even weakly bound non-covalent complexes can be detected intact and their mass analysed. MS methods such as ESI-MS have begun to be recently used for the determination of association constants of non-covalent interactions. One of the advantages of these methods is that signals due to protein, ligand, and protein–ligand ion signals can be detected separately. In addition higher order structure formation, e.g. dimerization may also be detected by MS methods, although analytical ultracentrifugation is probably better suited for this purpose.

Fluorescence spectroscopy is a technique that is very sensitive (picogram quantities of material can be detected) and well-suited for measurement in the real time domain. Its principal shortcomings, however, are that only the structure of the fluorescent probe and the immediate environment is reported and the data obtained is not easy to interpret. Nonetheless, fluorescence spectroscopy has found wide use in studying the physico-chemical properties of proteins, protein–ligand interactions, and protein dynamics. This is because almost all proteins contain naturally fluorescent amino acid residues such as tyrosine and tryptophan. In addition, a large number of fluorescent dyes have been developed that can be used to specifically probe the function and/or structure of macromolecules. In cases where only limited quantities of proteins are available (e.g. a recently expressed recombinant protein or a precious protein pharmaceutical), fluorescence spectroscopy is often the method of choice for studying properties such as stability, hydrodynamics, kinetics, or ligand binding, because of its exquisite sensitivity. Even in cases where the structure of proteins is 'well-known' (either from X-ray diffraction or NMR), *electron paramagnetic resonance* (EPR), *time-resolved fluorescence spectroscopy*, and *stopped-flow methods* can be particularly useful for investigating their dynamic behaviour.

Circular dichroism (CD) is an important, widely used and commonly available technique for secondary structure determination. It requires expertise in data collection/analysis to be used to its full potential, which is not always the case. Nevertheless it has the following advantages in that relatively small amounts of material are required (0.1–1 mg/ml can often suffice) and results can be obtained quickly (in hours rather than weeks or months as in the case of NMR and X-ray). The amino acid sequence of the protein need not be known—although obviously it helps. CD can be used to examine the effects of environmental conditions such as pH, temperature, etc., on the overall protein conformation, in the presence and absence of a ligand. It is in fact a very useful technique prior to X-ray crystallography and/or NMR for screening mutant proteins, obtained by molecular cloning methods, in relation to their secondary structure characteristics. CD, used in the stop-flow mode is particularly useful for studying protein–ligand interactions in real time. In order to minimize the signal to noise ratio, in CD studies, solution components should, ideally, be UV transparent; this limits the use of certain buffers, salts, and solvents. Unfortunately because of theoretical problems, the technique cannot be easily/reliably used to determine and interpret tertiary structure or changes therein.

Fourier transform infrared spectroscopy (FT-IR) is also very useful for secondary structure determination of proteins, where the information content is very similar to CD techniques. However there is greater flexibility in FT-IR, at least from the viewpoint of the type of samples that can be investigated: tissue slices, cells, solid state (powder; freeze dried) samples, crystals, thin films, and aqueous proteins and protein–ligand samples. FT-IR can be used to investigate the structure of proteins in the presence and absence of ligands, solubilized in D_2O by using the protein amide 1 band for analysis, provided that the absorption bands of the ligand do not interfere. Use of H_2O is a problem. Other pros and cons of this technique are similar to those for CD and Raman spectroscopy. It is, for example, useful for examining the effect of variations in solution conditions on proteins and protein–ligand interactions. In addition the presence and absence of hydrogen bonds can also be investigated. FT-IR is limited to the use of short path length cells and research grade instruments are not commonly available.

Raman spectroscopy has been used for many years to study structural and enzymatic proteins as well as protein–ligand (particularly enzyme–substrate/inhibitor) interactions, albeit in laboratories specializing in the technique. Again it is an excellent method for the study of variable solution state conditions. The use of resonance Raman is advantageous, since specificity and sensitivity are improved relative to off-resonance Raman. Information about electronic states can also be gained from resonance Raman spectra. Fluorescence effects *may* pose a problem but the FT-Raman technique can very often eliminate these. Its use is not limited by cell path lengths and as in FT-IR different sample states, e.g. cellular tissue can be examined.

4 Non-spectral methods

Equilibrium dialysis and pH titration represents the traditional approach—which dates back to the classical Scatchard analysis—to the study of ligand binding. Another titration probe, *ITC (isothermal titration calorimetry)* together with the related *DSC (differential scanning calorimetry)* technique are now widely and increasingly, used to examine proteins and protein–ligand interactions. The two techniques are complementary and are the only methods currently available which allow the direct determination of the enthalpy of an intra- or intermolecular reaction. Other thermodynamic parameters can also be determined directly, indirectly, or by using model-based assumptions. Both techniques provide invaluable information in relation to the energetics of a system and significant advances have been made, especially over the last two decades, in the theoretical basis of interpretation of experimental results. It has to be emphasized however that they are macroscopic techniques with attendant advantages and disadvantages which, should not be overlooked in terms of experimental design and interpretation of results. A similar complementary pair of techniques are *sedimentation velocity* and *sedimentation equilibrium analytical ultracentrifugation*. The latter, like calorimetry, provides an absolute thermodynamic probe of interactions (in terms of the stoichiometry and association or dissociation constants, K_d, and the related Gibbs free energy $\Delta G = RT.\ln K_d$) and molecular mass (oligomeric structure). The former (sedimentation velocity) is a good tool for the separation and analysis of heterogeneous mixtures of the various reaction components, as well as a highly useful solution conformation probe of the reactants and products. Another useful procedure—also based on a separation principle (but requiring a separation medium as opposed to a centrifugal field) is *affinity chromatography* (frontal and zonal procedures). Interaction stoichiometries can also be obtained by the technique of coupling size exclusion chromatography to multi-angle laser light scattering. This *SEC/MALLS* procedure (*Table 4*) is very much complementary to analytical ultracentrifugation and is considered in the earlier book by Creighton in the Practical Approach series (see Chapter 9 of ref. 6). Analytical ultracentrifugation and SEC are just two examples of hydrodynamic procedures. Others, based on the electrical properties of proteins and ligands, are *electo-optics* (which can provide valuable information on the effect of ligand binding on solution conformation of a protein) based—like fluorescence depolarization—on the measurement of rotational diffusion behaviour, and the rapidly emerging probe (with its great resolving power on very small quantities) of *capillary electrophoresis*. *Solution X-ray* and *neutron scattering*, also provide a very sensitive handle on the effects of ligand binding on the solution conformation of a protein, and the surface probe of *atomic force microscopy* provides the potential for individual complexes to be visualized (although very difficult to use and data interpretation can be problematic).

5 Computational methods/molecular modelling

Contemporary computing facilities and capacity have revolutionized virtually every aspect of scientific investigations related to protein–ligand interactions. This includes experimental protocol design as well as collection and analysis of data. Most people, unfortunately, associate molecular modelling with the now commonplace colourful complex two- and three-dimensional protein or protein–ligand docking images produced via graphic workstations. However this gives a totally erroneous picture of the importance of molecular modelling techniques for analysing and predicting the physical properties of molecules. It is perhaps, impossible to underestimate the usefulness of such techniques both currently and in the future.

Table 5 List of some of the protein servers on the World Wide Web

Server/software	Description	Web address (URL)
PDB	Database of experimentally determined protein and nucleic acid structures compiled at Brookhaven National Laboratory	http://www.pdb.bnl.gov/
SWIS-PROT	Database of protein sequences	http://expasy.hcuge.ch/www/tools.html
PIR	Database of protein sequences	http://www.gdb.org/Dan/proteins/pir.html
BioSCAN	A rapid search and sequence analysis	http://genome.cs.unc.edu/bioscan.html
BLITZ	Provides specific sequences most similar to or containing the most similar regions to a query sequence	http://www.ebi.ac.uk/searches/blitz_imput.html
NNPREDICT	Predicts secondary structure	http://www.cmpharm.ucsf.educ/~nomi/nnpredict.html
PHD	Automatic service for predicting a protein structure	http://embl-heidelberg.de/predictprotein/predictprotein.html
SWISS-MODEL (PoMod)	Experimental protein modelling server at the Glaxo Institute	http://expasy.hcuge.ch/swissmod/SWISS-MODEL.html
RASMOL (PC) and RASMAC (Mac)	Molecular graphics program intended for visualization of proteins and related molecules	http://www.unmass.edu/microbio/rasmol/

6 Information sources

Useful sources of information include the protein structural databases given in *Table 5*. In addition many scientific literature reference databases now exist. These include, but are not limited to, the following:

(a) Mimas (ISI®, Citation Indexes & ISTP® http://wos.mimas.ac.uk/). Access requires institutional affiliation.

(b) Chemical abstracts.

(c) MedLine.

(d) PubMed (www.ncbi.nlm.nih.gov/). A new facility called PubMed Central is due at the time of going to press.

7 Additional remarks

To date we have obtained, what may appear at first sight, an incredible amount of data and theoretical insight into the properties and behaviour underpinning the function of proteins in biological systems. However in reality it could, nay should, be argued that we have merely scratched the surface in our quest to understand and predict the intricate and interwoven complexities of protein–ligand interactions. There is no doubt that a multidisciplinary approach is required which entails intra- and/or inter-laboratory collaborations which in turn necessitate the collaborator or non-expert to have some level of understanding of the techniques/methods in which they themselves are not experts. It is to be hoped that the subject matter included in this volume will help, at least in part, to achieve this aim.

However the discerning reader will note that the foregoing overview has not dealt with any topic in detail but has merely attempted to wet their appetite for questioning current achievements and future exploration. There are an infinite number of unanswered questions compared to answers. For example, non-covalently controlled phenomena are still very poorly understood. It has not yet been possible to design, *from first principles*, even a small molecular weight molecule that binds with a 'designed' affinity and specificity to a binding site in a protein of known macroscopic structure, despite widespread interest in, e.g. rational drug design. Again at the expense of repetition, questions should be asked, above all by novice researchers as to the applicability of 'test-tube' experiments to biological systems. What are the effects for example, of the fact that biological systems do not operate under equilibrium thermodynamic conditions? Further what are the implications of the concentrations at which molecules occur, the packing density (as well as membrane association) of reactions and the activity of water, in cells compared to the manner and conditions under which *in vitro* experiments are conducted? In truth we are still at the first stage of a long quest. Nonetheless the techniques described in this volume are helping us at least make some inroads in the right direction.

References

1. Longhi, S., Czjzeck, M., and Cambillau, C. (1998). *Curr. Opin. Struct. Biol.*, **8**, 730.
2. Williams, D. H. and Westwell, M. S. (1998). *Chem. Soc. Rev.*, **27**, 57.
3. Philo, J. S. (1999). In *Current Protocols in Protein Science* (ed. P. T. Wingfield), pp. 20.1.1.–20.1.13. J. Wiley & Sons, NY.
4. Hensley, P. (1996). *Structure*, **4**, 367.
5. Harding, S. E. (1993). *Biotech. Gen. Eng. Rev.*, **11**, 317.
6. Creighton, T. E. (1997). *Protein structure: a Practical Approach*. Oxford University Press.

Chapter 2
Equilibrium dialysis and rate dialysis

Bent Honoré

Department of Medical Biochemistry, University of Aarhus, Ole Worms Allé, Bldg. 170, DK-8000 Aarhus C, Denmark.

1 Introduction

Studies on the interaction of low molecular mass molecules, i.e. ligands, with high molecular mass protein molecules play fundamental roles in biomedical research. Among important examples to be mentioned are drug binding to plasma proteins, especially albumin (1), vitamins to their transport proteins, as well as binding of various substances, e.g. hormones and transmitter substances to receptor proteins in their target cells. Ideally the questions to be answered from such studies are how many ligand molecules, N, may bind to a given macromolecule and with what affinities, $K_1, K_2, ..., K_N$ under the given set of conditions specified by, e.g. pH, temperature, ionic strength, and buffer composition. One way of approaching the answer is to measure several sets of concentrations of free ligand and protein bound ligand. The classical technique used to measure these parameters is termed *equilibrium dialysis* (2) and is based on the principle of separating the free ligand from the protein bound ligand by allowing the former to dialyse through a semi-permeable membrane until the concentration of ligand in the dialysate, i.e. the protein-free compartment, under ideal conditions, is equal to the concentration of unbound ligand in the retentate, i.e. the protein-containing compartment (*Figure 1A*). Although various other techniques based on different principles have been introduced to measure these parameters, equilibrium dialysis remains a sound and simple technique with the advantages that the interacting molecules are present under physiological conditions, i.e. in solution, and neither the ligand molecule nor the protein molecule are modified in any way. The procedures for setting up experiments and the necessary control experiments are described for ligands which absorb UV or visible light and ligands which can be obtained in radiolabelled form.

Although equilibrium dialysis is sound in its principle it possesses some inherent problems that under certain conditions may cause serious artefacts. These include, among many others, osmotic dilution of retentate and Donnan effects due to the presence of the protein on only one side of the membrane (3).

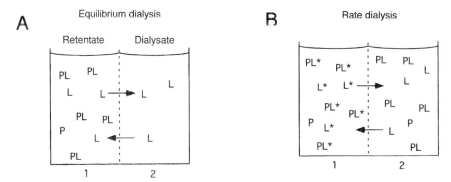

Figure 1 Principles of dialysis techniques. In order to measure the binding strength of a ligand, L, to a protein, P, various concentration values of free ligand and protein bound ligand should be measured. (A) Equilibrium dialysis is performed with the protein present in one compartment (compartment 1 or retentate) separated from another compartment (compartment 2 or dialysate) by a semi-permeable membrane. The membrane allows dialysis of the ligand, L, but not the protein, P. At equilibrium, under ideal conditions, the concentration of ligand in compartment 2 is equal to the concentration of free ligand in compartment 1. The free ligand may be measured by spectroscopy in compartment 2 or the ligand concentration may be measured in either compartment when a radiolabelled substance is used. (B) Rate dialysis is performed with solutions containing identical concentrations of protein and ligand in either compartment. At the start of dialysis the radiolabelled ligand is present in compartment 1 only and the rate of dialysis to compartment 2 is measured. The dialysis rate constant of the free ligand is measured without protein present.

Furthermore, reliable measurement of the free ligand concentration, which is a key element in determining the affinity of a ligand, is a problem with tightly bound ligands. This is partly due to the fact that the free concentration of a high-affinity ligand is very low and thereby very inaccurately determined and partly due to the fact that impurities, whether these are light absorbing, light scattering, or radiochemical may interfere strongly with the determination of the low free ligand concentration. However, if the ligand can be obtained in radioactive form a modified dialysis procedure using the very same equipment as in the case of equilibrium dialysis may be resorted to. The principle of this dialysis technique was developed more than 20 years ago in parallel, but independently, in two different European laboratories, in Aarhus (4-6) and in Nijmegen (7-9). The technique measures the rate of dialysis of a trace amount of a ligand between two compartments containing identical concentrations of ligand and protein (*Figure 1B*). Thereby, this technique very elegantly circumvents the problems of osmotic dilution and Donnan effects and minimizes the problem of measuring very low ligand concentrations—problems that are inherent and may be troublesome in equilibrium dialysis. The modified technique is known under various names; dialysis rate method/determination (4, 5, 7, 10), rate dialysis (11), dialysis exchange method (12), dialysis exchange rate determination (13), and symmetric dialysis (8, 9). Here we will refer to the technique as *rate dialysis* since it is a short name that simultaneously expresses the difference and the similarity to equilibrium dialysis.

Equilibrium dialysis as well as rate dialysis give binding data in the form of sets of concentrations of free ligand and the corresponding average number of ligand molecules bound to the protein. Once the binding data are in hand the next problem to be solved is the interpretation of the results. Such interpretation of the binding mechanism is not straightforward and may require computer analysis. The reader should consult the literature on the topic (14–20).

2 Equilibrium dialysis

2.1 Principle

With equilibrium dialysis two buffered solutions, one containing protein called retentate or compartment 1, and the other without protein called dialysate or compartment 2 are separated by a semi-permeable membrane, i.e. a membrane that allows diffusion of a low molecular mass ligand but not the high molecular mass protein molecule to be tested (*Figure 1A*). At the start of dialysis the ligand may be introduced into either one of the compartments or in both compartments in equal concentrations. In the latter case binding of the ligand to the protein in the retentate, compartment 1, results in a net diffusion of ligand from the dialysate into the retentate. The ligand diffuses until the free ligand concentration in compartment 1, $[L_F]_1$, is equal to the concentration of ligand in compartment 2, $[L]_2$. Depending on the method used the ligand concentration may then be measured in compartment 2, with light absorbing ligands, or in both compartments, with radiochemical ligands and we may calculate the concentration of bound ligand, $[L_B]_1$. When the protein concentration in compartment 1, $[P]_1$, is known the average number of bound ligand molecules per protein molecule, r, may be obtained from the following equation:

$$r = \frac{[L_B]_1}{[P]_1} = \frac{[L]_1 - [L]_2}{[P]_1} \qquad [1]$$

where $[L]_1$ is the sum of the free and bound ligand concentrations in compartment 1. We will now consider measurement of the free ligand concentration by light absorption spectroscopy and by scintillation counting of a radioactively labelled ligand.

2.2 Measurement of free ligand by UV or visible light absorption

If the ligand to be measured contains UV or visible light absorbing structures, e.g. conjugated double bonds, phenol rings, or quinoide groups, the free ligand concentration may then be spectroscopically measured in the dialysate. In order to reliably measure the concentration of the ligand in the experiment it is necessary first to record the absorption spectrum of the ligand in order to select

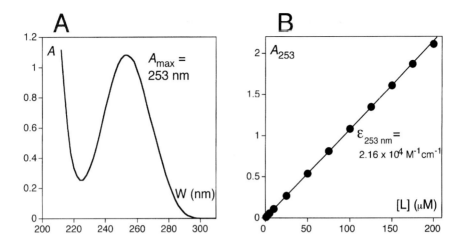

Figure 2 Spectroscopic determination of the concentration of biphenylacetate. (A) Light absorption spectrum of 100 µM biphenylacetate recorded in a quartz cuvette with a path length of 0.5 cm in the sample position and buffer in the reference position. Measurements were performed in a double beam Beckman Acta M V spectrophotometer. The spectrum shows an absorption maximum at 253 nm. (B) shows a standard curve of the absorption at 253 nm versus the concentration of biphenylacetate. The regression line for points obtained with an absorption below 1.5 has a slope of 1.079×10^4 corresponding to an absorption coefficient of 2.16×10^4 $M^{-1}cm^{-1}$. It is characteristic for a high quality spectrophotometer that points with high absorption (around 2) falls only slightly below the regression line. This deviation is much higher in spectrophotometers of lower quality. The experimental data were used with permission from ref. 6.

the wavelength of the absorption maximum, e.g. 253 nm for biphenylacetate (*Figure 2A*) and measure the absorption of the ligand at 253 nm, A_{253}, over a wide concentration range to determine the molar absorption coefficient and at the same time test the quality of the spectrophotometer (*Figure 2B*, see *Protocol 1*). If the substance, as in case of biphenylacetate, obeys the *Beer–Lambert law* we find a strict linear correlation between A_{253}, and the concentration of the ligand (*Figure 2B*). We can then calculate the molar absorption coefficient, ε_{253}, which is 2.16×10^4 $M^{-1}cm^{-1}$ in the present example. Some ligands do not obey the Beer–Lambert law, i.e. the absorption versus concentration relationship deviates from linearity. This may be caused by dimerization or precipitation of the ligand at high concentrations.

If the ligand itself does not significantly absorb light it may occasionally be possible to perform a chemical reaction that produces a light-absorbing substance, e.g. as described for sulfonamides which may be determined spectrophotometrically after diazotization (21).

Protocol 1

Measurement of drug concentration by UV or visible light absorption spectroscopy

Equipment and reagents

- Double beam UV/visible light absorption spectrophotometer
- Two matched quartz cuvettes with identical path lengths, e.g. 0.5 cm or 1 cm
- Buffer: 66 mM sodium phosphate buffer pH 7.4 or another appropriate for the experiments

Method

1. Weigh the ligand accurately and dissolve it in the appropriate buffer in concentrations between, e.g. 1 µM and 200 µM.
2. Zero adjust the spectrophotometer with both sample and reference cuvettes.
3. Replace the sample cuvette with a cuvette containing the drug at an appropriate concentration[a] and record the absorption spectrum over a wide wavelength range, e.g. 200–500 nm, or whatever is appropriate.
4. Determine the maxima of the spectrum and select a maximum at a convenient wavelength, e.g. between 250–300 nm, to be used for concentration measurements.
5. At the selected wavelength, λ, measure the absorbance, A_λ, at various concentrations of ligand and plot A_λ against the concentration of the ligand, [L]. The slope of the curve is determined by linear regression. The correlation coefficient is the molar extinction coefficient, ε_λ. A strict linear correlation should be obtained if the drug obeys the *Beer–Lambert law*, $A_\lambda = \varepsilon_\lambda \times [L] \times l$ where l is the path length of the cuvette.

[a] The appropriate concentration is determined on a trial and error basis.

2.3 Running an equilibrium dialysis experiment

A dialysis experiment can be set up according to *Protocol 2*. Various dialysis chambers may be used with various volumes from microlitres to millilitres. In each case the principle is the same; the two chambers are separated by a semi-permeable membrane. In the present example we use cylindrically formed home made polyacrylamide dialysis chambers suitable for introduction of 1 ml samples in each compartment (*Figure 3*). A chamber consists of two halves each with a compartment constructed by a cylinder formed bore 1.9 cm in diameter and 0.5 cm in depth. The two halves of the chamber are held together by four screws with a trapped dialysis membrane between the compartments. There are various types of membranes on the market. An important issue to consider is the chemical properties of the membrane. Thus, ideally the membrane should not bind the ligand or protein in question. Various types of membranes

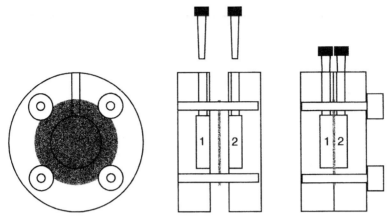

Figure 3 Example of a dialysis chamber. Two chamber halves each with a compartment (1 and 2) constructed by a cylinder formed bore 1.9 cm in diameter and 0.5 cm in depth. The two chamber halves are held together by four screws with a trapped dialysis membrane between the compartments (grey). The compartment volumes are 1.4 ml, convenient for introduction of 1 ml samples. An air bubble is left and serves as a stirrer when the chambers are put on a rotating device in a thermostatted cabinet.

may be tried in order to find one suitable for a given set of experimental conditions (*Protocol 3*). Furthermore, the membrane molecular cut-off should be high enough to allow the ligand molecule to pass through and low enough to retain the protein molecule. Checks should be made to ensure that the protein does not leak through the membrane (*Protocol 3*). The volume in each compartment in the present example is 1.4 ml, convenient for introduction of a 1 ml sample volume. An air bubble is left and stirs the solution when the chambers are put on a rotating plate in an air thermostat set at a constant temperature. Maintaining a constant temperature is important since the strength of many binding reactions depends upon the temperature. In general, the interaction affinity decreases with the increasing temperature.

Protocol 2

Setting up a dialysis experiment with spectroscopic measurement of the ligand[a]

Equipment and reagents

- Double beam UV/visible light absorption spectrophotometer
- Dialysis chambers with 1.4 ml compartments suitable for 1 ml sample volumes[b]
- Dialysis membrane[c]
- Thermostatted rotating equipment for home-made apparatus or water-bath for, e.g. Dianorm apparatus
- Two matched quartz cuvettes with identical path lengths, e.g. 0.5 cm or 1 cm

Protocol 2 continued

- Buffer: 66 mM sodium phosphate buffer pH 7.4 or another appropriate for the experiments
- Known concentration of pure protein dissolved in the appropriate buffer
- Ligand dissolved in the appropriate buffer and diluted to various concentrations

Method

1. Cut the membranes from the dialysis tube to an appropriate size for positioning in the dialysis chambers.[d]
2. Soak the membranes in distilled water with three changes of water over about 1 h.
3. Soak the membranes in the appropriate buffer for the experiment with two or three buffer changes for about 30 min to 1 h.[e]
4. Dry the membranes softly with paper tissue before mounting in the chambers. Avoid too much buffer remaining on the membrane since this will dilute the samples.
5. Introduce 0.5 ml of the ligand solution into all compartments 1 and 2.
6. Introduce 0.5 ml of the protein solution into all compartments 1.
7. Introduce 0.5 ml buffer with no ligand into all compartments 2.[f]
8. Place the chambers on a rotating plate in a thermostatted cabinet and allow dialysis to proceed for the time necessary to achieve equilibrium. This time is determined in separate experiments as described in *Protocol 3*.
9. Measure the absorption in all the solutions from compartments 2 at the selected absorption maximum, A_λ, and calculate the concentration of the ligand from the equation $[L]_2 = A_\lambda/(\varepsilon_\lambda \times l)$ using the molar absorption coefficient, as determined in *Protocol 1*.[g]

[a] Together with the experiment several control experiments should be performed for proper measurements as detailed in Section 2.4 and in *Protocol 3*.

[b] Home-made apparatus of polyacrylamide or various commercial equipment can be used from, e.g. Amika Corp, Dianorm GmbH or Spectrum Medical Industries.

[c] The chemical properties and molecular cut-off depends upon the protein and ligand in question, e.g. cellophane or acetylcellulose with molecular cut-off at 10–20 kDa. Membranes are available from Dianorm GmbH and Spectrum Medical Industries.

[d] 3 cm in diameter in the present case.

[e] If necessary, the membranes may be boiled for 15 min, in a buffer, e.g. sodium bicarbonate in order to extract light absorbing or light scattering impurities before they are put in buffer.

[f] All compartments thus contain 1 ml solutions. Compartments 1 contain the protein in half of the original concentration and all the compartments contain the drug in half of the original concentrations.

[g] It is important to include chambers with no protein and no ligand in order to check for leaching of light absorbing or light scattering impurities from the membrane or protein preparation during dialysis since the presence of these will give an overestimation of the free ligand concentration and thereby an underestimate of the affinity. If the problem cannot be circumvented by boiling the membranes, the ligand may be purified by HPLC before measurements.

At each measured free ligand concentration in compartment 2, $[L]_2$, the bound ligand concentration in compartment 1, $[L_B]_1$, is calculated from:

$$[L_B]_1 = 2 \times ([L_i]_2 - [L]_2) \qquad [2]$$

where $[L_i]_2$ is the initial ligand concentration present in compartment 2 at the start of the dialysis. The measured average number of bound ligand molecules per protein molecule, r, at that concentration of free ligand is then obtained from Equation 1. A binding isotherm for the specific set of ligand and protein is thus characterized by a set of values of free ligand concentrations, $[L]$, and their corresponding r values. From the law of mass action it follows that there is only one value of r for any given free ligand concentration no matter which protein concentration is present. The set of values obtained from the dialysis experiments together represent the experimentally determined binding isotherm for the ligand–protein interaction under the chosen conditions specified by pH, temperature, ionic strength, and buffer composition used.

The results may be plotted in several ways. The average number of bound ligand molecules per protein molecule, r, can be plotted against the free ligand concentration $[L]$ on a linear scale (linear plot, *Figure 4A*) or on a logarithmic scale (logarithmic plot, *Figure 4B*). The latter plot gives a more even distribution of the experimental points especially at low ligand concentration. Alternatively $r/[L]$ can be plotted versus r (Scatchard plot, *Figure 4C*). The various plots each have their advantages and disadvantages. However, all of them may be used to obtain an overview of the quality of the experimental data. The experimental reproducibility generally tends to decrease as the free ligand concentration approaches zero. This is not obvious in the linear plot (*Figure 4A*). The logarithmic plot shows this slightly better but it is seen especially in the Scatchard plot

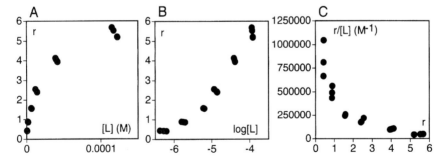

Figure 4 Various plots of experimental data from binding of flurbiprofen to human serum albumin measured by equilibrium dialysis. (A) A linear plot of the average number of bound ligand molecules per protein molecule, r, versus the free concentration of ligand, $[L]$. This plot may erroneously indicate saturation of the protein molecule with ligand at high ligand concentrations. (B) A logarithmic plot. This plot does not erroneously indicate saturation at high ligand concentrations. (C) A Scatchard plot. This plot clearly shows that at low ligand concentrations (close to the y-axis) the experimental accuracy decreases. The intercept with the y-axis gives the first stoichiometric binding constant. However, the Scatchard plot may also tend to erroneously suggest saturation of the protein molecule at high ligand concentrations. The experimental data were used with permission from ref. 6.

(*Figure 4C*), where the measurements tend to scatter close to the y-axis when the ligand concentration approaches zero.

2.4 Running control experiments for equilibrium dialysis

In order to obtain reliable measurements of the binding affinity of a ligand to a purified protein it is necessary to run several control experiments (detailed in *Protocol 3*) and to consider other sources of errors.

(a) The purified protein may contain bound ligands which may interfere with binding of the ligand in question and should be removed before analysis.

(b) Determination of the time taken for establishment of binding equilibrium in the system used.

(c) Light absorbing or light scattering impurities may accumulate during dialysis and interfere with measurement of the ligand.

(d) The ligand may bind to the membrane and/or chamber walls during dialysis.

(e) Protein may leak through the dialysis membrane.

(f) Retentate may be diluted due to the osmotic pressure from the presence of a high protein concentration.

(g) Charged ligands may be unequally distributed across the membrane due to the Donnan effect.

(h) Ligand may be unstable during dialysis.

(i) Bacterial growth.

2.4.1 The protein may initially contain ligands that interfere with measurement of the analysed ligand

The first thing to ascertain is whether the purified protein, initially, contains the ligand that is to be analysed because this will disturb the measurements, as the total concentration of ligand will be unknown. Also, the presence of other types of ligands may interfere with the measurements since these other ligands could compete for binding with the analysed ligand. Albumin purified from serum, contains e.g. fatty acids (1) that may be removed by a defatting procedure. EF-hand Ca^{2+} binding proteins may carry Ca^{2+} as well as other divalent cations which should be removed by passage through a chelating agent like Chelex 100 (22), etc. It is up to the investigator to suggest which types of ligands the protein preparation may contain, from accumulated knowledge about the protein.

2.4.2 Check for establishment of binding equilibrium

A certain time is necessary for the system to establish binding equilibrium. The time needed to reach this may be determined as follows (*Protocol 3A*). A set of chambers are set up with identical concentrations of ligand and protein. The ligand concentration is measured in compartment 2 at various times after the

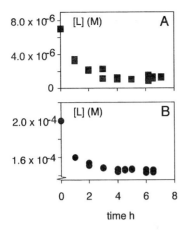

Figure 5 Check for establishment of equilibrium with dialysis of phenylbutazone bound to human serum albumin. Initial concentrations of phenylbutazone in compartment 2 are 7 μM (A, ■) and 200 μM (B, ●), respectively. Equilibrium is established within 5–6 h. The experimental data were used with permission from ref. 6.

start of dialysis. The time chosen for dialysis should be sufficient to attain equilibrium and not significantly longer due, possibly, to difficulties with bacterial growth (see later). *Figure 5* shows the results of an experiment for the binding of phenylbutazone to human serum albumin. The initial concentrations in compartments 1 and 2 were 7 μM (*Figure 5A*) and 200 μM (*Figure 5B*), respectively, with 30 μM albumin at 37 °C pH 7.4, 66 mM phosphate buffer. In this system equilibrium is established within 5–6 h. Equilibrium may be established more quickly by adding the ligand to the protein side, the retentate (23).

2.4.3 Check for extraction of light absorbing or light scattering impurities from the dialysis membrane or protein preparation

During dialysis a certain degree of turbidity may accumulate in the solutions due to extraction of impurities from the dialysis membrane or from the protein preparation. Turbidity of the solution will give a certain light absorption or light scattering that disturbs the measurement of the ligand. Therefore experiments should include a few chambers with only buffer in both compartments and a few chambers with buffer and protein but no ligand in order to check that the light absorption, at the given wavelength in these compartments, is negligible compared to the absorption in the chamber with the lowest ligand concentration analysed (*Protocol* 3B). If the absorption, however, is not negligible try to extract the impurities from the dialysis membrane by boiling it in a buffer. However, the impurities may also come from the protein sample and it may thus be necessary to purify the solution in compartment 2 after dialysis and before measurement of the ligand, e.g. by HPLC. Otherwise the ligand can be measured by scintillation counting if available in radiolabelled form (see Section 2.5).

2.4.4 Check for binding of the ligand to chamber walls and membrane

Test whether the ligand binds to the chamber walls or to the membrane (24) by introducing known concentrations of ligand into the compartments with mounted membrane and dialyse for the same time as in the actual experiment. At the end of dialysis the concentration of ligand measured in the compartments should be the same, within experimental limits, as introduced into the compartments (*Protocol 3C*).

2.4.5 Check for protein leakage through the membrane

It is important to check the dialysis membrane for its ability to retain the protein in compartment 1. Leakage of only a few per cent of the protein may lead to significant disturbances. To test for leakage, known concentrations of protein are introduced into compartment 1 and the concentration in compartments 1 and 2 are measured after dialysis (*Protocol 3D*). Measurements can be done by UV light absorption around 280 nm or in the case of turbid samples by using a dye binding assay (25, 26).

2.4.6 Osmotic dilution of retentate

Osmosis may be a problem if high concentrations of protein are used due to volume changes during dialysis (27–29). This may be monitored by measuring the concentration of protein in the retentate at various times. If osmosis is a problem the protein concentration will decrease during dialysis. Lowering the protein concentration may help to solve the problem. Theoretical expressions that correct for osmotic dilution effects have been published (3).

2.4.7 Donnan effects

In general proteins possess a certain net charge at the given pH at which the experiment is performed. As an example albumin has about 20 negative charges at pH 7.4 (30). This will give an unequal distribution of a charged ligand across the membrane due to the Donnan equilibrium. At an albumin concentration of 30 µM in a 66 mM sodium phosphate buffer the ratio between the concentrations in each of the compartments for a ligand with one negative charge is 1.002. This unequal distribution will increase with increasing protein concentration and decreasing buffer concentration. Thus, the Donnan effect may be minimized by lowering the protein concentration, increasing the buffer concentration or, if possible, by changing the pH to values near the isoelectric point of the protein. Expressions that correct for the Donnan effect have been presented (3). The magnitude of the Donnan effect may be measured by analysing the distribution of a charged ligand that is known not to bind to the protein.

2.4.8 Instability of ligand during dialysis

It is important that the ligand is stable under the conditions of the experiment. Instability may be a problem with ligands that contain several conjugated double bonds as these compounds can be unstable in daylight. Protection from light may solve the problem. Usually structures with phenol rings like those in, e.g. many anti-inflammatory drugs are stable (6). Some ligands may participate in chemical reactions e.g. acetylsalicylic acid (aspirin), which acetylates albumin (31) and thereby is converted to salicylic acid during dialysis.

2.4.9 Bacterial growth

Bacterial growth may be a problem if dialysis is performed for a very long time at 37 °C. However, the system can be checked for the time taken to adjust to equilibrium and the shortest possible time compatible with achievement of equilibrium should be chosen. This time may be 5-6 h where bacterial growth usually presents no major problem. Otherwise the solution can be sterilized by filtration or antibiotics may be added, e.g. gentamicin or kanamycin (32). In the latter case, check that the antibiotic does not interfere with binding of the ligand.

Protocol 3

Control experiments for equilibrium dialysis using spectroscopic measurement of the ligand

Equipment and reagents

- See *Protocol 2*

A Check for time to obtain binding equilibrium

1. Set up several dialysis chambers with the highest and lowest ligand concentrations used in the experiment and stop the dialysis at various time intervals, e.g. at each hour for 6-8 h.
2. Measure the ligand concentration in compartments 2 as described in *Protocol 2*.
3. Plot the ligand concentrations in compartments 2 *versus* time. Equilibrium is established when the concentration does not change further with time. It is important to note whether turbidity interferes with the measured absorbances.

B Check for extraction of light absorbing or light scattering substances from the membrane or protein preparation during dialysis (turbidity)

1. Set up a few chambers with pure buffer in compartments 1 and 2 and a few chambers with protein present in compartments 1 at the same concentration as used in the experiments and buffer in compartments 2.
2. Dialyse for the same time as used for the main experiment (as determined above, *Protocol* 3A).
3. Measure the absorption, A_λ, in compartments 2 and compare them with the measured absorbances obtained in the main experiment.[a]

Protocol 3 continued

C Check for binding of ligand to chamber walls and membrane

1. Dilute the ligand to appropriate concentrations, e.g. the same concentrations as used in the experiments (*Protocol 2*).
2. Set up dialysis experiments as described in *Protocol 2* with 0.5 ml ligand present in all compartments 1 and 2 and 0.5 ml pure buffer introduced into all compartments 1 and 2. No protein is present. Dialysis is performed for the same time as in the experiment (as determined above, *Protocol 3A*).
3. Measure the ligand concentrations in compartments 2 after dialysis. Ideally they should be the same as introduced into the compartments.

D Check for protein leakage through the membrane

1. Set up chambers with a known protein concentration in compartments 1 and pure buffer in compartments 2.
2. Measure the protein concentration in both compartments after dialysis for the time used in the experiments. Measurement of the protein concentration may be performed by UV light absorption spectroscopy near 280 nm; or for turbid samples, by using a protein assay kit based on dye binding as described by, e.g. Lowry (25) or Bradford (26). The protein concentration in compartment 2 after dialysis should be negligible.

[a] The contribution to the absorbance from the turbid solution should be negligible otherwise it is necessary to purify the ligand by HPLC or if possible extract impurities from the membrane before dialysis by boiling the membranes in a buffer.

2.5 Measurement of ligand concentration by using a radiolabelled ligand

Spectroscopic measurement of the ligand concentration can only be used when the ligand does contain light absorbing structures or reacts with a substance that gives a compound that may be spectroscopically determined. If this is not the case another way of measuring the ligand concentration should be used. If the ligand can be obtained radiolabelled the concentration can be measured by scintillation counting. The use of radiolabelled ligands possesses advantages as well as disadvantages. First of all it is very important that the radiochemical purity of the ligand is as high as possible. The concentrations of free and bound ligand may then be measured by using the following relationships:

$$[L_F] = \frac{A_2}{A_1 + A_2} [L_T] \text{ and } [L_T] = [L_i]_1 + [L_i]_2 \qquad [3]$$

$$[L_B] = \frac{A_1 - A_2}{A_1 + A_2} [L_T] \text{ and } [L_T] = [L_i]_1 + [L_i]_2 \qquad [4]$$

where A_1 and A_2 are the measured radioactivities in compartments 1 and 2 and $[L_i]_1$ and $[L_i]_2$ are the initial concentrations introduced into compartments 1 and

2, respectively. In the case of very high affinity ligands it is advisable to check the results at low concentrations of ligand with rate dialysis since this technique is especially suited for analysing high affinity ligands (see Section 3).

Protocol 4

Setting up dialysis experiments with a radiolabelled ligand

Equipment and reagents[a]

- Scintillation counter
- Scintillation liquid
- Radiolabelled ligand[b]
- Chromatographic equipment for purification of radiolabelled ligand[c]

Method

1. Set up dialysis experiments as described in *Protocol 2*, steps 1–8. Add a trace amount of the radiolabelled substance to the ligand solution. Enter the protein and ligand into the chambers as described in *Protocol 2*, steps 5–7. Dialysis is allowed to proceed for the time necessary to achieve equilibrium.

2. Withdraw equal volumes, e.g. 0.8 ml from each of the compartments at the end of dialysis and place in scintillation vials. Make a background sample by using 0.8 ml of buffer.

3. Add 2 ml of scintillation fluid to each sample and to the background.

4. Count each sample until the standard deviation is at the desired level, e.g. below 1%, i.e. with total count figures above 10 000.[d]

5. Subtract the background from each measurement and calculate the free and bound ligand concentrations from Equations 3 and 4.

[a] Dialysis chambers, dialysis membranes, buffer solution, protein solution, ligand solution, and thermostatted equipment are described in *Protocol 2*.

[b] If dissolved in an organic solvent this should be removed before use, e.g. under a stream of nitrogen, and the ligand can then be redissolved in the appropriate buffer.

[c] In order to obtain good results with equilibrium dialysis it is extremely important that the substance is very pure. The necessary purity depends upon the affinity. For high affinity ligands a purity around 98–99% may not be sufficient. The presence of unbound radiochemical impurities may be checked by dialysing the radiolabelled substance in the presence of increasing protein concentrations as described in Section 2.5.1. For purification thin layer chromatographic procedures as well as column chromatographic procedures may apply but this depends on the specific substance used. If purification is difficult or impossible the rate dialysis technique is highly recommended (see Section 3)

[d] Check that the quenching of radioactivity is equal in compartment 1 and 2 by counting equal amounts of radioactivity in the presence and absence of protein. If unequal quenching occurs the count figures should be corrected accordingly.

2.5.1 Control experiments for equilibrium dialysis with a radiolabelled ligand

The control experiments performed with a radiolabelled ligand should be similar to those described for a light absorbing ligand in *Protocol 3*. With equilibrium dialysis using a radiolabelled ligand it is very important to determine the radiochemical purity of the ligand and purify it if necessary, although this may be difficult (33). Even a few per cent of a radiochemical impurity may be detrimental for the analysis of high-affinity ligands. If a given ligand is 99% bound to a protein under a given set of concentrations even 1–2% of radiochemical impurity may surpass the low free ligand concentration unless the impurity binds very strongly to the protein (10). In such cases purification of the ligand is necessary in order to obtain reliable results by equilibrium dialysis. This may be done by chromatographic procedures, the nature of which depends on the specific type of substance that is measured. The amount of unbound radiochemical impurity can be checked by dialysing the radiolabelled ligand from compartment 1 to compartment 2 in the presence of varying but high concentrations of the protein in each of the compartments in order to avoid errors occurring from osmotic dilution and Donnan equilibria. At infinitely high protein concentration the amount of dialysed radiolabelled ligand should be zero. Any dialysis at this high protein concentration occurs from dialysis of an unbound radiolabelled impurity. *Figure 6* sets out an example using dialysis of decanoate with human serum albumin present in each of the chambers. A

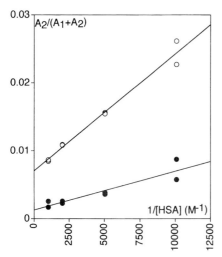

Figure 6 Measurement of unbound radiochemical impurity of radiolabelled decanoate. Dialysis is performed for a given time at various high concentrations of protein (human serum albumin, HSA) present in each compartment. At infinitely high protein concentration a fraction (0.7%) has dialysed from compartment 1 to compartment 2 (○) and since a similar fraction is present in compartment 1 the total amount of impurity is 1.4%. After chromatographic purification the impurity amounts to 0.3% (●). The experimental data were used with permission from refs 10 and 13.

fraction of 0.7% has dialysed into compartment 2 at infinitely high albumin concentration. An equal amount is present in compartment 1 so that the total amount of unbound impurity accounts for 1.4% of the total radioactivity. After purification of the substance by thin layer chromatography the impurity accounts for 0.3% of the total. In such a case it is better to use the rate dialysis technique, Section 3.

3 Rate dialysis

3.1 Principle and theory

Although equilibrium dialysis is a sound technique for measuring binding of ligands the technique may give false results. It fails, especially when high affinity ligands are examined since the free ligand concentration approaches zero with increasing affinity of the ligand. With the use of radiolabelled ligands the presence of radiochemical impurities is a serious problem. Furthermore, the sources of errors occurring from osmosis and Donnan effects may become troublesome at high protein concentrations. In these cases another excellent dialysis procedure may be resorted to: rate dialysis (4–13) also called symmetric dialysis (8, 9). With this technique errors from osmosis and Donnan effects are avoided and errors due to the presence of radiochemical impurities are also minimized (10). Finally, the reproducibility at low ligand concentrations is better since with this technique we do not measure the free ligand concentration directly but a parameter that is proportional to the free ligand concentration, i.e. the dialysis rate of the free ligand. The technique uses the same equipment as used for equilibrium dialysis. The only difference is that in rate dialysis, solutions with identical concentrations of protein and ligand are introduced into both compartments with the radioactive ligand initially present only in compartment 1. Binding equilibrium is thus established in both of the compartments before dialysis has begun. The dialysis of the ligand is allowed to proceed for a given time such that it is insufficient for establishment of equal radioactivity in the compartments. The dialysis is then stopped and the radioactivity present in each of the compartments is measured.

Consider firstly, dialysis of a radioactive ligand in the system without protein present. *Figure 7* shows the dialysis of decanoate in this system. The ligand dialyses strictly according to a first order process. We may thus express the concentration difference between compartment 1 and 2 in the following way:

$$\frac{d([L^*_T]_1 - [L^*_T]_2)}{dt} = -k \times ([L^*_T]_1 - [L^*_T]_2) \qquad [5]$$

where $[L^*_T]_1$ and $[L^*_T]_2$ are the total radioactive ligand concentrations in compartments 1 and 2, respectively. If we presume that, in the presence of a protein, the equilibrium between the free ligand and the protein bound ligand is adjusted much faster than the exchange of ligand between compartments 1 and 2 and that the dialysis rate constant of the ligand, k, is independent of the concentration of the protein, we have the following expressions:

$$\frac{[L]}{[L_T]} = \frac{[L]_1}{[L_T]_1} = \frac{[L^*]_1}{[L^*_T]_1} = R \; \wedge \; \text{for } t > 0 \; \frac{[L^*]_2}{[L^*_T]_2} = R \qquad [6]$$

where $[L^*]_1$ and $[L^*]_2$ are the free radioactive ligand concentrations in compartments 1 and 2, respectively. In the presence of a protein the dialysis of the radioactive ligand is then presumed to follow a first order process dependent upon the concentration difference between the free radioactive ligand concentrations in compartments 1 and 2:

$$\frac{d([L^*_T]_1 - [L^*_T]_2)}{dt} = -k \times ([L^*]_1 - [L^*]_2) \qquad [7]$$

By entering the relations described in Equations 6 we have,

$$\frac{d([L^*_T]_1 - [L^*_T]_2)}{dt} = -k \times R \times ([L^*_T]_1 - [L^*_T]_2) \qquad [8]$$

By integrating Equation 8 in the time interval from start of dialysis $t = 0$ to end of dialysis $t = t$ we obtain:

$$\int_{([L^*_T]_1 - [L^*_T]_2)_{t=0}}^{([L^*_T]_1 - [L^*_T]_2)_{t=t}} \frac{1}{([L^*_T]_1 - [L^*_T]_2)} \, d([L^*_T]_1 - [L^*_T]_2) = \int_{t=0}^{t=t} -k \times R \, dt \qquad [9]$$

Since the radioactive difference between the two compartments at $t = 0$ may be calculated from the sum of radioactivities at the end of dialysis we then have the following relation for the dialysis:

$$\frac{[L]}{[L_T]} = \frac{1}{k \times t} \times \ln \frac{[L^*_T]_1 - [L^*_T]_2}{[L^*_T]_1 + [L^*_T]_2} \qquad [10]$$

When A_1 and A_2 represent the radioactive counts in equal sized samples from compartments 1 and 2, respectively Equation 10 may be written as:

$$\frac{[L]}{[L_T]} = -\frac{1}{k \times t} \times \ln \frac{A_1 - A_2}{A_1 + A_2} \qquad [11]$$

In order to determine the rate constant, k, we have that in the absence of a binding protein $[L]$ is equal to $[L_T]$. Equation 11 then reduces to:

$$\ln \frac{A_1 - A_2}{A_1 + A_2} = -k \times t \qquad [12]$$

By plotting the left side of Equation 12 on the y-axis *versus* time on the x-axis we should obtain a straight line with the slope $-k$. The inset in *Figure 7* shows the measurements obtained with dialysis of decanoate in the dialysis chambers as described in *Figure 3*. In this system decanoate dialyses with a rate constant of 0.035 min^{-1}. Protocol 5 describes how k is measured. Alternatively, some authors standardise their experiment by results from equilibrium dialysis (9).

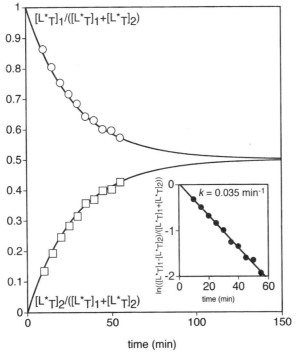

Figure 7 Rate dialysis. With no protein present radiolabelled decanoate is initially introduced into compartment 1 and is allowed to dialyse into compartment 2 with no protein present. The relative concentrations of radioactive ligand present in compartment 1, $[L^*_T]_1/([L^*_T]_1 + [L^*_T]_2)$ (○), and in compartment 2, $[L^*_T]_2/([L^*_T]_1 + [L^*_T]_2)$ (□), are measured at various times. The concentration of radioactive ligand present in each compartment versus time strictly follows exponential functions (solid lines). The rate constant for the process may be obtained by plotting the natural logarithm to the relative concentration difference between the compartments *versus* time (shown in the inset). The rate constant, k, as obtained from the slope of the regression line is 0.035 min^{-1}. The experimental data were used with permission from ref. 10.

Protocol 5

Determination of the dialysis rate constant, *k*, for the ligand

Equipment and reagents

- Dialysis chambers and membranes as in Protocol 2
- Scintillation counter
- Radioactive ligand dissolved in appropriate buffer
- Buffer: 66 mM sodium phosphate buffer pH 7.4 or another appropriate for the experiments
- Scintillation liquid

Protocol 5 continued

Method

1. Incubate solutions with radioactive ligand and buffer at the appropriate temperature.
2. Mount the buffer-soaked and lightly dried membranes in the dialysis chambers.
3. Introduce all solutions with radioactive labelled ligand into compartments 1 and incubate the chambers in the air thermostat at the rotating plate for temperature equilibration.
4. Introduce temperature equilibrated buffer into all compartments 2. The specific time for introduction into each chamber is noted. In practice the chambers may be started with intervals of 1 min. Several chambers are set up.
5. Allow dialysis to proceed for various time intervals. Stop the dialysis by pipetting the solutions from compartments 1 and 2 into scintillation vials.[a]
6. Stop chambers with 5 or 10 min intervals depending on the nature of the ligand and the system used.
7. Add 2 ml scintillation fluid to each vial and count the radioactivity in a scintillation counter until the standard deviation of the count figures is at the desired level, e.g. below 1%, i.e. total count figures above 10 000.
8. Subtract the background radioactivity from each count figure. A_1 represents the measurements from compartments 1 while A_2 represents the measurements from compartments 2.
9. Plot $\ln((A_1 - A_2)/(A_1 + A_2))$ versus time. A straight line should be obtained. Determine the slope of the line, $-k$, by linear regression. The dialysis rate constant is then obtained by multiplying by -1.

[a] If 1 ml volumes are present in either compartment it may be convenient to pipette a smaller volume for counting, e.g. 0.8 ml. The same volume should be used from both sides. Use also 0.8 ml of buffer to count as background.

3.2 Running experiments using rate dialysis

For rate dialysis, experiments are set up in virtually the same manner as used for equilibrium dialysis. In one set of experiments similar concentrations of protein may be present in each compartment and the concentration of ligand can vary. Before dialysis a trace amount of radioactivity is added to the solutions intended for compartments 1. Optionally a similar concentration of unlabelled substance may be added to the solutions intended for compartments 2. The time of dialysis is an important parameter. It may be necessary to run pilot experiments in order to determine which time should be chosen. Several factors should be taken into consideration. The dialysis time should be sufficiently long so that the time of dialysis can be accurately measured. Also, if a radiochemical impurity is present it is advisable to dialyse a certain fraction of radioactivity before stopping the experiment otherwise the impurity may account for too large a fraction of the dialysed substance (10). It is reasonable to dialyse a fraction of more than about 0.2 of the initial radioactivity present in compartment 1. On the other hand the dialysis should not proceed for such a long time that the radioactive concentrations in each compartments are close to being equal. As a rule of thumb the dialysis should be stopped before a fraction of about 0.3 of the radioactivity initially present in compartment 1 has dialysed to compartment 2. *Protocol 6* outlines the details of the procedure.

Protocol 6
Setting up a ligand binding experiment using rate dialysis

Equipment and reagents
- Dialysis chambers and membranes as in Protocol 2
- Radioactive ligand dissolved in appropriate buffer
- Protein dissolved in buffer at known concentration
- Buffer: 66 mM sodium phosphate buffer pH 7.4 or another appropriate for the experiments
- Ligand dissolved in buffer at various known concentrations

Method
1. Make solutions with varying concentrations of ligand. It may be convenient to have a fixed protein concentration and then vary the ligand concentration. For each experimental point more than 2 ml solutions should be made, e.g. 2.5 ml.
2. Split the solutions from step 1 in two tubes with 1.2 ml in each tube.
3. Add a trace amount of radioactive ligand to all the solutions for compartments 1.
4. Add a similar volume with a similar amount of unlabelled ligand to the solutions for compartments 2.[a]
5. Incubate the solutions at the temperature at which the measurements are undertaken.
6. Mount a set of dialysis chambers with dialysis membranes exactly as described for equilibrium dialysis, *Protocol 2*.
7. Introduce all solutions with radiolabelled ligand into compartments 1 and mount the cells on the rotating plate in the thermostatted cabinet. Allow an equilibration time of about 10 min.
8. Initiate the dialysis in a given chamber by introducing the corresponding solution for compartment 2. The dialysis in various chambers may be started with 1 min intervals.
9. Dialyse a fraction of between about 0.2 and 0.3 of the radioactivity from compartment 1 into compartment 2.[b]
10. Stop the dialysis by pipetting similar volumes, e.g. 0.8 ml from compartment 1 and compartment 2 into scintillation vials. Make a background with 0.8 ml of buffer.

Protocol 6 continued

11. Add 2 ml scintillation solution to each vial and count the radioactivity in a scintillation counter until the standard deviation of the count figures are at the desired level, e.g. below 1%, i.e. total count figures above 10 000.
12. Subtract the background radioactivity from each count figure. A_1 represents the measurements from compartments 1 while A_2 represents the measurements from compartments 2.
13. Calculate the free ligand concentrations in each experiment from Equation 11 where the total ligand concentration, the rate constant, and the time of dialysis are known. Also calculate the bound ligand concentration [L_B] by subtracting the free from the total ligand concentration and calculate the average number of bound ligand molecules per protein molecule, r, by dividing with the total protein concentration.
14. Plot the results in one of the plots that shows the relation between [L] and r^c.

[a] This is especially important if the amount of added radioactive ligand is more than just a trace amount. For each experimental point we thus have two tubes containing identical concentrations of ligand and protein. However, one of the tubes contain the ligand radioactively labelled whereas the other contains the ligand unlabelled.

[b] The time for the dialysis thus vary with the given set of conditions used and can only be estimated from pilot experiments. In general, at the same protein concentration, the dialysis times should be longer for lower ligand/protein ratios.

[c] If points scatter seriously at low ligand concentrations (especially seen in a Scatchard plot) it may be necessary to run experiments at a higher protein concentration.

Rate dialysis inherently possesses significant advantages compared with equilibrium dialysis as previously mentioned. In order to obtain more precise determinations the protein concentration may be increased as seen in *Figure 8*.

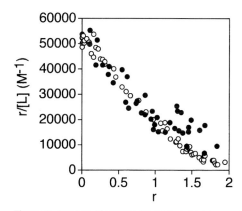

Figure 8 Binding of hexanoate to human serum albumin determined by rate dialysis. The concentration of human serum albumin is 300 μM (○) and 30 μM (●), respectively. The reproducibility increases with the increasing protein concentration. The experimental data were used with permission from ref. 13.

The increase in protein concentration does not affect the other sources of errors otherwise affecting equilibrium dialysis, i.e. osmosis and Donnan equilibria.

3.3 Rate dialysis in microchambers

Rate dialysis may also be performed in microchambers using 50 µl of solution for one experiment, i.e. 25 µl of solution for each compartment. Dialysis equipment suitable for this has been previously described (34). The principle is exactly the same as used for the larger chambers *Figure 3* and *Protocol 6*. Solutions identical with respect to concentrations are introduced into compartments 1 and 2 and only the solution intended for compartment 1 contains the ligand radiolabelled. The procedure used for these chambers is described in *Protocol 7*. The solutions are not stirred but remains stationary and the chamber geometry is different from the larger dialysis chambers. This means that the dialysis in practice proceeds slightly different from that in the chambers described in *Figure 3*. Empirically, the dialysis of the ligand in the microchambers may be described by a modification of Equation 11 (6):

$$\frac{[L]}{[L_T]} = -\frac{1}{k \times (t + t_0)} \times \ln \frac{A_1 - A_2}{A_1 + A_2} \qquad [13]$$

Apparently the dialysis proceeds strictly according to a first order process with a time shift, t_0, that depends upon the procedure of filling, withdrawal, and rinsing of chambers and varies with the substance dialysed (6).

Protocol 7
Setting up rate dialysis using microchambers

Equipment and reagents

- Microdialysis chambers (25 µl volumes on either side of the membrane) and microdialysis apparatus
- Radioactive ligand dissolved in appropriate buffer
- Protein dissolved in buffer at known concentration
- Buffer: 66 mM sodium phosphate buffer pH 7.4 or another appropriate for the experiments
- Ligand dissolved in buffer at various, known, concentrations

Method

1. Mount the dialysis membranes in microdialysis chambers and introduce the appropriate buffer into each of the compartments.
2. Make solutions, essentially as described in *Protocol 6*.[a]
3. Withdraw buffer from each of the compartments before dialysis by pipetting. The residue of buffer is removed with a strip of filter paper.

Protocol 7 continued

4 Introduce the solution intended for compartment 1 into the compartment and allow it to equilibrate with the membrane for 1–1.5 min.

5 Start the dialysis by introducing the solution for compartment 2 into the compartment.

6 Allow dialysis to proceed for a given time, e.g. 10–20 min and stop the dialysis by withdrawing the solution from compartment 1 first. Note the exact time. Rinse the chamber once by introducing and withdrawing buffer. The buffer is combined with the first solution taken out. Then withdraw the solution from compartment 2 and introduce buffer for 1 min into compartment 2 for extraction of the radioactivity in the membrane. Withdraw the buffer from compartment 2 and combine it with the first solution taken out from compartment 2.

7 Add 2 ml of scintillation solution to each of the vials and count the radioactivity by scintillation counting. Also establish a background by counting on equal volume of buffer and scintillation solution.

8 Subtract the background radioactivity from each count figure. A_1 represents the measurements from compartments 1 while A_2 represents the measurements from compartments 2.

9 Calculate the free ligand concentration from Equation 13 by using the values of k and t_0 as determined below.

10 Measure the rate constant for dialysis of the ligand without protein present by dialysing the radioactive ligand at various times with no protein present. Plot $\ln((A_1 - A_2)/(A_1 + A_2))$ versus time. A straight line should be obtained. Determine the slope of the line, $-k$, by linear regression. The dialysis rate constant is then obtained by multiplying by -1. From the regression line t_0 may be determined from the intercept with the x-axis. This intercept is around -1 min or -2 min. The numerical value used is t_0.

^a One determination requires 50 µl, i.e. 25 µl for each compartment. Make well beyond 25 µl per compartment for one determination, e.g. 100 µl.

In practice this microchamber method has proven its value. *Figure 9* shows results for dialysis of salicylate by rate dialysis in microchambers (*Protocol 7*) compared with results from equilibrium dialysis performed in the macrochambers (*Protocol 2*). As can be seen there is very good agreement between the techniques under conditions where both of the techniques work.

4 Rate dialysis for measurement of free ligand concentrations in complex mixtures, e.g. serum

In many cases the biologically active form of a given drug (35) or a hormone (36) is the non-protein bound concentration rather than the total concentration. For

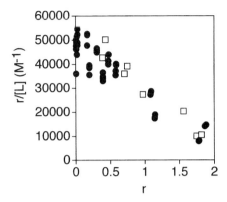

Figure 9 Comparison of salicylate binding to human serum albumin. Binding experiments were performed with rate dialysis in microchambers using protein concentrations between 100–1000 µM (●) and with equilibrium dialysis in macrochambers using 30 µM protein (□). The experimental data were used with permission from ref. 6.

monitoring drug therapy or for analysis of hormone action it is therefore reasonable to measure the free concentration rather than the total concentration in serum. Such measurements are usually performed by equilibrium dialysis where the serum sample is dialysed against a buffer. However, as previously stated equilibrium dialysis inherently possesses some major methodological problems.

(a) The high concentration of proteins in serum gives an osmotic pressure that may lead to dilution of the serum sample.

(b) The high protein concentration is the basis for an unequal distribution of charged ligands due to the Donnan equilibrium.

(c) Ligands that may compete with or even increase the protein binding of the tested ligand are dialysed away from the serum sample and could thus change the binding of the ligand in the serum sample.

(d) Low free ligand concentrations equate with high affinity ligands. Low free ligand concentrations are usually difficult to measure due to a bad signal/noise ratio.

Of these points (a) has been dealt with thoroughly (27–29, 37–39) and expressions that correct for the errors in (a) as well as in (b) have been presented (3). However, the problems mentioned in (c) and (d) are impossible or difficult to deal with. With the availability of rate dialysis all of the problems listed are circumvented or minimized since the same concentration of protein is present in both compartments, i.e. both compartments contain the same osmotic pressure with no osmotic dilution of the serum sample. Likewise the effect of the Donnan equilibrium is omitted so that a charged ligand will be equally distributed between the two compartments. Interacting ligands will remain in the serum and not be dialysed away. Finally, measurements of high affinity ligands can be performed more conveniently since the dialysis time within a certain frame is flexible and can be prolonged.

For measurements in serum a macro- as well as a micromethod may be applied. The principle is the same as used for measurements in pure buffer as detailed in *Protocol 6* for macrochambers and in *Protocol 7* for microchambers. The radioactive ligand may be added to the serum sample without dilution by entering the radioactive ligand, which usually is dissolved in an organic solvent like ethanol, in a test-tube and then evaporate the solvent. Serum is added to the test-tube and equilibrated for a few hours and then used for dialysis. The dialysis times depend upon the binding strength. A fraction of about 0.2–0.3 of the amount of radioactivity initially present in compartment 1 should be dialysed for the reasons given previously (see Section 3.2). The rate constant of the dialysed substance should also be measured by dialysing the substance various times in pure buffer without protein present. It is then possible to determine the fraction of free ligand from Equation 11 (macromethod) or Equation 13 (micromethod). Alternatively, the measurements may be standardized with analysis using equilibrium dialysis (9) or by measuring standard solutions with known concentrations of free and bound ligand (5, 11). If measured, the total ligand concentration may be multiplied with the fraction of free ligand and thereby give the concentration of free ligand. *Protocol 8* describes the procedure for the macrochambers and microchambers.

Protocol 8

Measurement of free ligand concentrations in serum using rate dialysis

Equipment and reagents

- Dialysis chambers as in *Protocol 2* or microchambers and apparatus as in *Protocol 7*
- Membranes as in *Protocol 2* and *Protocol 7*
- Radioactive ligand in organic solvent
- Buffer: 66 mM sodium phosphate buffer pH 7.4
- Scintillation counter
- Scintillation liquid
- Serum sample

A Preparation of serum sample

1 Place aliquots of the radioactive ligand into small test-tubes and evaporate the organic solvent over a stream of nitrogen.

2 Introduce the serum sample into the tube and incubate for a few hours at 37 °C under gentle shaking. For the macromethod use 1.2 ml per analysis, for the micromethod use well beyond 25 µl, e.g. 100 µl per analysis. A similar volume of serum sample without radioactive ligand is incubated as well at 37 °C.

B Macromethod

1. Mount the buffer-soaked and lightly dried membranes in the dialysis chambers.

2. Introduce serum samples with radioactive labelled ligand into compartments 1 and incubate the chambers in the air thermostat at the rotating plate for temperature equilibration.

3. Introduce the corresponding temperature equilibrated serum sample into compartment 2. The specific time for introduction is noted. In practice for analysis of several serum samples the chambers may be started with intervals of 1 min.

4. Allow dialysis to proceed for a specific time interval. Preferably a fraction of between 0.2–0.3 should be dialysed.

5. Stop the dialysis by pipetting serum from compartments 1 and 2 into scintillation vials. If 1 ml volumes are present in either compartment it may be convenient to pipette a smaller volume for counting, e.g. 0.8 ml. The same volume should be used from both sides. Make a sample of 0.8 ml serum for background counting.

6. Add 2 ml scintillation fluid to each vial and count the radioactivity in a scintillation counter until the standard deviation of the count figures is at the desired level, e.g. below 1%, i.e. total count figures above 10 000.

7. Subtract the background radioactivity from each count figure. A_1 represents the measurements from compartments 1 while A_2 represents the measurements from compartments 2.

8. Calculate the fraction of free ligand, $[L]/[L_T]$, from Equation 11 using the dialysis times and the rate constant.[a]

9. Measure the total ligand concentration by an appropriate method for the given ligand.

10. Calculate the free ligand concentration by multiplying the fraction of free ligand with the total ligand concentration.

C Micromethod

1. Mount the dialysis membranes in microdialysis chambers as described in *Protocol 7*, steps 1–3.

2. Introduce the radioactive serum intended for compartment 1 into the compartment and allow it to equilibrate with the membrane for 1–1.5 min.

3. Start the dialysis by introducing the corresponding serum sample for compartment 2 into the compartment.

4. Allow dialysis to proceed for a given time, e.g. 10–20 min and stop the dialysis by withdrawing the solution from compartment 1 first. Note the exact time. Rinse the chamber once by introducing and withdrawing buffer. The buffer is combined with the first solution taken out. Then withdraw the solution from compartment 2 and introduce buffer for 1 min into compartment 2 for extraction of the radioactivity in

Protocol 8 continued

the membrane. Withdraw the buffer from compartment 2 and combine it with the first solution taken out from compartment 2. Then add 2 ml of scintillation solution to each of the vials and count the radioactivity by scintillation counting. A background is also counted.

5 Subtract the background radioactivity from each count figure. A_1 represents the measurements from compartments 1 while A_2 represents the measurements from compartments 2.

6 Calculate the fraction of free ligand, $[L]/[L_T]$, from Equation 13 using the dialysis times, the rate constant, and t_0.[b]

7 Measure the total ligand concentration by an appropriate method for the given ligand.

8 Calculate the free ligand concentration by multiplying the fraction of free ligand with the total ligand concentration.

[a] The dialysis rate constant can be measured as described in *Protocol 5* for macrochambers.
[b] The dialysis rate constant and t_0 can be measured as described in *Protocol 7* for microchambers.

By using rate dialysis it is thus possible to measure the fraction of free ligand, $[L]/[L_T]$, or the free concentration of ligand, $[L]$, from *Protocol 8* in serum samples without having to take care of osmotic dilution and Donnan effects.

Acknowledgements

I am indebted to the late Professor Rolf Brodersen for having introduced me into the field of thermodynamics and ligand binding studies using dialysis techniques. I thank Dr. Anders O. Pedersen for reading and commenting on the manuscript. Also, I thank the late Signe Andersen and Nina Jørgensen for technical assistance. The dialysis equipment was constructed at the Department of Medical Biochemistry workshop, University of Aarhus. Research in the author's laboratory is supported by grants from the Danish Medical Research Council.

References

1. Peters, T., Jr. (ed.) (1996). *All about albumin. Biochemistry, genetics, and medical applications.* Academic Press, San Diego.
2. Davis, B. D. (1943). *J. Clin. Invest.*, **22**, 753.
3. Bowers, W. F., Fulton, S., and Thompson, J. (1984). *Clin. Pharmacokinet.*, **9 Suppl 1**, 49.
4. Brodersen, R. (1978). *Acta Pharmacol. Toxicol.*, **42**, 153.
5. Honoré, B., Brodersen, R., and Robertson, A. (1983). *Dev. Pharmacol. Ther.*, **6**, 347.
6. Honoré, B. and Brodersen, R. (1984). *Mol. Pharmacol.*, **25**, 137.
7. Ross, H. A. (1978). *Experientia*, **34**, 538.
8. Swinkels, L. M., Ross, H. A., and Benraad, T. J. (1987). *Clin. Chim. Acta*, **165**, 341.

9. van Hoof, H. J., Swinkels, L. M., Ross, H. A., Sweep, C. G., and Benraad, T. J. (1998). *Anal. Biochem.*, **258**, 176.
10. Honoré, B. (1987). *Anal. Biochem.*, **162**, 80.
11. Willcox, D. L., McColm, S. C., Arthur, P. G., and Yovich, J. L. (1983). *Anal. Biochem.*, **135**, 304.
12. Pedersen, A. O., Hust, B., Andersen, S., Nielsen, F., and Brodersen, R. (1986). *Eur. J. Biochem.*, **154**, 545.
13. Honoré, B. and Brodersen, R. (1988). *Anal. Biochem.*, **171**, 55.
14. Klotz, I. M. and Hunston, D. L. (1971). *Biochemistry*, **10**, 3065.
15. Klotz, I. M. and Hunston, D. L. (1975). *J. Biol. Chem.*, **250**, 3001.
16. Klotz, I. M. and Hunston, D. L. (1979). *Arch. Biochem. Biophys.*, **193**, 314.
17. Klotz, I. M. and Hunston, D. L. (1984). *J. Biol. Chem.*, **259**, 10060.
18. Rodbard, D., Lutz, R. A., Cruciani, R. A., Guardabasso, V., Pesce, G. O., and Munson, P. J. (1986). *NIDA Res. Monogr.*, **70**, 209.
19. Brodersen, R., Nielsen, F., Christiansen, J. C., and Andersen, K. (1987). *Eur. J. Biochem.*, **169**, 487.
20. Brodersen, R., Honoré, B., Pedersen, A. O., and Klotz, I. M. (1988). *Trends Pharmacol. Sci.*, **9**, 252.
21. Bratton, A. C. and Marshall, E. K. (1939). *J. Biol. Chem.*, **128**, 537.
22. Vorum, H., Madsen, P., Rasmussen, H. H., Etzerodt, M., Svendsen, I., Celis, J. E., *et al.* (1996). *Electrophoresis*, **17**, 1787.
23. Øie, S. and Guentert, T. W. (1982). *J. Pharm. Sci.*, **71**, 127.
24. Briggs, C. J., Hubbard, J. W., Savage, C., and Smith, D. (1983). *J. Pharm. Sci.*, **72**, 918.
25. Lowry, O. H., Rosebrough, N. J., Farr, A. L., and Randall, R. J. (1951). *J. Biol. Chem.*, **193**, 265.
26. Bradford, M. M. (1976). *Anal. Biochem.*, **72**, 248.
27. Tozer, T. N., Gambertoglio, J. G., Furst, D. E., Avery, D. S., and Holford, N. H. (1983). *J. Pharm. Sci.*, **72**, 1442.
28. Huang, J. D. (1983). *J. Pharm. Sci.*, **72**, 1368.
29. Lima, J. J., MacKichan, J. J., Libertin, N., and Sabino, J. (1983). *J. Pharmacokinet. Biopharm.*, **11**, 483.
30. Tanford, C. (1950). *J. Am. Chem. Soc.*, **72**, 441.
31. Walker, J. E. (1976). *FEBS Lett.*, **66**, 173.
32. Ilett, K. F., Hughes, I. E., and Jellett, L. B. (1975). *J. Pharm. Pharmacol.*, **27**, 861.
33. Werblin, T., Kim, Y. T., Smith, C., and Siskind, G. W. (1973). *Immunochemistry*, **10**, 3.
34. Brodersen, R., Andersen, S., Jacobsen, C., Sønderskov, O., Ebbesen, F., Cashore, W. J., *et al.* (1982). *Anal. Biochem.*, **121**, 395.
35. Rowland, M. (1980). *Ther. Drug Monit.*, **2**, 29.
36. Mendel, C. M. (1989). *Endocr. Rev.*, **10**, 232.
37. Moll, G. W. J. and Rosenfield, R. L. (1978). *J. Clin. Endocrinol. Metab.*, **46**, 501.
38. Lockwood, G. F. and Wagner, J. G. (1983). *J. Pharm. Pharmacol.*, **35**, 387.
39. Curry, S. H. and Hu, O. Y. (1984). *J. Pharmacokinet. Biopharm.*, **12**, 463.

Chapter 3
Quantitative characterization of ligand binding by chromatography

Donald J. Winzor

Centre for Protein Structure, Function and Engineering, Department of Biochemistry, University of Queensland, Brisbane, Queensland 4072, Australia.

1 Introduction

The power of chromatography as a means of separating molecules on the basis of charge (ion-exchange chromatography), size (exclusion or gel chromatography), and biospecificity (affinity chromatography) is recognized universally. Less well known is the scope of the technique for the quantitative characterization of ligand binding in terms of the stoichiometry and strength of the equilibrium phenomenon. Although biochemists are usually disinclined to embark upon such a venture into physical chemistry, their experience and expertise with the technique as a fractionation procedure render the quantitative characterization of interactions by chromatography a far less daunting prospect than studies by (say) analytical ultracentrifugation. Indeed, only slight modifications of the classical chromatographic techniques are required to accommodate their application to the evaluation of equilibrium constants.

A biologically important feature of ligand binding is the non-covalence of the interaction, the equilibrium nature of which allows the extent of complex formation to vary with prevailing reactant concentrations. From the viewpoint of gaining information about the distribution of reactants between free and complexed states, the ability of chromatography to separate molecules on the basis of size, charge, or biospecificity is a decided asset. However, the obvious advantage of this resolving power is offset to a considerable extent by the fact that a complexed state has no independent existence. Any chromatographic experiment aimed at quantitative characterization of an equilibrium phenomenon must therefore be devised in such a way that the measured concentration of a reactant may be related unequivocally to the composition of an unperturbed equilibrium mixture of reactants and complex(es).

2 General experimental aspects

Because rapidity of equilibrium attainment is a feature of many macromolecular interactions, the chromatographic behaviour cannot reflect directly the composition of the original equilibrium mixture. In that regard the use of gradient techniques is obviously precluded by an inability to specify the environment to which the elution profile relates. Furthermore, even when isocratic elution is used to generate the chromatographic pattern, there is continual re-equilibration in response to the attempted separation of complex(es) from reactants by differential migration. Traditional interpretation of elution profiles thus needs to be replaced by analyses that take into account this complication and thereby render possible the quantitative characterization of rapid equilibria by chromatography (1–3).

2.1 Zonal and frontal affinity chromatographic techniques

Most isocratic chromatographic studies employ zonal analysis, which entails the application of a small zone of solute mixture onto a column, and subsequent elution with more of the buffer with which the column has been pre-equilibrated. As the zone migrates through the chromatographic bed, it undergoes continual dilution because of (i) axial dispersion and (ii) resolution into separate zones (*Figure 1a*). For any system undergoing reversible complex formation both of these consequences are deleterious from the viewpoint of quantitative characterization of the interaction: the dilution gives rise to an ever-decreasing proportion of complex, the concentration of which tends to zero as the resolution of the two reactants is effected. Thus, although the elution profile shown in *Figure 1a* does reflect the gel chromatographic behaviour of a non-interacting mixture, the pattern is not diagnostic inasmuch as a system undergoing reversible complex formation would yield a similar elution profile.

A more rewarding approach entails the switch to frontal chromatography (4, 5), in which a sufficient volume of mixture is applied to the column to ensure the existence in the elution profile of a region where the composition matches that of the applied solution (*Figure 1b*). Thereafter the column is eluted with buffer to generate a second elution profile (*Figure 1c*). The advantage of the frontal technique is its provision of advancing and trailing elution profiles, which are independent and related unequivocally to the chromatographic characteristics of the mixture applied to the column.

2.2 Comparison of advancing and trailing elution profiles: the concept of non-enantiography

Whereas zonal chromatography provides no obvious diagnostic for the existence of a rapidly equilibrating interaction, the frontal procedure affords a ready means of detecting reversible complex formation. Consider, for example, the gel chromatographic behaviour of a system involving 1:1 complex formation

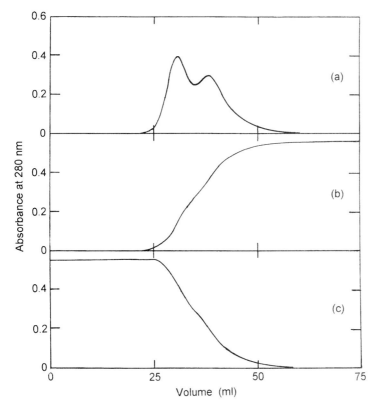

Figure 1 Chromatographic analysis of a non-interacting mixture of ovalbumin and soybean trypsin inhibitor on a column of Sephadex G-75 (2.0 × 23.5 cm) equilibrated with Tris–chloride buffer pH 8.0, I 0.15. (a) Zonal elution profile at 280 nm obtained by applying 1 ml of a mixture comprising 180 μM soybean trypsin inhibitor and 90 μM ovalbumin. (b) Advancing elution profiles obtained in frontal chromatography of a 10-fold dilution of the same mixture. (c) Trailing elution profile generated by the application of buffer to the column equilibrated with mixture.

between two dissimilar proteins, A and S. Because the complex, AS, is the fastest migrating species, there is a tendency for the complex to migrate ahead of the two reactants in the advancing elution profile; but such separation is countered by the consequent dissociation of complex to re-establish equilibrium. The corresponding effect in the leading edge of the trailing elution profile is less profound because the preferential migration of AS leads to its congregation in a region where A, S, and AS had already coexisted. The non-equivalence of circumstances responsible for generation of the advancing and trailing elution profiles gives rise to observable differences in their chromatographic characteristics (*Figure 2*). In this frontal gel chromatographic study of an electrostatic interaction between soybean trypsin inhibitor (A) and cytochrome c (S) on Sephadex G-75 (6), the presence of a maximum in the advancing elution profile for the A constituent is not mirrored in the trailing pattern. Similarly, the

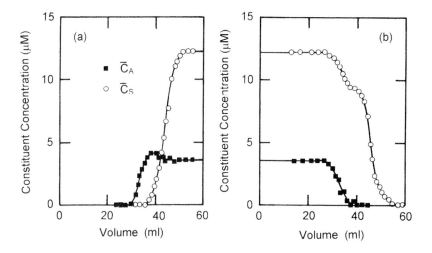

Figure 2 Elution profiles obtained in frontal chromatography of a mixture of soybean trypsin inhibitor ($\bar{C}_A$ = 3.5 µM) and cytochrome c ($\bar{C}_S$ = 12.2 µM) on a column of Sephadex G-75 (1.5 × 32 cm) equilibrated with phosphate buffer pH 6.8, I 0.01. (a) Advancing elution profiles for both constituents; (b) corresponding trailing elution profiles. Data are taken from ref. 6.

decidedly biphasic nature of the trailing elution profile for the cytochrome c constituent has no counterpart in the advancing elution profile. Such non-enantiography (non-identity) of chromatographic behaviour in the advancing and trailing elution profiles of a frontal experiment is a characteristic by which the existence of rapid equilibria may be diagnosed.

2.3 Elution volume: an equilibrium parameter

The elution volume obtained by frontal chromatography is an equilibrium parameter, despite being derived from a mass-migration experiment (7–9). Indeed, the product of elution volume (V_A) and the applied solute concentration in a frontal chromatographic experiment (C_A) defines the amount of solute retained by the column; and is thus governed by the volume of the mobile (liquid) phase and the coefficient describing partition of solute between mobile and stationary phases. Furthermore, because it is merely reflecting the partition coefficient describing solute distribution, the elution volume is independent of flow rate—except in instances where an increase in flow rate changes the proportion of liquid phase as the result of column compaction. Although the boundary spreading exhibits a flow rate dependence under conditions where partition is not effectively instantaneous on the time-scale of solute migration (*Figures 3a* and *b*), identification of the elution volume as the first moment, $V_A = (1/C_A) \Sigma (V \Delta C_A)$ where V is the mean effluent volume corresponding to each increment in solute concentration ΔC_A, of the elution profile (advancing or trailing) allows its

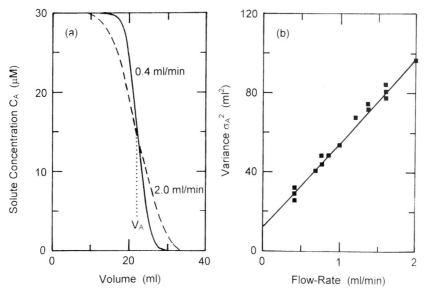

Figure 3 Effect of flow rate on the trailing elution profiles for methyl orange (30 μM) in frontal chromatography on a column of Sephadex G-25 (0.9 × 9.0 cm) equilibrated with 0.05 M phosphate buffer pH 6.8. (a) Elution profiles obtained at the indicated flow rates (ml/min), together with the elution volume (V_A) obtained as the first moment (centroid) of the profile. (b) Flow rate dependence of the boundary spreading, as monitored by the second moment (variance σ_A^2) of the boundary. Data in (b) are taken from ref. 10.

consideration as an equilibrium parameter related to the composition of the applied solution (10, 11). Indeed, the theoretical expressions for the characterization of interactions by quantitative affinity chromatography have been developed on the basis of partition equilibrium considerations; and then converted to column chromatographic format by taking advantage of the fact that the total amount of solute on the column, $V_A C_A$, may also be regarded as the product of the column volume accessible to solute, V_A^*, and a corresponding total concentration, ($\bar{\bar{C}}_A$), that includes solute associated with the stationary phase (12–15). In many respects a frontal chromatographic experiment may be regarded as a simple partition equilibrium procedure in which the distribution of solute between the two phases, $\bar{\bar{C}}/(C_A)$, is being monitored by the corresponding ratio of elution volumes, V_A^*/V_A (14, 15).

2.4 Measurement of binding constants

In studies of the interaction between two solutes in solution the smaller reactant is designated as the ligand, S, and the larger ligand as the acceptor, A. The ligand is considered to be univalent in its interaction with A; but the acceptor can possess several sites for interaction with ligand. Provided that the acceptor sites are equivalent and independent, the difference between the total concentration (in mol.l^{-1} or M) of ligand, $\bar{C}_S = C_S + i\Sigma C_{ASi}$, and the free ligand concentration, C_S, exhibits a rectangular hyperbolic dependence upon free ligand concentration that is governed by the expression

$$(\bar{C}_S - C_S) = nK_{AS}\bar{C}_A/(1 + K_{AS}C_S) \qquad [1]$$

where $\bar{C}_A$ is the total concentration of acceptor in the mixture and K_{AS} is the intrinsic binding constant for the interaction of ligand with n sites on the acceptor (16). Experimentally, the existence of equivalent and independent acceptor sites is often demonstrated by plotting the results in terms of the Scatchard (17) linear transform of *Equation 1*, namely,

$$(\bar{C}_S - C_S)/C_S = nK_{AS}\bar{C}_A - K_{AS}(\bar{C}_S - C_S) \qquad [2]$$

which predicts a linear dependence of $(\bar{C}_S - C_S)/C_S$ upon $(\bar{C}_S - C_S)$ with slope of $-K_{AS}$ and an ordinate intercept of $nK_{AS}\bar{C}_A$: the abscissa intercept thus yields $n\bar{C}_A$, the total concentration of acceptor sites.

The goal of a study aimed at the evaluation of K_{AS} is therefore to determine C_S for mixtures with known composition, $\bar{C}_A$, $\bar{C}_S$. Alternatively, in instances where the magnitude of n is known, measurement of the free concentration of acceptor, C_A, suffices on the grounds that $(n\bar{C}_A - C_A)$, the concentration of acceptor sites occupied by ligand, may be substituted for $(\bar{C}_S - C_S)$ in *Equations 1* and *2*; and that the free ligand concentration is therefore $(\bar{C}_S - n\bar{C}_A + C_A)$.

3 Frontal gel chromatographic studies of ligand binding

A major breakthrough in the development of chromatography as a means of studying biological interactions was the introduction of cross-linked dextran (Sephadex) gels as chromatographic media (16). For these gels and their subsequent counterparts (BioGel, Sepharose, Controlled-Pore-Glass, Fractogel, etc.), chromatographic migration reflects a partition equilibrium that is (i) very rapidly established, (ii) dependent upon molecular size, and (iii) effectively insensitive to solute concentration. These three characteristics have rendered gel (size exclusion) chromatography an extremely versatile technique for studying a whole range of macromolecular interactions. In that regard most of the studies to be described entail conventional gel chromatography; but it should be understood that the experiments can now be done more rapidly and on a much smaller scale by adapting the methodology to HPLC.

3.1 Evaluation of the binding constant for a 1:1 interaction

In gel chromatography of a mixture comprising two reactants (A, S) in association equilibrium with 1:1 complex AS there are three combinations of elution volumes that may be encountered, the most general situation being that in which the elution volumes of all three species differ. For systems in which A and S are both macromolecular, the larger size of the complex renders likely the combination $V_{AS} < V_A < V_S$ (6, 19, 20); but for systems involving small reactants the combination $V_A < V_{AS} < V_S$ is also possible (21, 22). The situation in which complex and faster-migrating reactant co-migrate ($V_{AS} = V_A < V_S$) is frequently encountered when S is a small ligand and A is macromolecular, because the similarity in sizes of the acceptor and acceptor–ligand complex ensures the identity of their elution volumes. Alternatively, the selection of a gel matrix that excludes the larger reactant can ensure the identity of V_A and V_{AS} even under circumstances where the ligand is also macromolecular. The general case is considered first.

Theoretical considerations of mass migration (2, 23, 24) have shown that the trailing elution profile for a situation with $V_{AS} < V_A < V_S$ comprises a reaction boundary, across which the concentrations of both constituents change, and a boundary corresponding to either pure S or pure A: for the situation where V_{AS} is intermediate between V_A and V_S, the pure solute boundary is invariably S. This feature of the trailing elution profile for a system with AS the fastest-migrating species is evident from *Figure 2*, in which the frontal gel chromatographic behaviour of a mixture of soybean trypsin inhibitor (A) and cytochrome *c* was used to illustrate the non-enantiography of advancing and trailing elution profiles that is symptomatic of rapid, reversible complex formation. To facilitate explanation of the means by which such data may be analysed, the trailing elution profile is reproduced in *Figure 4*, which also contains the nomenclature used for quantitative description.

Although frontal gel chromatography has yielded a region (β) corresponding to pure ligand, the concentration of cytochrome *c* in this pure ligand phase (C_S^β)

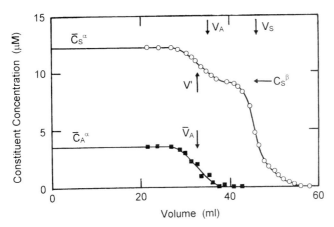

Figure 4 Elaboration of the parameter nomenclature used for the characterization of 1:1 complex formation between reactants by frontal gel chromatography (*Equation 3*, divided throughout by $\bar{C}_A$). The diagram, which is an enlargement of *Figure 2b*, depicts the volume-dependence of the constituent concentrations of soybean trypsin inhibitor ($\bar{C}_A$) and cytochrome *c* ($\bar{C}_S$) in the trailing elution profile obtained for a mixture thereof.

does not match its concentration in the plateau of original composition (C_S^α). Furthermore, C_S^α is not readily obtained. However, considerations of mass conservation allow the equilibrium concentration of pure A in the equilibrium mixture, C_A^α, to be obtained from the expression (25):

$$C_A^\alpha = [\bar{C}_A^\alpha (\bar{V}_A - V_S) - \bar{C}_S^\alpha (V' - V_S) + C_S^\beta (V' - V_S)]/(V_A - V_S) \quad [3]$$

where $\bar{V}_A$ denotes the centroid (first moment) of the gradient in A constituent concentration and V' that of S constituent in the region of the trailing reaction boundary: V_A and V_S are the elution volumes of the individual reactants. In instances where separate recording of the two constituent elution profiles is not possible, V' and $\bar{V}_A$ must both be approximated by the first moment of the observed reaction boundary ($\alpha\beta$) in the elution profile reflecting total solute concentration. Knowledge of C_A^α for a mixture with known constituent composition ($\bar{C}_A^\alpha$, $\bar{C}_S^\alpha$) then allows the equilibrium constant to be calculated from its definition, namely:

$$K_{AS} = C_{AS}^\alpha/(C_A^\alpha C_S^\alpha) = (\bar{C}_A^\alpha - C_A^\alpha)/[C_A^\alpha (\bar{C}_S^\alpha - \bar{C}_A^\alpha + C_A^\alpha)] \quad [4]$$

An alternative procedure in instances where separate elution profiles for the A and S constituents are available (as in *Figure 2*) entails combination of the measured constituent elution volumes, $\bar{V}_A$ and $\bar{V}_S$, with the expressions for their theoretical description,

$$\bar{V}_A = (V_A C_A^\alpha + V_{AS} C_{AS}^\alpha)/\bar{C}_A^\alpha \quad [5a]$$

$$\bar{V}_S = (V_S C_S^\alpha + V_{AS} C_{AS}^\alpha)/\bar{C}_S^\alpha \quad [5b]$$

which enables C_A^α to be evaluated from the relationship (6, 20):

$$C_A^\alpha = [\bar{C}_A^\alpha (\bar{V}_A - V_S) - \bar{C}_S^\alpha (\bar{V}_S - V_S)]/(V_A - V_S) \quad [6]$$

As before, Equation 4 is then used for the evaluation of K_{AS}. We can now describe the practical steps required for determining the binding constant in a 1:1 interaction (*Protocol 1*).

Protocol 1

Determination of the binding constant for 1:1 complex formation between acceptor and macromolecular ligand by frontal gel chromatography

Equipment and reagents

- Gel (size exclusion) chromatography column
- Ligand-free acceptor solution
- Acceptor-free ligand solution
- Acceptor–ligand mixture
- Pre-equilibrating buffer

Protocol 1 continued

Method

1. Equilibrate an appropriate gel (size exclusion) chromatography column with buffer at the flow rate and temperature to be used for the binding study.

2. Load a sufficient volume of ligand-free acceptor solution onto the column to ensure the existence in the elution profile of a plateau region with the applied concentration, C_A^α. Elute with pre-equilibrating buffer.

3. Obtain the elution volume (V_A) as the first moment of the elution profile, which is effectively the effluent volume at which $C_A = C_A^\alpha/2$ if the boundary is reasonably symmetrical.

4. Repeat steps 2 and 3 with an acceptor-free ligand solution to obtain V_S.

5. Apply a sufficient volume of acceptor–ligand mixture with defined composition ($\bar{C}_A^\alpha$, $\bar{C}_S^\alpha$) to ensure the existence of a plateau region with the applied composition in the elution profile. Record the elution profile for each constituent (as in *Figure 2*) if possible. Otherwise, record a single elution profile in terms of total solute concentration (or total absorbance).

6. Generate the trailing elution profile by switching to elution with the buffer used for column pre-equilibration until the concentrations of acceptor and ligand constituents have both decreased to zero.

7. Measure the concentration of ligand in the pure solute phase (C_S^β) and also the elution volumes $\bar{V}_A$ and V' (see *Figure 4*): $\bar{V}_S$ is obtained more readily from the centroid of the ligand boundary in the advancing elution profile (*Figure 2a*).

8. Determine the concentration of free acceptor (C_A^α) via *Equation 3*, using the first moment of the boundary separating the α and β phases as both $\bar{V}_A^\alpha$ and V' if separate elution profiles for the two constituents are unavailable.

9. Compare the value of C_A^α so obtained with that determined by applying *Equation 6* in instances where separate elution profiles for the two constituents are available.

10. Calculate the equilibrium constant (K_{AS}) by means of *Equation 4*.

On the grounds that the application of *Equation 6* to obtain C_A^α requires the estimation of only two parameters ($\bar{V}_A$, $\bar{V}_S$) in addition to the elution volumes of the individual reactants (V_A, V_S), this procedure should provide a more accurate estimate of the free acceptor concentration than that obtained from *Equation 3*: the latter requires additionally a magnitude for C_S^β. However, *Equation 3* has provided the greatest boost to the determination of equilibrium constants by frontal gel chromatography—because of the simplification that results when the complex and faster-migrating reactant co-migrate ($V_{AS} = V_A$). Under those circumstances the identity of $\bar{V}_A$ and V_A also ensures the identity of V' and V_A, whereupon *Equation 3* simplifies to

$$C_A^\alpha = \bar{C}_A^\alpha - \bar{C}_S^\alpha + C_S^\beta \qquad [7]$$

which in turn signifies the identity of C_S^α and C_S^β. In other words, by employing a gel matrix that excludes the larger reactant (A) and hence acceptor–ligand complex from the stationary phase, the equilibrium concentration of ligand in the applied mixture is given directly by C_S^β. Furthermore, whereas the application of *Equations 3–6* is restricted to the analysis of gel chromatographic elution profiles reflecting acceptor–ligand interactions with 1:1 stoichiometry, the direct measurement of C_S^α as C_S^β for systems with co-migration of acceptor and acceptor–ligand complexes is independent of the valence of either the acceptor or the ligand.

3.2 Interaction of a univalent ligand with a multivalent acceptor

Although equilibrium dialysis is the classical method for investigating the interaction of a small ligand with a macromolecular acceptor, it has been shown above that frontal gel chromatography also has the potential for direct measurement of the equilibrium concentration of ligand, C_S^α, provided that acceptor and acceptor–ligand complexes co-migrate (26–28). Results from the trailing elution profile may therefore be analysed in terms of *Equation 1* or its linear transform (*Equation 2*) to obtain the stoichiometry (n) as well as the strength (K_{AS}) of the interaction. Furthermore, the availability of gel (molecular sieve) matrices with different porosities allows considerable latitude in the designation of a ligand as being small. Thus, frontal gel chromatography (*Protocol 2*) on matrices such as Sephadex G-25 and BioGel P-2, which exclude acceptors with molecular weights in excess of 2000, is suitable for studying the interactions of proteins with small metabolites and metal ions. However, selection of a chromatographic matrix with larger pores (e.g. Sepharose 6B) allows the same technique to be used with a protein as ligand for the characterization of a DNA–protein interaction. Consequently, gel chromatography is a more versatile technique than equilibrium dialysis for the characterization of ligand binding: the only requirement is a large difference between the sizes of acceptor and ligand, there being much less restriction on the maximum size of the smaller reactant (S).

Protocol 2

Determination of the stoichiometry and strength of an acceptor–ligand interaction by frontal gel chromatography

Equipment and reagents

- Column
- Gel (molecular sieve) matrix
- Acceptor–ligand mixture
- Elution buffer

Method

1 Select a gel (molecular sieve) matrix that excludes the larger reactant (A) but not the ligand (S) from the stationary phase.

Protocol 2 continued

2. Prepare a column with dimensions (bed volume) appropriate to the means of assaying the column effluent; and equilibrate with buffer at the flow rate and temperature to be used for the binding study.

3. Load a sufficient volume of acceptor–ligand mixture with defined composition ($\bar{C}_A^\alpha$, $\bar{C}_S^\alpha$) to ensure the existence of a plateau region with the applied composition in the elution profile.

4. Generate the trailing profile by switching to elution with buffer until the concentration of ligand has decreased to zero. Monitor this trailing elution profile by a method commensurate with measurement of the concentration of ligand in the pure solute (β) phase.

5. Determine the concentration of ligand in the β-phase, which corresponds to its equilibrium concentration, C_S^α, in the plateau of original composition.

6. Repeat steps 3–5 for a series of mixtures with the same $\bar{C}_A^\alpha$ but a range of total ligand concentrations, $\bar{C}_S^\alpha$.

7. Analyse the results of the series of experiments by non-linear regression analysis in terms of *Equation 1* to obtain n and K_{AS} as the two curve-fitting parameters. Alternatively, employ the Scatchard linear transform of the binding equation (*Equation 2*) to obtain the two parameters from the slope ($-K_{AS}$) and abscissa intercept ($n\bar{C}_A$).

Aspects of *Protocol 2* are illustrated in *Figures 5a* and *b*, which refer to characterization of the stoichiometry and strength of the enzyme–coenzyme interaction between rabbit muscle lactate dehydrogenase and NADH by frontal gel chromatography on Sephadex G-25 equilibrated with imidazole–chloride buffer pH 7.4 (29). From the trailing elution profile in an experiment with 2.5 μM enzyme ($\bar{C}_A^\alpha$) and 10.3 μM coenzyme ($\bar{C}_S^\alpha$), the equilibrium concentration of NADH ($C_S^\alpha = C_S^\beta$) is 5.6 μM (*Figure 5a*). Results of this experiment and a series of others with the same lactate dehydrogenase concentration (2.5 μM) and 3–30 μM NADH are presented as a Scatchard plot in *Figure 5b*, which signifies an intrinsic binding constant of 1.5 ($\pm$ 0.2) $\times$ 10^5 M^{-1} for the interaction of coenzyme with four equivalent and independent sites on the tetrameric lactate dehydrogenase.

3.3 Allowance for multivalence of the ligand

In the event that the ligand is multivalent in its interaction with acceptor sites the concentration of ligand in the pure solute phase of the trailing elution profile (C_S^β) continues to describe its free concentration in the applied mixture. However, *Equations 1* and *2* no longer provide a valid means of analysis, because ligand univalence is implicit in their derivation. For an *f*-valent ligand the general counterpart of the Scatchard equation becomes (30):

$$(\bar{C}_S^{1/f} - C_S^{1/f})/C_S^{1/f} = nK_{AS}\bar{C}_A - fK_{AS}\bar{C}_S^{(f-1)/f}(\bar{C}_S^{1/f} - C_S^{1/f}) \qquad [8]$$

which indicates that a linear dependence of $(\bar{C}_S^{1/f} - C_S^{1/f})/C_S^{1/f}$ upon $\bar{C}_S^{(f-1)/f}(\bar{C}_S^{1/f} - C_S^{1/f})$ is the requirement for equivalence and independence of acceptor sites for a

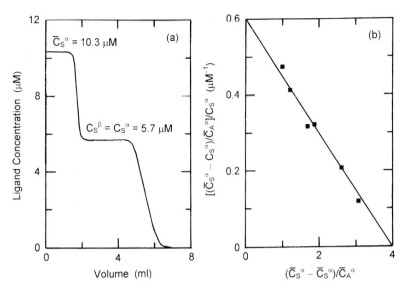

Figure 5 Determination of the stoichiometry (n) and strength (K_{AS}) of the interaction between lactate dehydrogenase (A) and NADH (S) by frontal gel chromatography on Sephadex G-25 (imidazole–chloride buffer pH 7.4, I 0.15). (a) Estimation of the free nucleotide concentration (C_S^α) from the trailing elution profile in an experiment with $\bar{C}_A^\alpha$ = 2.5 µM and $\bar{C}_S^\alpha$ = 10.3 µM. (b) Scatchard plot (*Equation 2*) of results for a series of mixtures with the same enzyme concentration (2.5 µM) and 3–30 µM NADH. Data in (b) are taken from ref. 29.

multivalent ligand. Such graphical evaluation of the binding constant from the slope ($-fK_{AS}$) of the dependence is open to criticism because of its reliance upon transformed experimental parameters; but that objection can now be avoided by the expression of binding in terms of the rectangular hyperbolic relationship (31):

$$f\bar{C}_S^{(f-1)/f}(\bar{C}_S^{1/f} - C_S^{1/f}) = nfK_{AS}\bar{C}_A\bar{C}_S^{(f-1)/f}C_S^{1/f}/[1 + fK_{AS}\bar{C}_S^{(f-1)/f}C_S^{1/f}] \quad [9]$$

Non-linear regression analysis of the dependence of $f\bar{C}_S^{(f-1)/f}(\bar{C}_S^{1/f} - C_S^{1/f})$ upon $f\bar{C}_S^{(f-1)/f}C_S^{1/f}$ thus yields K_{AS} and $n\bar{C}_A$, the total concentration of acceptor sites, as the two curve-fitting parameters.

Because the application of either *Equation 8* or *Equation 9* is clearly conditional upon assignment of a magnitude to f, the ligand valence, there has been a tendency amongst experimenters to opt for analysis in terms of *Equation 1* or *Equation 2*—an action taken to obviate this requirement. However, such action merely signifies that unity has been selected as the most appropriate value of the ligand valence.

The analysis of results for a multivalent ligand is illustrated in *Figures 6a* and *b*, which refer to a study of a mixture of Dextran T-2000 and concanavalin A on glyceryl-CPG 170 (30). Despite the size of the ligand (M_S = 52 000), exclusion chromatography continues to provide a means of evaluating C_S^α (*Figure 6a*). On the grounds that concanavalin A is dimeric under the conditions of the experiment, f has been assigned a value of 2 in the analysis of results by multivalent

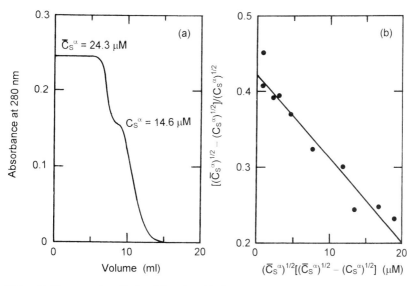

Figure 6 Allowance for effects of ligand multivalence in the characterization of an acceptor–ligand interaction by frontal gel chromatography. (a) Trailing elution profile obtained with a mixture of Dextran T-2000 (1 mg/ml) and concanavalin A ($\bar{C}_S^\alpha$ = 24.3 μM) in phosphate–chloride buffer pH 5.5. (b) Multivalent Scatchard analysis (*Equation 8* with $f = 2$) of results obtained for a series of such mixtures with constant Dextran T-2000 concentration and $\bar{C}_S^\alpha$ = 1–100 μM. Data in (b) are taken from ref. 30.

Scatchard analysis (*Figure 6b*). An equilibrium constant of 5600 M^{-1} is obtained from the slope ($-2K_{AS}$) for the interaction of Dextran T-2000 with concanavalin A (30).

3.4 Evaluation of K_{AS} from the dependence of the constituent elution volume of acceptor upon ligand concentration

Mention has already been made of the use of constituent elution volumes of acceptor and ligand to characterize 1:1 interactions by means of *Equations 4* and *6*. However, situations are frequently encountered in which the only information available from a gel chromatographic experiment is an elution profile for one constituent. It has therefore been necessary to devise a procedure that allows quantitative characterization of an acceptor-ligand interaction from such data.

In gel chromatography on a porous matrix the interaction of ligand with acceptor can often be detected by a change in constituent elution volume of the acceptor. Provided that chemical equilibrium is rapidly attained on the time-scale of acceptor and ligand separation, the constituent elution volume of acceptor ($\bar{V}_A$) is given by *Equation 5a*. If interaction is restricted to a single acceptor site ($n = 1$), *Equation 5a* may be rewritten as:

$$\bar{V}_A = [V_A C_A^\alpha + V_{AS}(\bar{C}_A^\alpha - C_A^\alpha)]/\bar{C}_A^\alpha \qquad [10]$$

on the grounds that C_{AS}^α is simply the difference between total and free concentrations of acceptor. Simple algebraic manipulation of *Equation 10* shows that:

$$C_A^\alpha = \bar{C}_A^\alpha (\bar{V}_A - V_{AS})/(V_A - V_{AS}) \quad [11]$$

The binding constant may thus be determined from the consequent value of C_A^α on noting that:

$$(\bar{C}_A^\alpha - C_A^\alpha)/C_A^\alpha = K_{AS} C_S^\alpha = K_{AS}(\bar{C}_S^\alpha - \bar{C}_A^\alpha + C_A^\alpha) \quad [12a]$$

Alternatively, specific evaluation of C_A^α can be avoided by its elimination from *Equation 12a* on the basis of *Equation 11* to obtain:

$$(V_A - \bar{V}_A)/(\bar{V}_A - V_{AS}) = K_{AS}[\bar{C}_S^\alpha - \bar{C}_A^\alpha (V_A - \bar{V}_A)/(V_A - V_{AS})] \quad [12b]$$

Provided that the accessible range of $\bar{C}_S^\alpha$ permits the evaluation of V_{AS} as the acceptor elution volume in the presence of a saturating concentration of ligand ($1/\bar{C}_S^\alpha \to 0$), the measurement of $\bar{V}_A$ as a function of $\bar{C}_S^\alpha$ affords a means of evaluating the binding constant from *Equation 12*.

Use of this approach to evaluating binding constants is illustrated with a study of the interaction between thiamin diphosphate (A) and magnesium ion (S) by frontal gel chromatography on Sephadex G-10 (22). Formation of the acceptor–ligand complex is accompanied by a decrease in size of the coenzyme and hence by an increase in the constituent elution volume of acceptor ($\bar{V}_A$) with increasing ligand concentration (*Figure 7a*). Analysis of the results from experiments with a range of ligand concentrations in accordance with *Equation 12b* (*Figure 7b*) yields a binding constant of 3000 M^{-1} for the Mg^{2+}-coenzyme interaction under the conditions examined (0.1 M Tris–HCl pH 7.6).

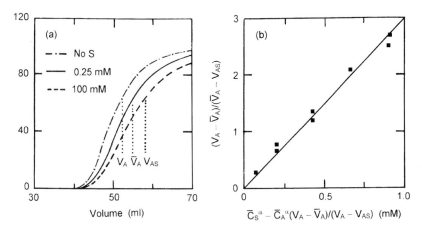

Figure 7 Evaluation of the binding constant for the interaction between thiamin diphosphate (A) and Mg^{2+} (S) in 0.1 M Tris–chloride pH 7.6, from the dependence of acceptor elution volume ($\bar{V}_A$) upon ligand concentration ($\bar{C}_S^\alpha$). (a) Frontal elution profiles for thiamin diphosphate ($\bar{C}_A^\alpha = 100$ μM) on a column of Sephadex G-10 (2.5 × 20 cm) in the presence of the indicated concentrations of MgCl$_2$. (b) Plot of results in accordance with *Equation 12b* to obtain the binding constant, K_{AS}. Data are taken from ref. 22.

4 Zonal gel chromatographic studies of ligand binding

As indicated in Section 2.1, the conventional zonal technique of column chromatography is not as rewarding as its frontal counterpart in providing information for the quantitative characterization of ligand binding. However, there is one zonal gel chromatographic procedure that has found widespread application in studies of interactions between proteins and small ligands.

4.1 The Hummel and Dreyer technique

A disadvantage of frontal gel chromatography for studying ligand binding is the relatively large amount of acceptor–ligand mixture (at least one column volume) that is required to generate the plateau of original composition. A more popular technique for characterizing protein–small ligand interactions has been the Hummel and Dreyer procedure (32), described in *Protocol 3* and which involves the application of a small zone (volume V_a) of acceptor with concentration $(C_A)_a$ to a column of tightly cross-linked gel (Sephadex G-25, BioGel P-2) pre-equilibrated with a known concentration of ligand, $(C_S)_p$. The resultant elution profile exhibits a peak of ligand concentration ($\bar{C}_S$) coincident with elution of acceptor (at $\bar{V}_A = V_A$) as the result of complex formation during acceptor zone migration through a concentration $(C_S)_p$ of free ligand. Trapezoidal integration of this peak area is used to calculate the amount of ligand bound to the amount of acceptor applied to the column. *Equation 1* then assumes the form:

$$[\Sigma\{\bar{C}_S - (C_S)_p\}(\Delta V)]/[V_a(C_A)_a] = nK_{AS}(C_S)_p/[1 + K_{AS}(C_S)_p] \tag{13}$$

which allows n and K_{AS} to be determined from the rectangular hyperbolic dependence of the left-hand side of *Equation 13* upon $(C_S)_p$ that is obtained from a series of experiments with a range of pre-equilibrating ligand concentrations.

Protocol 3

Characterization of ligand binding by the Hummel and Dreyer procedure

Equipment and reagents

- Column
- Gel matrix
- Pre-equilibration buffer

Method

1 Select a gel matrix that excludes the larger reactant (A) but not the ligand (S) from the stationary phase.

2 Prepare a column with dimensions that ensure complete separation of A and S. Equilibrate with buffer that has been supplemented with a known concentration of ligand, $(C_S)_p$.

Protocol 3 continued

3. Apply a small volume (V_a) of acceptor solution with known concentration, $(C_A)_a$, and elute with the ligand-supplemented buffer used for column pre-equilibration.

4. Either record the elution profile at a wavelength suitable for monitoring the total ligand concentration C_S, or divide the eluate into fractions with known volume for subsequent determination of C_S.

5. Repeat steps 2–4 for a range of pre-equilibrating ligand concentrations, $(C_S)_p$.

6. For each elution profile determine the amount of ligand bound as $\Delta V \Sigma [\bar{C}_S - (C_S)_p]$, where ΔV is the volume increment used to obtain the area of the ligand peak by trapezoidal integration.

7. Determine n and K_{AS} by non-linear regression analysis of the results in accordance with *Equation 13*. Alternatively, analyse the results in accordance with the Scatchard linear transformation thereof.

Use of the Hummel and Dreyer procedure to quantify the interaction between a small ligand and a protein is illustrated in *Figures 8a* and *b*, which refer to the binding of calcium ion to the first two modules of low-density lipoprotein receptor (33). In this FPLC adaptation of the Hummel and Dreyer procedure a PC3.2/10 fast-desalting column (Sephadex G-25) was equilibrated at 20 °C and 0.2 ml/min with Tris-chloride buffer pH 7.5 containing 5–365 µM CaCl$_2$ (supplemented with ^{45}CaCl$_2$ tracer). The ligand elution profile obtained by applying a zone (0.1 ml) of receptor (90 µM) to the column pre-equilibrated with 40 µM CaCl$_2$ is shown in *Figure 8a*. Clearly evident are the peak of total ligand concentration at V_A and the consequent negative peak at V_S reflecting depletion of the free ligand concentration as the result of complex formation during acceptor migration down the column. Analysis of the results from the series of experiments according to *Equation 13* is summarized in *Figure 8b*, which signifies an intrinsic binding constant of 2×10^4 M^{-1} for two equivalent and independent binding sites on the first two modules of low-density lipoprotein receptor (33).

4.2 A limitation of the Hummel and Dreyer technique

Implicit in the extraction of binding data from Hummel–Dreyer elution profiles is an assumption that acceptor and acceptor–ligand complex(es) co-migrate (21). Because most applications of the technique have entailed studies of interactions between proteins and small ligands by means of a gel such as Sephadex G-25 (as in *Figure 8*), that requirement has been met by the exclusion of acceptor and acceptor and complexes from the gel phase. However, the validity of this assumption does not extend to the interpretation of Hummel–Dreyer elution profiles obtained in gel chromatographic studies of nucleotide–metal ion interactions on Sephadex G-10 (34)—systems for which $V_A < V_{AS} < V_S$ (21). Numerical simulations of profiles for the ATP–Mg^{2+} system on Sephadex G-10 have shown (21) that the free ligand concentration within the acceptor zone of the elution

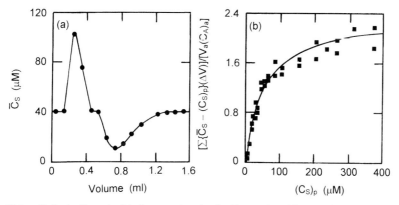

Figure 8 Evaluation of a binding constant by the Hummel and Dreyer (32) procedure. (a) Ligand elution profile obtained by FPLC of a zone (0.1 ml, 90 μM) of acceptor (first two modules of low-density-lipoprotein receptor) on a PC3.2/10 fast-desalting column pre-equilibrated with Tris–chloride buffer pH 7.5 containing 40 μM CaCl$_2$ supplemented with 0.5 nCi/μl ^{45}CaCl$_2$. (b) Determination of the stoichiometry (n) and binding constant (K_{AS}) from the plot of results according to *Equation 10*. Data are taken from ref. 33.

profile is less than $(C_S)_p$ for this combination of elution volumes (*Figure 9a*). The trapezoidal integration used in *Equation 13* thus underestimates the value of $\Sigma (\bar{C}_S - C_S)(\Delta V)$, the amount of complexed ligand. On the other hand, faster migration of the complex ($V_{AS} < V_A < V_S$) would give rise to a local peak in free ligand concentration within the acceptor zone (*Figure 9b*), whereupon the trapezoidal integration would overestimate the amount of ligand bound. Under either circumstance a better estimate of the binding constant for 1:1 complex formation is obtained by substituting the apparent value, $(K_{AS})_{app}$, determined from *Equation 13* into the expression (21):

$$K_{AS} = (K_{AS})_{app}(V_S - V_A)/(V_S - V_{AS}) \qquad [14]$$

However, the procedure for characterizing such systems from the dependence of the constituent elution volume of acceptor ($\bar{V}_A$) upon ligand concentration (*Figure 7*) provides a readier means of evaluating the binding constant for a 1:1 interaction.

5 Evaluation of binding constants by quantitative affinity chromatography

Thus far this description of chromatographic methods for the characterization of acceptor–ligand interactions has centred on the use of gel chromatography for such purposes. Attention is now turned to the characterization of interactions by quantitative affinity chromatography, for which the experimentalist can find a detailed description in two recent articles (15, 35). Despite its extensive coverage therein, the technique affords one of the most versatile methods of evaluating binding constants for acceptor–ligand interactions; and therefore warrants inclusion here, albeit in a condensed form. Inasmuch as the emphasis

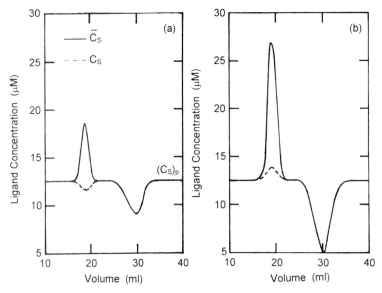

Figure 9 Effect of the non-identity of the elution volumes of acceptor and acceptor–ligand complex on computer-simulated Hummel and Dreyer elution profiles. These simulated experiments for a system with 1:1 interaction between acceptor (A) and ligand (S) that is governed by an association constant (K_{AS}) of 8×10^4 M^{-1} entailed the application of a zone (0.825 ml) of a mixture comprising 75 μM A and 12.5 μM S to a column (0.97 × 45 cm) pre-equilibrated with the same concentration of ligand [$(C_S)_p$ = 12.5 μM]. (a) Simulated profiles for total and free ligand concentration in an experiment where V_A = 17.7 ml, V_{AS} = 20.5 ml, and V_S = 30.0 ml. (b) Corresponding profiles for a system with V_{AS} = 17.7 ml, V_A = 20.5 ml, and V_S = 30.0 ml. Data are taken from ref. 21.

of earlier sections has been confined to the characterization of acceptor–ligand interactions in solution, this summary of quantitative affinity chromatography will also focus on that aspect.

5.1 General theoretical aspects

The initial step in the study of an acceptor–ligand interaction by quantitative affinity chromatography is to select a matrix that interacts specifically with acceptor (A). This is often accomplished by covalently attaching the ligand of interest to a gel such as Sepharose to create a matrix-associated reactant (X) with affinity for A. For example, the interaction of *Staphylococcus* nuclease (A) with 3'-(p-aminophenylphosphate)-5'-phosphate (S) has been studied by using the same ligand immobilized on Sepharose (36). Ligand-facilitated elution of acceptor (A) occurs because its interactions with matrix sites (X) and ligand (S) are mutually exclusive (*Figure 10a*). The competitive effect is illustrated in *Figure 10b*, where benzamidine has been used to displace β-trypsin from a column of gly-gly-D-arg-Sepharose (37). Inasmuch as such displacement of A from affinity sites is a function of the binding constants for both interactions (K_{AS}, K_{AX}), rigorous characterization of an acceptor–ligand interaction requires prior evaluation of the acceptor–matrix equilibrium constant, K_{AX}.

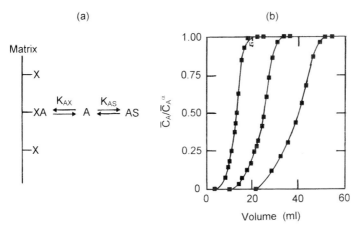

Figure 10 Characterization of acceptor–ligand interactions by quantitative affinity chromatography. (a) Schematic representation of the situation in which the interactions of acceptor (A) with matrix sites (X) and competing ligand (S) are mutually exclusive. (b) Advancing elution profiles showing benzamidine-facilitated elution of β-trypsin from a column of gly–gly–D-arg–Sepharose equilibrated with ligand-supplemented Tris–maleate buffer pH 6.0. Data in (b) are taken from ref. 37.

The evaluation of K_{AX} emanates from the concentration dependence of the constituent elution volume of acceptor, $\bar{V}_A$, in the absence of ligand. As noted in Section 2.3, the amount of acceptor retained by the column during the advancing stage of a frontal chromatographic experiment is $\bar{V}_A \bar{C}_A$, where $\bar{V}_A$ is the constituent elution volume and $\bar{C}_A$ the total concentration of A in the liquid phase (including acceptor–ligand complexes if S is also present). However, this amount of acceptor can also be described as the product of V_A^*, the column volume accessible to A, and an effective total concentration of acceptor, $\bar{\bar{C}}_A$, that includes acceptor bound to affinity sites. There is also an effective total concentration of matrix sites, $\bar{\bar{C}}_X$, the magnitude of which must evolve from the analysis.

Because the acceptor (A) has n sites for interaction with ligand (S), it also exhibits the same multivalence towards X. By analogy with Equation 8, the multivalent Scatchard (17) expression for the acceptor–matrix interaction is:

$$(\bar{\bar{C}}_A^{1/n} - \bar{C}_A^{1/n})/\bar{C}_A^{1/n} = K_{AX}\bar{\bar{C}}_X - nK_{AX}\bar{\bar{C}}_A^{(n-1)/n}(\bar{\bar{C}}_A^{1/n} - \bar{C}_A^{1/n}) \quad [15]$$

where K_{AX} is the intrinsic binding constant. Elimination of $\bar{\bar{C}}_A$, which is not a measurable quantity in column chromatography, by substituting $\bar{V}_A/V_A^*$ for $\bar{\bar{C}}_A/\bar{C}_A$ in Equation 15 then gives:

$$[(\bar{V}_A/V_A^*)^{1/n} - 1] = K_{AX}C_X - nK_{AX}(\bar{V}_A/V_A^*)^{(n-1)/n}\bar{C}_A[(\bar{V}_A/V_A^*)^{1/n} - 1] \quad [16]$$

K_{AX} and $\bar{\bar{C}}_X$ may thus be evaluated from the linear dependence of $[(\bar{V}_A/V_A^*)^{1/n} - 1]$ upon $(\bar{V}_A/V_A^*)^{(n-1)/n}\bar{C}_A[(\bar{V}_A/V_A^*)^{1/n} - 1]$, after assigning a value to n, the acceptor valence.

Situations can be encountered where a conventional Scatchard plot of affinity chromatographic results is linear despite multivalence of the acceptor (12, 38)—a finding rationalized intuitively at the time on the grounds that physical restrictions precluded multiple interaction of acceptor with immobilized affinity sites. Confirmation that the binding constant deduced under such circumstances is the product nK_{AX} has been provided by a subsequent formal explanation of such behaviour (39), which arises when the total concentration of acceptor greatly exceeds that of matrix sites ($n\bar{\bar{C}}_A \gg \bar{C}_X$). Under conditions where the concentration of bound acceptor, ($\bar{\bar{C}}_A - \bar{C}_A$), is small relative to the acceptor concentration in the liquid phase, $\bar{C}_A$, the total acceptor concentration $\bar{\bar{C}}_A$ may be written as:

$$\bar{\bar{C}}_A = \bar{C}_A + (\bar{\bar{C}}_A - \bar{C}_A) = \bar{C}_A(1 + \delta) \qquad [17]$$

where $\delta = (\bar{\bar{C}}_A - \bar{C}_A)/\bar{C}_A$ is much smaller than unity. It then follows from the binomial theorem that:

$$\bar{\bar{C}}_A^{1/n} = \bar{C}_A^{1/n}(1 + \delta/n + \ldots) \qquad [18]$$

whereupon *Equation 15* simplifies to:

$$(\bar{\bar{C}}_A - \bar{C}_A)/(n\bar{C}_A) = K_{AX}\bar{C}_X - K_{AX}(\bar{\bar{C}}_A - \bar{C}_A) \qquad [19]$$

Its chromatographic counterpart (*Equation 16*) thus becomes:

$$[(\bar{V}_A/V_A^*) - 1] = nK_{AX}\bar{C}_X - nK_{AX}\bar{C}_A[(\bar{V}_A/V_A^*) - 1] \qquad [20]$$

which signifies a slope of $-nK_{AX}$ for the dependence of $[(\bar{V}_A/V_A^*) - 1]$ upon $\bar{C}_A[(\bar{V}_A/V_A^*) - 1]$.

Although the presence of competing ligand, S, introduces a second interaction governed by binding constant K_{AS}, results obtained in the presence of a free ligand concentration C_S may also be analysed according to *Equation 16*, the only difference being that the solute-matrix equilibrium constant is a constitutive parameter, $\tilde{K}_{AX}$, defined by the relationship (14, 40):

$$\tilde{K}_{AX} = K_{AX}/(1 + K_{AS}C_S) \qquad [21]$$

Under conditions where the total ligand concentration, $\bar{C}_S$, is the quantity of known magnitude C_S is eliminated from *Equation 21* by means of the expression:

$$C_S = \bar{C}_S - (Q - 1)n\bar{C}_A/Q \qquad [22]$$

where $Q = K_{AX}/\tilde{K}_{AX}$. The required ligand binding constant (K_{AS}) is thus obtained from the slope of the linear dependence of $K_{AX}/\tilde{K}_{AX}$ upon either C_S or $[\bar{C}_S - (Q - 1)n\bar{C}_A/Q]$.

5.2 Evaluation of ligand binding constants by frontal affinity chromatography

The above theoretical section sets the scene for outlining the steps for determining a binding constant by frontal affinity chromatography (*Protocol 4*).

Protocol 4

Determination of binding constants by frontal affinity chromatography

Equipment and reagents

- Affinity column
- Buffer

Method

1. Equilibrate the affinity column with buffer at the flow rate and temperature to be used for the binding study.

2. For each of a series of solutions of A with different acceptor concentrations, load a sufficient volume of solution to ensure that the elution profile contains a plateau region with the applied concentration $\bar{C}_A^\alpha$.

3. Obtain the elution volume ($\bar{V}_A$) for each applied acceptor concentration as the first moment (centroid) of the elution profile (effectively the eluate volume at which $\bar{C}_A = \bar{C}_A^\alpha/2$ when the boundary is reasonably symmetrical.

4. Determine the acceptor-matrix binding constant (K_{AX}) and the effective total concentration of matrix sites ($\bar{\bar{C}}_X$) from a Scatchard plot in accordance with *Equation 16*, which requires prior assignment of a value of n, the number of binding sites on the acceptor.

5. Perform frontal experiments on a range of solutions of A (with different $\bar{C}_A$) that are either in dialysis equilibrium with ligand-supplemented buffer for which the free ligand concentration (C_S) is known, or are supplemented with a known total concentration, $\bar{C}_S$, of ligand. In the latter instance care must be taken to ensure that application of mixture is continued until the plateau region reflects $\bar{C}_S^\alpha$ as well as $\bar{C}_A^\alpha$.

6. Measure the constituent elution volumes ($\bar{V}_A$) as the first moment of the elution profiles for acceptor constituent; and evaluate the constitutive binding constant ($\bar{K}_{AX}$) by multivalent Scatchard analysis (*Equation 16*). Note that the abscissa intercept has the same magnitude as that determined in step 4.

7. Determine the ligand binding constant ($\bar{K}_{AS}$) from the slope of the dependence of $K_{AX}/\bar{K}_{AX}$ upon either C_S (*Equation 21*) or $\bar{C}_S - (Q - 1)n\bar{C}_A/Q$ (*Equation 22*).

Evaluation of the binding constant for an acceptor-ligand system is illustrated (*Figures 11a* and *b*) by characterization of the interaction between NADH and tetrameric rabbit muscle lactate dehydrogenase ($n = 4$) by frontal affinity chromatography of the enzyme on trinitrophenyl-Sepharose (41). Multivalent Scatchard analysis of results in the absence of coenzyme (*Figure 11a*) yields a binding constant (K_{AX}) of 1.5×10^4 M^{-1} for the enzyme-matrix interaction, and a total matrix site concentration ($\bar{\bar{C}}_X$) of 28 μM. Although the slopes of the plots

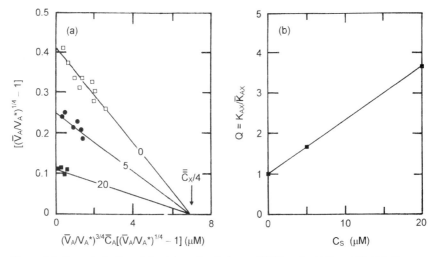

Figure 11 Characterization of the interaction between NADH and rabbit muscle lactate dehydrogenase by frontal affinity chromatography of the enzyme on trinitrophenyl–Sepharose. (a) Multivalent Scatchard plot (*Equation 20* with $n = 4$) of results obtained in the presence of the indicated concentrations (μM) of NADH (C_S). (b) Evaluation of the binding constant for the enzyme–coenzyme interaction (K_{AS}) from the dependence of the resultant $\tilde{K}_{AX}$ values upon C_S (*Equation 21*).

reflecting competitive binding in the presence of 5 and 20 μM NADH (C_S) are not as well defined as that of the plot for enzyme alone, advantage is taken of the fact that the same total concentration of matrix sites (i.e. the same abscissa intercept) applies to the three sets of experiments. The slope of the linear dependence of the resultant $\tilde{K}_{AX}$ values upon C_S (*Figure 11b*) signifies a K_{AS} of 1.3×10^5 M^{-1} for the interaction of NADH with rabbit muscle lactate dehydrogenase under these conditions (41).

5.3 Zonal quantitative affinity chromatography

Zonal affinity chromatography has been used extensively for characterizing acceptor–ligand interactions (36, 42, 43). However, the continual dilution of acceptor during zonal chromatography clearly precludes the assignment of a magnitude to its concentration, $\bar{C}_A$. Under those circumstances the only option is to assume that the $\bar{C}_A$ term contributes negligibly to the right-hand side of *Equation 16*, which thus needs to be truncated as:

$$[(\bar{V}_A/V_A^*)^{1/n} - 1] \cong K_{AX}\bar{\bar{C}}_X \qquad [23]$$

Inasmuch as the left-hand side of Equations 16 and 23 describes the product of the binding constant and the free concentration of matrix sites (C_X), the assumption inherent in zonal quantitative affinity chromatography is that the total concentration of matrix sites is an acceptable approximation of its free counterpart (41). Introduction of *Equation 21* to cover systems involving the chromatography of acceptor in the presence of a concentration C_S of free ligand then leads to the expression:

$$[(\bar{V}_A/V_A^*)^{1/n} - 1] = K_{AX}\bar{\bar{C}}_X/(1 + K_{AS}C_S) \qquad [24]$$

for the evaluation of K_{AS} from a series of experiments in which a zone of acceptor is applied to an affinity column pre-equilibrated with a range of ligand concentrations.

Protocol 5
Evaluation of binding constants by zonal affinity chromatography

Equipment and reagents
- Affinity column
- Buffer

Method
1. Equilibrate the affinity column with buffer containing a known concentration, C_S, of ligand.
2. Subject a small zone of acceptor to chromatography on the column. Measure the elution volume by assuming that $\bar{V}_A$ is given with sufficient accuracy by the effluent volume corresponding to the peak of the eluted zone.
3. Re-equilibrate the affinity column with buffer containing a range of concentrations of ligand. Determine $\bar{V}_A$ for each C_S by step 2.
4. Obtain V_A^* as the limiting value of $\bar{V}_A$ as $C_S \to \infty$ (i.e. as $1/C_S \to 0$).
5. Determine K_{AS} and the product $K_{AX}\bar{\bar{C}}_X$ as the two curve-fitting parameters to emanate from non-linear regression analysis of the $(C_S, \bar{V}_A)$ data set in terms of Equation 24.

The zonal procedure as described in Protocol 5 and which is clearly more economical in regard to acceptor requirements, is rendered simpler than its frontal counterpart by elimination of the need to evaluate a binding constant for the acceptor–matrix interaction (K_{AX}). However, that simplification has been achieved by invoking an assumption that $C_X \cong \bar{\bar{C}}_X$—an approximation that can be rendered valid by employing a highly substituted affinity matrix (41). The fact that highly substituted affinity matrices were initially deemed unsuitable for zonal quantitative affinity chromatography (36, 42) reflected the use of a reciprocal linear transform of Equation 24, namely:

$$1/[(\bar{V}_A/V_A^*)^{1/n} - 1] = 1/(K_{AX}\bar{\bar{C}}_X) + K_{AS}C_S/(K_{AX}\bar{\bar{C}}_X) \qquad [25]$$

to obtain K_{AS} as the ratio of the slope and ordinate intercept of the dependence of $1/[(\bar{V}_A/V_A^*)^{1/n} - 1]$ upon C_S. This transform thus precludes analysis of results for which the ordinate intercept, $1/(K_{AX}\bar{\bar{C}}_X)$, is indistinguishable from zero; and hence of results for which the implicit assumption in Equations 24 and 25 is demonstrably validated. The solution to this dilemma is either to employ the suggested non-linear regression analysis in terms of *Equation 24* or to take advantage of an alternative linear transform:

$$[(\bar{V}_A/V_A^*)^{1/n} - 1] = K_{AX}\bar{\bar{C}}_X - K_{AS}C_S[(\bar{V}_A/V_A^*)^{1/n} - 1] \quad [26]$$

which allows K_{AS} to be determined from the slope of the dependence of $[(\bar{V}_A/V_A^*)^{1/n} - 1]$ upon $C_S[(\bar{V}_A/V_A^*)^{1/n} - 1]$ (14, 41).

Although the zonal technique of quantitative affinity chromatography lacks the rigour of the frontal procedure, it can play a useful role in the characterization of interactions where the competing ligand is also macromolecular. Acceptance of an approximate characterization of the acceptor–ligand interaction allows *Equation 24* (or one of its linear transforms) to be used for the analysis of acceptor elution volumes ($\bar{V}_A$) obtained by zonal gel chromatography of acceptor on an affinity column pre-equilibrated with defined concentrations, C_S, of macromolecular ligand. However, it is important to realize that *Equation 24*, as well as *Equation 17* for that matter, entails an inherent assumption that V_A^* describes not only the accessible volume for acceptor but also for acceptor–ligand complex(es). In situations where S is also macromolecular, this requirement can be met by choosing an affinity matrix that excludes acceptor and hence all acceptor–ligand complexes from the matrix phase. This approach is illustrated by quantitative characterization of the interaction between dimeric concanavalin A and a glycoprotein, ovalbumin, by zonal affinity chromatography (44).

Results of affinity chromatography experiment involving the application of a small zone of lectin solution to a Sephadex G-50 column pre-equilibrated with 1.2–14.0 µM ovalbumin are plotted according to *Equations 25* and *26* in *Figures 12a* and *b* respectively. Although the results clearly exhibit the predicted linear dependencies, the fact that the interaction of concanavalin A with affinity sites is confined to saccharide residues on the Sephadex G-50 bead surface makes it necessary to justify experimentally the use of analytical expressions based on the premise that $n\bar{C}_A \ll \bar{\bar{C}}_X$. An independent estimate of the ordinate intercepts, $1/(K_{AX}\bar{\bar{C}}_X)$ and $K_{AX}\bar{\bar{C}}_X$, was therefore sought by frontal affinity chromatography of the lectin alone on the same Sephadex G-50 column (44). Concordance of the zonal affinity chromatographic data with the rigorously determined estimate of $K_{AX}\bar{\bar{C}}_X$ instils confidence in their analysis according to the relevant transform of *Equation 24*. The consequent intrinsic binding constant of 5.3×10^5 M^{-1} is therefore considered to describe the interaction of ovalbumin with two equivalent and independent sites on dimeric concanavalin A under the conditions of the investigation (44).

5.4 Other forms of quantitative affinity chromatography

Inasmuch as the fundamental quantitative expressions for quantitative affinity chromatography are based on partition equilibrium considerations (14, 35), it can be argued that direct measurement of the acceptor distribution between liquid and matrix phases is more straightforward than its deduction from

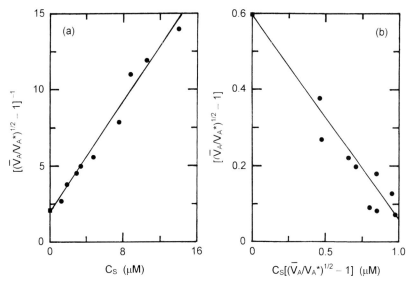

Figure 12 Characterization of the interaction between concanavalin A and ovalbumin by zonal affinity chromatography of the lectin on a column of Sephadex G-50 (0.9 × 12.5 cm) equilibrated with phosphate–chloride buffer pH 5.5 containing 1.2–14 µM glycoprotein. (a) Plot of results according to *Equation 25*. (b) Plot of results according to *Equation 26*. In both cases the point at the ordinate intercept corresponds to the product $K_{AX}\bar{C}_X$ obtained by frontal affinity chromatography of concanavalin A alone on the same column.

elution volumes. This discussion of quantitative affinity chromatography therefore concludes with a brief indication of those alternative procedures.

Application of the quantitative affinity chromatography expressions to simple partition equilibrium measurements is exemplified by characterization of the biospecific adsorption of glycolytic enzymes to muscle myofibrils (45–47). In studies where an acceptor–ligand reaction in solution is the interaction of interest, the need for precise control of a fixed amount of affinity matrix in each reaction mixture can be met by a recycling partition equilibrium procedure (13, 48), which allows all partition measurements to be made with the same slurry of affinity matrix. Its passage through a porous membrane allows the liquid phase of the slurry to be monitored continuously for acceptor content before being returned to the stirred slurry via a peristaltic pump. An obvious advantage of this recycling technique is the facility for conducting stepwise titrations by sequential additions of acceptor (or ligand) solution to supplement a reaction mixture that is already at equilibrium.

An alternative way of quantifying acceptor partition between liquid and adsorbed states is to measure the concentration of matrix-bound acceptor in the equilibrium mixture by biosensor technology (49, 50). Although solid phase immunoassays (RIA, ELISA) also have the potential to provide similar information, the measurement of matrix-bound acceptor entails the prior removal of uncomplexed reactants. Measurement of the amount of acceptor bound to a biosensor surface thus has the advantage of quantifying the acceptor distribution in an unperturbed equilibrium mixture—an important consideration in studies of rapidly attained chemical equilibria. The study of ligand binding by means of the BIAcore instrument (49) bears a strong resemblance to frontal affinity chromatography in the sense that the equilibrium response reflects the concentration of bound acceptor in equilibrium with a liquid phase that has the composition ($\bar{C}_A$, $\bar{C}_S$) of the applied solution—a consequence of the flow of acceptor solution across the affinity matrix attached to the biosensor surface. On the other hand, the cuvette design of the IAsys instrument renders such measurements of bound acceptor more akin to those obtained by recycling partition equilibrium studies. Indeed, the feasibility of performing stepwise titrations by this technology has been demonstrated (50).

These various procedures add considerably to the versatility of quantitative affinity chromatography for the characterization of acceptor–ligand interactions; but are accorded only brief mention here because of their more detailed consideration in a recent volume of this series (35).

6 Concluding remarks

Chromatography has much to offer as a method for quantitative characterization of acceptor–ligand interactions. Of the various gel chromatographic methods described, those entailing the use of matrices conducive to co-migration of acceptor and acceptor–ligand complexes are undoubtedly the simplest to apply. For example, frontal gel chromatography under such circumstances provides a direct means of evaluating the equilibrium concentration of free ligand (C_S) in the applied mixture with known composition ($\bar{C}_A$, $\bar{C}_S$). Furthermore, the availability of gel (exclusion) matrices with markedly different pore sizes to achieve that combination of elution volumes ($V_{ASi} = V_A < V_S$) allows the characterization of many acceptor–ligand interactions in which the ligand is also macromolecular. Quantitative affinity chromatography is also an extremely versatile technique —not only in that regard but also in the sense that the same basic expressions can be applied to results from a variety of partition equilibrium procedures as well as to column chromatographic data. As noted elsewhere (35), a particular asset is its ability to accommodate an extraordinarily wide range of binding affinities (from less than 10^2 to greater than 10^9 M^{-1}. Thus, whereas chromatography is usually considered solely in terms of its undoubted prowess for solute separation and purification, the technique also has powerful potential for the quantitative characterization of macromolecular interactions in terms of binding affinity and stoichiometry.

References

1. Cann, J. R. (1970). *Interacting macromolecules: the theory and practice of their electrophoresis, ultracentrifugation, and chromatography.* Academic Press, New York.
2. Nichol, L. W. and Winzor, D. J. (1972). *Migration of interacting systems.* Clarendon Press, Oxford.

3. Winzor, D. J. (1981). In *Protein-protein interactions* (ed. C. Frieden and L. W. Nichol), p. 129. Wiley, New York.
4. Claesson, S. (1946). *Ark. Kemi Mineral. Geol.*, **24A**, No. 7.
5. Winzor, D. J. and Scheraga, H. A. (1963). *Biochemistry*, **2**, 1263.
6. Cann, J. R. and Winzor, D. J. (1987). *Arch. Biochem. Biophys.*, **256**, 78.
7. Gilbert, G. A. (1966). *Nature*, **212**, 296.
8. Nichol, L. W., Ogston, A. G., and Winzor, D. J. (1967). *J. Phys. Chem.*, **71**, 726.
9. Hogg, P. J. and Winzor, D. J. (1984). *Arch. Biochem. Biophys.*, **234**, 55.
10. Winzor, D. J., Munro, P. D., and Cann, J. R. (1991). *Anal. Biochem.*, **194**, 54.
11. Munro, P. D., Winzor, D. J., and Cann, J. R. (1993). *J. Chromatogr.*, **646**, 3.
12. Nichol, L. W., Ogston, A. G., Winzor, D. J., and Sawyer, W. H. (1974). *Biochem. J.*, **143**, 435.
13. Nichol, L. W., Ward, L. D., and Winzor, D. J. (1981). *Biochemistry*, **20**, 4856.
14. Winzor, D. J. and Jackson, C. M. (1993). In *Handbook of affinity chromatography* (ed. T. Kline), p. 253. Marcel Dekker, New York.
15. Winzor, D. J. (1998). In *Quantitative analysis of biospecific interactions* (ed. P. Lundahl, E. Greijer, and A. Lundqvist), p. 33. Harwood Press, Chur, Switzerland.
16. Klotz, I. M. (1946). *Arch. Biochem.*, **9**, 109.
17. Scatchard, G. (1949). *Ann. N. Y. Acad. Sci.*, **51**, 660.
18. Porath, J. and Flodin, P. (1959). *Nature*, **183**, 1657.
19. Nichol, L. W., Ogston, A. G., and Winzor, D. J. (1967). *Arch. Biochem. Biophys.*, **121**, 517.
20. Gilbert, G. A. and Kellett, G. L. (1971). *J. Biol. Chem.*, **246**, 5079.
21. Cann, J. R., Appu Rao, A. G., and Winzor, D. J. (1989). *Arch. Biochem. Biophys.*, **270**, 173.
22. Booth, C. K., Nixon, P. F., and Winzor, D. J. (1992). *J. Chromatogr.*, **609**, 83.
23. Gilbert, G. A. and Jenkins, R. C. L. (1959). *Proc. R. Soc. London, Ser. A*, **253**, 420.
24. Nichol, L. W. and Ogston, A. G. (1965). *Proc. R. Soc. London, Ser. B*, **163**, 343.
25. Nichol, L. W. and Ogtson, A. G. (1965). *J. Phys. Chem.*, **69**, 1754.
26. Nichol, L. W. and Winzor, D. J. (1964). *J. Phys. Chem.*, **68**, 2455.
27. Cooper, P. F. and Wood, G. C. (1968). *J. Pharm. Pharmacol.*, **20**, 150S.
28. Nichol, L. W., Jackson, W. J. H., and Smith, G. D. (1971). *Arch. Biochem. Biophys.*, **144**, 438.
29. Ward, L. D. and Winzor, D. J. (1983). *Biochem. J.*, **215**, 685.
30. Hogg, P. J. and Winzor, D. J. (1985). *Biochim. Biophys. Acta*, **843**, 159.
31. Harris, S. J., Jackson, C. M., and Winzor, D. J. (1995). *Arch. Biochem. Biophys.*, **316**, 20.
32. Hummel, J. P. and Dreyer, W. J. (1962). *Biochim. Biophys. Acta*, **63**, 530.
33. Bieri, S., Atkins, A. R., Lee, H. T., Winzor, D. J., Smith, R., and Kroon, P. A. (1998). *Biochemistry*, **37**, 10994.
34. Colman, R. F. (1972). *Anal. Biochem.*, **46**, 358.
35. Winzor, D. J. (1997). In *Affinity separations: a practical approach* (ed. P. Matejtschuk), p. 39. IRL Press, Oxford.
36. Dunn, B. M. and Chaiken, I. M. (1974). *Proc. Natl. Acad. Sci. USA*, **71**, 2382.
37. Kasai, K. and Ishii, S. (1978). *J. Biochem. (Tokyo)*, **84**, 1051.
38. Brinkworth, R. I., Masters, C. J., and Winzor, D. J. (1975). *Biochem. J.*, **151**, 631.
39. Kalinin, N. L., Ward, L. D., and Winzor, D. J. (1995). *Anal. Biochem.*, **228**, 238.
40. Winzor, D. J., Munro, P. D., and Jackson, C. M. (1992). *J. Chromatogr.*, **597**, 37.
41. Bergman, D. A. and Winzor, D. J. (1986). *Anal. Biochem.*, **153**, 380.
42. Abercrombie, D. M. and Chaiken, I. M. (1985). In *Affinity chromatography: a practical approach* (P. D. G. Dean, W. S. Johnson, and F. A. Middle), p. 380. IRL Press, Oxford.
43. Swaisgood, H. E. and Chaiken, I. M. (1987). In *Analytical affinity chromatography* (ed. I. M. Chaiken), p. 65. CRC Press, Boca Raton, FL.

44. Hogg, P. J. and Winzor, D. J. (1987). *Anal. Biochem.*, **163**, 331.
45. Kuter, M. R., Masters, C. J., and Winzor, D. J. (1983). *Arch. Biochem. Biophys.*, **225**, 384.
46. Harris, S. J. and Winzor, D. J. (1989). *Arch. Biochem. Biophys.*, **275**, 185.
47. Harris, S. J. and Winzor, D. J. (1989). *Biochim. Biophys. Acta*, **999**, 95.
48. Hogg, P. J., Jackson, C. M., and Winzor, D. J. (1991). *Anal. Biochem.*, **192**, 303.
49. Ward, L. D., Howlett, G. J., Hammacher, A., Weinstock, J., Yasukawa, K., Simpson, R. J., *et al.* (1995). *Biochemistry*, **34**, 2901.
50. Hall, D. R. and Winzor, D. J. (1997). *Anal. Biochem.*, **244**, 152.

Chapter 4
Sedimentation velocity analytical ultracentrifugation

Stephen E. Harding
NCMH Physical Biochemistry Laboratory, University of Nottingham, School of Biosciences, Sutton Bonington, Leicestershire LE12 5RD, UK.

Donald J. Winzor
Centre for Protein Structure, Function and Engineering, Department of Biochemistry, University of Queensland, Brisbane, Queensland 4072, Australia.

1 Introduction: sedimentation velocity and sedimentation equilibrium

There has been a general misconception amongst biochemists that the analytical ultracentrifuge does not provide an absolute means of molecular mass determination and hence of the characterization of interaction phenomena. This has arisen from a lack of awareness about the difference between the two types of experiment that can be performed in the analytical ultracentrifuge.

(a) **Sedimentation velocity**: an experiment performed at sufficiently high speed for the centrifugation of solute away from the centre of rotation to be monitored as the rate of movement of a sedimenting boundary. For a given rotor speed, solvent viscosity and solvent density the rate of migration depends upon the overall size and shape of the macromolecule or macromolecule–ligand complex.

(b) **Sedimentation equilibrium**: an experiment performed at a lower speed so that the sedimentation and back-diffusion forces are of comparable magnitudes and therefore give rise to an equilibrium distribution of solute concentration. Because there is no net transport at equilibrium, shape effects do not come into play and the distribution becomes an absolute function of molecular mass for a single solute. For an interacting system the distribution is an absolute reflection of the mass action relationship between the species participating in the chemical equilibrium reaction (concentrations as well as molecular masses of the participating species).

The criticism that analytical ultracentrifugation does not provide an absolute determination of molecular mass thus only applies to sedimentation velocity, which can nevertheless be used to great effect in the identification and characterization of solute–ligand interactions. In such studies one needs to allow for the effects of shape on the sedimentation coefficients of putative solute–ligand complexes—allowances that become less equivocal for the interaction of a protein with a small ligand ($M < 500$). Indeed, sedimentation velocity studies have been crucial to detection of the conformational changes associated with the allosteric regulation of aspartate transcarbamylase (1) and pyruvate kinase (2). On the grounds that biochemists tend to be more familiar with the technique of sedimentation velocity, this variant of analytical ultracentrifugation for the study of acceptor–ligand systems is addressed in the current chapter: the following chapter considers the application of sedimentation equilibrium for the same purpose. Although treated separately, a combination of the two types of ultracentrifuge measurement can often provide an even greater inroad into the understanding of interactions between macromolecular acceptors and a wide range of ligand types.

2 Basic principles of sedimentation velocity

Before the user embarks on the analysis of interacting systems—which can present a number of difficulties—he/she needs to have a grasp for the basic principles of sedimentation velocity: we start by outlining the original (and still used) procedure for determining the sedimentation coefficient of a non-interacting solute, which for the purposes of illustration is taken to be a homogeneous protein.

2.1 Measurement of a sedimentation coefficient

A solution of protein is placed in a specially designed cell in which the sector-shape of the channels in the centrepiece (*Figure 1*) allows unimpeded migration of protein molecules in a radially outward direction in response to the applied centrifugal field. One sector is filled with protein solution and the other with buffer to provide a reference cell for the absorption and Rayleigh optical systems. At the commencement of a sedimentation velocity experiment the concentration of solute is uniform throughout the cell, but subjection of the solution column to a high centrifugal field (typically 50 000–60 000 r.p.m. for a protein with a molecular mass of 10–100 kDa) leads to progressive removal of solute from the inner region of the cell (*Figure 2*). Migration of the moving boundary of solute is recorded optically, and the sedimentation coefficient, s_A, then determined from its definition (rate of migration per unit field), namely:

$$s_A = (dr_p/dt)/(\omega^2 r) = (d \ln r_p/dt)/\omega^2 \qquad [1]$$

where r_p denotes the radial position of the protein boundary after centrifugation for time t at angular velocity ω, which is expressed in radians per second (1 revolution = 2π radians, and ω = r.p.m. $\times 2\pi/60$). The linear dependence of

Figure 1 Centrepieces commonly used in an ultracentrifuge cell (Photograph courtesy of Beckman Instruments, Palo Alto, USA). (a) Standard 12 mm optical path length double sector. Used in XL-A, XL-I, and Model E ultracentrifuges. A range of centrepiece materials are available: users should check the inertness of the solvent in their solutions. One sector used for solution, the other for reference solvent. (b) Six channel (12 mm). Three solution/solvent pairs. These are generally used only for sedimentation equilbrium measurements because (i) the shorter solution column length requirements; (ii) a speed limitation of ~ 40 000 r.p.m. (c) Single channel (12 mm). For the Model E ultracentrifuge only, and not suitable for sedimentation equilibrium experiments. Users should familiarize themselves with the difference between 'optical path length' and 'solution column length'. Standard cell centrepieces are 12 mm optical path length for the XL centrifuges. Shorter path lengths are available to attenuate optical signals (e.g. lower the UV absorbance). Long (30 mm) path length cells for amplification of optical signals can be used in Model E but not in XL ultracentrifuges. Solution column lengths are typically 10 mm for sedimentation velocity (corresponding to ~ 0.4 ml in a 12 mm path length cell) and ~ 3 mm for sedimentation equilibrium (corresponding to ~ 0.1 ml).

ln r_p upon t thus has a slope of $\omega^2 s_A$. The sedimentation coefficient has units of time, which are usually reported in Svedberg units S (1 S = 10^{-13} sec).

For a spherical solute with molecular mass M_A and a radius a the value of the sedimentation coefficient measured at temperature T in buffer b, $(s_A)_{T,b}$, is related to molecular parameters by the expression:

$$(s_A)_{T,b} = M_A(1 - \bar{v}_A \rho_{T,b})/(N 6\pi \eta_{T,b} a) \quad [2]$$

where N is Avogadro's number and $\bar{v}_A$ is the partial specific volume of the protein (effectively the reciprocal of the solute density): the other two parameters refer to the density ($\rho_{T,b}$) and viscosity ($\eta_{T,b}$) of the buffer medium in which the solute is migrating. For a non-spherical solute the same expression is used except that a now refers to the radius of the equivalent hydrodynamic sphere—the source of the dependence of sedimentation coefficient upon shape of the solute. To take into account the dependence of $(s_A)_{T,b}$ upon solvent parameters, sedimentation coefficients are corrected to values for migration in a solvent with the density and viscosity of water. From *Equation 2* the corrected value, $(s_A)_{20,w}$, is therefore:

$$(s_A)_{20,w} = (s_A)_{T,b}(\eta_{T,b}/\eta_{20,w})[(1 - \bar{v}_A \rho_{20,w})/(1 - \bar{v}_A \rho_{T,b})] \quad [3]$$

where $\eta_{20,w}$ and $\rho_{20,w}$ are the viscosity and density respectively of water at 20 °C.

Because its derivation is based on the premise of unhindered migration, *Equation 2* refers to the sedimentation coefficient of solute in infinitely dilute solution, $(s_A^o)_{20,w}$—a parameter that needs to be obtained from the dependence

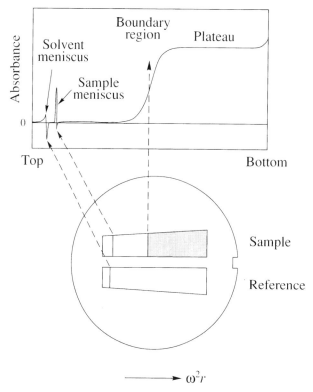

Figure 2 Plan view of a double sector centrifuge cell during a sedimentation velocity experiment and corresponding UV absorption optical record. The sample solution is placed in one sector and a sample of the solvent in the reference sector. The reference sector is usually filled slightly more than the sample sector, so that the reference meniscus does not obscure the sample profile. For simplicity the boundary in the schematic cell is shown as infinitely sharp: because of diffusion effects this will not be the case, as reflected in the scan. From ref. 80, and reproduced courtesy of Beckman Instruments. (NB. For sedimentation equilibrium experiments (Chapter 5), the reference channel should contain solvent that has been in dialysis equilibrium with the sample solution.)

of $(s_A)_{20,w}$ upon the weight-concentration of solute, c_A. For proteins this dependence is of the form:

$$(s_A)_{20,w} = (s_A^o)_{20,w}(1 - k_s c_A) \qquad [4]$$

where the Gralén coefficient k_s is in the vicinity of 0.007 ml/g. For nucleic acids and polysaccharides the concentration dependence is expressed more appropriately in the form:

$$(s_A)_{20,w} = (s_A^o)_{20,w}/(1 + k_s c_A + ...) \qquad [5]$$

2.2 Measurement of molecular mass by sedimentation velocity

Unequivocal determination of M_A from sedimentation velocity experiments requires replacement of the $(6\pi\eta_{T,b}a)$ term in *Equation 2* by an independent measure of the frictional coefficient, $f_A = 6\pi\eta_{T,b}a$: the diffusion coefficient, $(D_A)_{T,b}$, provides such a means. Combination of the description of the diffusion coefficient in molecular terms:

$$(D_A)_{T,b} = RT/(Nf_A) = RT/(6\pi\eta_{T,b}a) \qquad [6]$$

where R is the universal gas constant, with *Equation 2* gives rise to the Svedberg equation,

$$M_A = RT(s_A)_{T,b}/[(D_A)_{T,b}(1 - \bar{v}_A\rho_{T,b})] \qquad [7]$$

The diffusion coefficient can be measured independently in a separate experiment; but advantage is frequently taken of the Lamm equation for centrifugal migration:

$$r(dc_A/dt) = -d[s_A\omega^2 rc_A - D_A(dc_A/dr)]/dr \qquad [8]$$

to obtain estimates of the diffusion coefficient from the extent of boundary spreading in a sedimentation velocity experiment. Any such value is, of course, an apparent diffusion coefficient because its elucidation is based on the premise that diffusion is the sole cause of boundary spreading, i.e. on the premise that the solute is homogeneous.

The lack of an analytical solution to this differential equation prompted the use of approximate solutions, the most notable of which is that obtained by Fujita (3) for the situation in which s_A varies linearly with solute concentration but D_A is constant (4–7). Currently, however, the requirement of an analytical solution to *Equation 8* is being obviated by employing numerical integration—a procedure which has the potential to allow the incorporation of concentration dependence of the diffusion coefficient as well as the sedimentation coefficient (8–14).

Protocol 1

Sedimentation velocity: basic operation and measurement of a sedimentation coefficient

Equipment and reagents

- Ultracentrifuge
- Optical system
- Protein
- Buffer

Method

1. Concentration requirements of the protein. This depends on the interaction being investigated. If it is a self-association and interaction strengths are being probed, the initial cell-loading concentrations chosen should be such that there are measurable amounts of reactants and products present.

2. Optical system: this depends on the concentration range and the protein. For absorption optics a minimum cell loading concentration equivalent to 0.1 absorbance units is required. An absorbance of 1.4 is the likely upper limit for strict adherence with the Lambert–Beer proportionality between absorbance and concentration—a limit

Protocol 1 continued

that is more critical in sedimentation equilibrium than in sedimentation velocity studies. For solutions with absorbance values greater than 3 shorter path length cells need to be employed (the minimum is about 3 mm). Although the absorbance can be decreased by a change to a less sensitive wavelength, the preferred alternative is a switch to interference or schlieren optics. Conventional cells (pathlength 12 mm) are usable down to about 0.1 mg/ml and up to 5 mg/ml if shorter cells are used. Above 5 mg/ml, schlieren optics are the only real option: consult an advanced user.

3 Choose the appropriate buffer/solvent. If possible, work with an aqueous solvent of sufficiently high ionic strength (> 0.05 M) to provide adequate suppression of non-ideality phenomena deriving from macromolecular charge effects. If denaturing/dissociating solvents are used, appropriate centrepieces need to be used (e.g. of the Kel-F type from Beckman instruments).

4 Load the sample into the cell. Double sector cells are used with the protein solution or protein–ligand solution (0.2–0.4 ml) in one sector and the reference buffer or solvent in the other. The latter is filled to a slightly higher level to avoid complications caused by the signal coming from the solvent meniscus; the scanning system subtracts the absorbance of the reference buffer from that of the sample. Electronic multiplexing allows multiple hole rotors to be used, so that several samples can be run at a time (see text above).

5 Choose the appropriate temperature. The modern XL ultracentrifuges can measure comfortably between 4 °C and 40 °C. For higher temperatures one of the authors (S. E. H.) has a specially adapted Model E ultracentrifuge which will measure up to 85 °C.

6 Choose the appropriate rotor speed. For a small globular protein of sedimentation coefficient ~ 2 Svedbergs (S, where $1S = 10^{-13}$ sec), a rotor speed of 50 000 r.p.m. gives rise to a measurable set of optical records after some hours. For larger protein systems (e.g. 12S seed proteins, 30S ribosomes) speeds below 30 000 r.p.m. can be employed.

7 Measure the sedimentation coefficient, s of the sedimenting component(s) (denoted s_A for the protein 'acceptor'). The sedimenting coefficient is defined by the rate of movement of the (protein) boundary (radial position r_p) per unit centrifugal field (*Equation 1*). Commercial software is available for identifying the centre of the sedimenting boundary (strictly the '2nd moment' of the boundary is more appropriate; practically there is no real difference). Personal choices vary, but the following options are available.

   (a) *Simple boundary analysis*: Plot out the boundaries from the $c(r)$ vs r plots from the absorbance or interference optical records (recorded at appropriate time intervals) using a high resolution printer or plotter and graphically draw a line through the user-identified boundary centres. Then use a graphics digitizing tablet to recapture the central boundary positions as a function of radial position. Routines such as *XLA-PLOT* (15) work out dr_p/dt and hence s, and also a correction of the loading concentration for average radial dilution during the run (caused by the sector shape of the cell channels).

Protocol 1 continued

(b) *Analysis of the entire concentration distribution {c(r) vs r} and its change with time.*

The on-line capture of data from the centrifuge into the computer now makes this type of analysis feasible. There are several routines currently available: popular ones include *SVEDBERG* (16) based on the Lamm (17) equation and *DCDT (or the more recent version, 'DCDT+')* (18) based on Rinde's concept of a sedimentation concentration distribution. For monodisperse systems, besides providing an accurate measure of s, these routines provide also an estimate for the translational diffusion coefficient, D_A. For polydisperse systems, a weighted-average sedimentation coefficient is returned for each boundary or component resolved. With *DCDT* a genuine distribution of sedimentation coefficient $g(s)$ is *not* returned directly, (i) because of the complication of diffusion: rather it is an 'effective' distribution, $g^*(s)$. However, extrapolation to infinite time using a procedure developed by Van Holde and Weichet (9) and incorporated by B. Demeler into the algorithm *ULTRASCAN* provides a way around this problem; (ii) The 's' itself is an apparent sedimentation coefficient, affected by non-ideality {sometimes this is denoted by 's^*'—so the true notation is $g^*(s^*)$, i.e 'an 'apparent' distribution of 'apparent' sedimentation coefficients', although most workers quote it as either $g^*(s)$ or $g(s^*)$.}

8 For each protein concentration used, correct s_A to standard conditions using *Equation 3* (and a similar equation for D_A if measured by *Equation 6*): $(D_A)_{20,w} = D_A(\eta/\eta_{20,w})(293/T)$, where T is the temperature at which D_A was measured. In *Equation 3* knowledge of $\bar{v}_A$, a parameter known as the partial specific volume (essentially the reciprocal of the anhydrous macromolecular density), is needed. This parameter can usually be obtained for proteins from amino acid composition data; for most proteins v_A is in the range 0.73–0.74 ml/g. Programmes such as *SEDNTERP* (20) perform this operation, and from provided amino acid (and carbohydrate) composition data estimates $\bar{v}_A$, as well as ε_{278} and the 'hydration' δ (see Section 4). For glycoproteins the carbohydrate composition also has to be considered ($\bar{v}_A \approx 0.6$ ml/g for carbohydrate); and a similar situation pertains to proteins containing prosthetic groups, which also affect the magnitude of $\bar{v}_A$. Where there is doubt, the partial specific volume should be measured experimentally by precision densimetry (21).

9 For a reversible interaction of the type:

$$\text{protein} + \text{ligand} \longleftrightarrow (\text{protein-ligand}) \qquad [9]$$

the concentration of protein (and ligand) affects the position of the equilibrium, and hence separate experiments with different loading concentrations are necessary to take into account the effect of concentration upon the sedimentation coefficient. This is discussed in detail later in this chapter. A complication is non-ideality (deriving from the exclusion volume and charge of the macromolecule and/or complex), which is also considered later. The non-ideality is incorporated into the 'Gralén' parameter, k_s, which is related to $s_{20,w}$ and c for dilute solutions of a non-interacting system by *Equations 4* and *5*: this also applies to protein–ligand interacting systems where there is no change in the extent of ligand binding over the concentration range considered.

3 General experimental aspects

Historically, many biochemists have shirked away from ultracentrifuge measurements because of the impression that analytical ultracentrifuges were large bulky instruments which were difficult to operate (correctly) and which yielded photographic records which were tedious to interpret. Such impressions have now changed with the appearance since 1990 of instruments about half of the size of the old traditional ones and with automatic or semi-automatic data capture of the optical records produced via photomultipliers or diode-array camera into a computer. Nonetheless, even with the new generation instruments, for measurements other than simple molecular weight or sedimentation coefficient determination, the general user is still advised to consult the design and interpretation of his/her data with an advanced user since there are many pitfalls awaiting the unwary. Additionally it is worth stressing that there are certain applications where consultation with the advanced user is mandatory. Examples include measurements at high solute concentration that require schlieren (i.e. refractive index gradient optics) or those measurements at low concentration requiring a long optical path length cell. Both types of measurement can only be performed on older instruments still in active use: these remaining few have generally themselves been upgraded with automatic data capture systems.

It is worth stressing that ultracentrifuges generally allow *multiplexing*: that is the analysis of two or more solutions *almost* (i.e. after allowance for the finite time for each scan) simultaneously. This is made possible by multi-hole rotors (four or eight hole with the Beckman XL-A and XL-I instruments, allowing three or seven ultracentrifuge cells respectively—the remaining hole being taken up by the reference counterbalance cell). In addition, special multichannel cells are available which permit more than one solution to be analysed, but these have a rotation speed limit of approximately 40 000 r.p.m. and give data of lower accuracy. *Full advantage should be taken of this opportunity for analysis, under identical experimental conditions, of protein–ligand systems compared against the appropriate controls.*

3.1 Optical systems for sedimentation velocity and sedimentation equilibrium

There are three types of optical detection of centrifuge records (22, 23).

(a) **UV/visible absorbance optics**. The aromatic amino acids tryptophan (Trp) and tyrosine (Tyr) both absorb radiation strongly in the near UV with a maximum at a wavelength of 278 nm. The extinction coefficient, ε_λ, at wavelength λ will depend on the proportion of Trp and Tyr in the amino acid composition: serine will also add slightly to the extinction at 278 nm, and phenylalanine will give a protein some absorbance at 256 nm. The far-UV (190–230 nm) can also be used (where the peptide bond absorbs) *so long as the solvent buffer does not also absorb appreciably*. Detection of the concentration $c(r)$ in terms of absorbance $A_\lambda(r)$ of light of wavelength λ (cm) at a radial position

r (cm) in a centrifuge cell of optical path length l (cm) is based on the Lambert–Beer law:

$$c(r) = A_\lambda(r)/\varepsilon_\lambda l \qquad [10]$$

The two techniques of sedimentation velocity and sedimentation equilibrium have different restrictions with regard to maximum absorbance (considered below).

(b) **Rayleigh interference (refractive index) optics**. All macromolecular solutions have a refractive index, n, greater than that of water (n_o), i.e. they have a positive refraction increment, $n_c = n - n_o$, to an extent which depends on the concentration, c (g/ml) of the macromolecule and the nature of the macromolecule itself, as manifested by the specific refractive increment, dn/dc (ml/g). dn/dc is a parameter in some ways analogous to the extinction coefficient, although unlike ε_{278} it is not heavily dependent on aromatic amino acid content. For proteins dn/d$c \approx 0.19$ ml/g, for carbohydrates it is approximately 0.15 ml/g. It can be measured accurately by refractometry (the accuracy being limited by the accuracy in concentration measurement) or by use of extensive tabulations (24). For a given radial position r in the ultracentrifuge cell $c(r) = n_c(r)/(dn/dc)$. $n_c(r)$ is registered by interference optics with monochromatic (normally laser) light in terms of absolute fringe numbers $J(r) = n_c(r)l/\lambda$: we thus end up with an equation for interference optics analogous to the Lambert–Beer expression, namely:

$$c(r) = J(r)\lambda/[(dn/dc)l] \qquad [11]$$

In practice what is actually measured is the absolute fringe number $J(r)$ relative to the absolute fringe number at the meniscus $J(r_a)$. This relative fringe number, termed $j(r)$, equals $J(r) - J(r_a)$. To obtain $J(r)$ therefore, an estimate of $J(r_a)$ is required. For sedimentation velocity this is normally trivial because $J(r_a) = 0$ after the boundary leaves the meniscus; but for sedimentation equilibrium it is generally not (a matter for the following chapter).

(c) **Schlieren optics**. This optical system records the refractive index (refraction increment) gradient as a function of radial position r:

$$dc(r)/dr = \{1/(dn/dc)\}.dn_c(r)/dr \qquad [12]$$

The choice of optical system depends on whether or not the protein has sufficient absorbing chromophore, the concentration range selected, and the type of experiment (sedimentation velocity or sedimentation equilibrium).

3.2 Sedimentation velocity optical records

Figure 3 shows the type of optical record for monodispersed non-interacting systems from the three types of optical system. The simplest record to visualize and interpret is the UV/visible absorption system (*Figure 3a*) which gives a direct record of concentration $c(r)$ as a function of radial position r, with the concentration expressed in absorbance units, $A_\lambda(r)$, and within the constraints of

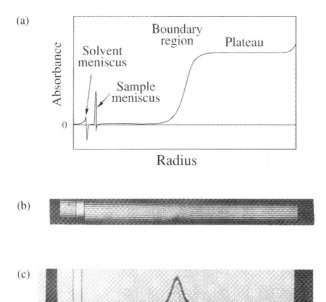

Figure 3 Optical records for sedimentation velocity on homogeneously sedimenting systems. (a) UV absorption (scanned). (b) Rayleigh interference. (c) Schlieren. The optical record in (a) is reprinted from ref. 80, courtesy of Beckman Instruments. (b) and (c) are reprinted from ref. 22, courtesy of Academic Press.

the Lambert–Beer law $\{A_\lambda(r) \propto c(r)\}$. In the interference system (*Figure 3b*) each fringe is a record of concentration $c(r)$ relative to the meniscus, expressed as relative fringe number displacement $j(r)$. The multiple fringes are effectively averaged by a Fourier transform (done automatically by the software coming with the Beckman XL-I) to produce an accurate record of the radial dependence of $j(r)$. On the other hand the schlieren optical record (*Figure 3c*) is a plot of refractive index gradient, $dn_c(r)/dr$, versus radial distance r. Since $n_c(r)$ is proportional to $c(r)$, a concentration gradient is accurately produced. It is possible by integration to produce a plot of $c(r)$ versus r, although for many applications, particularly those involving liganded systems, it is an advantage to have a direct record of the concentration gradient distribution.

3.3 Data capture

There are several options available, as explained in *Protocol 1*. Visual inspection of the $c(r)$ vs r records (absorption/ interference optics) or $dc(r)/dr$ (schlieren optics) vs r can give a rapid idea of the heterogeneity of the system (*Figure 4a,b*) from the number and shape of the boundaries. However, the components need to have quite different sedimentation coefficients; and casual inspection cannot distinguish between a non-interacting mixture of species (a heterogeneous system) and a mixture of species undergoing chemical re-equilibration (a chemically interacting system). Such analysis can be enhanced by transforming the *c(r)* vs *r* plots via the *DCDT* routine (18) into a plot of the apparent distribution of sedimentation coefficient, $g^*(s)$ versus *s* (see *Protocol 1*) which takes on the appearance of a schlieren diagram (*Figure 4c*) even though the functions describing them are different. The plot of $g^*(s)$ vs *s* can also be produced from schlieren records (25).

By model-fitting Gaussian distributions to either the $dc(r)/dr$ vs *r* or $g^*(s)$ vs *s* diagrams using standard computer packages such as *PRO-FIT* (26), the user can:

(a) Resolve sedimenting components present.

(b) Provide an accurate estimate for the sedimentation coefficient(s) *s*.

(c) Estimate the amount of each component present.

An important requirement is a minimum number of scans: the helpfiles accompanying the above computer routines should help

3.4 Two complications

There is a complication known as the Johnston–Ogston effect (27) that arises in the analysis of simple mixtures. Because of the inverse dependence of sedimentation coefficient upon solute concentration, the boundary of slower solute is migrating faster than slow solute in the mixture. This leads to a pile-up of slower solute in the region immediately behind the faster-migrating boundary, and hence to overestimation of the proportion of slower-migrating solute: the proportion of faster-migrating solute is correspondingly underestimated.

Another complication is that for a rapid self-association or interaction between solutes of similar size, only a single symmetric boundary may be evident: the sedimentation coefficient obtained in this case is a weighted average of the reactants and product.

3.5 Co-sedimentation diagrams

A useful way of assaying for interactions (other than self-associations) is possible if the reacting species exhibit optical absorption at different regions of the UV/visible spectrum. Optical records of solute distribution are taken at wavelengths where successively one of the reacting species is visible but the other(s) is/are transparent, after which these records are compared with controls of the reactants by themselves at the same concentration (absorbance). This method is particularly useful for monitoring the interaction of a small ligand with a protein. *Figure 5a* illustrates the situation where interaction of the ligand (cofactor B12) with acceptor (methylmalonyl mutase) is stoichiometric (i.e. complete), whereas *Figure 5b* presents a situation involving reversible equilibrium between ligand (methyl orange) and acceptor (bovine serum albumin).

3.6 Concentration dependence of the sedimentation coefficient

For reversible interactions involving protein the concentration of protein and ligand is important. In order to probe the reversible interaction in terms of

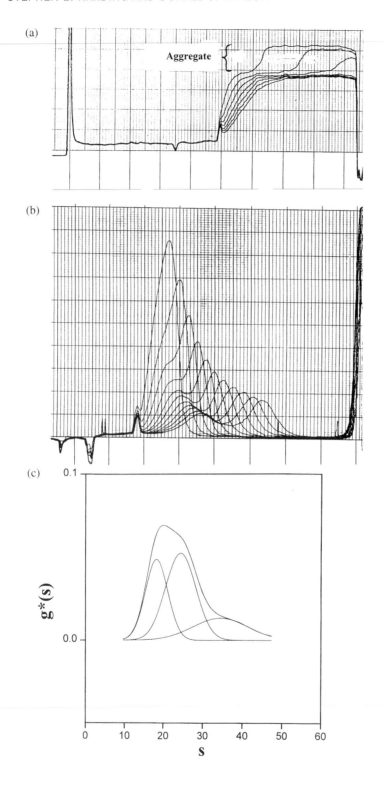

Figure 4 Optical records for mixed solute systems. (a) Scanning UV absorption optical records of the Gene5 protein with aggregate. 0.7 mg/ml, monochromator wavelength 278 nm; scan interval 8 min; rotor speed 40 000 rev/min; temperature 20.0 °C; measured $s_{20,w}$ = (35.3 ± 1.4)S (faster boundary) and (2.6 ± 0.1)S (slower boundary). (From ref. 81.) (b) Scanning schlieren optical records for rat IgE solution with a low molecular weight impurity. 4.52 mg/ml, monochromator wavelength 546 nm; scan interval 8 min; rotor speed 40 000 rev/min; temperature 20.0 °C; measured $s_{20,w}$ = (7.53 ± 0.15) S (IgE) and (3.8 ± 0.1)S (slower impurity). (Davis, K. G., Burton, D. R., and Harding, S. E., unpublished data.) (c) $g*(s)$ vs s (Svedbergs) plots for a trp-mutant GroEL chaperonin system. Upper profile the direct transform from the Rayleigh interference optical record. Lower three profiles from a three component Gaussian fit to these data. Peak maxima, areas respectively are: 1st peak (18.3S, 0.352 units); 2nd peak (24.4S, 0.503 units); 3rd peak (34.5S, 0.229 units). 0.7 mg/ml, number of scans 18; rotor speed 40 000 rev/min; temperature 20.0 °C. Relative peak areas do not change with differing loading concentration, implying the three observed components are NOT in reversible interaction equilibrium. (Walters, C., Clarke, A., and Harding, S. E. unpublished data.)

stoichiometry and strength it is necessary to make measurements over a range of different loading concentrations, since the position of the equilibrium will depend on the concentration of protein (and ligand): higher concentrations will favour the equilibrium towards the right-hand side of *Equation 9*.

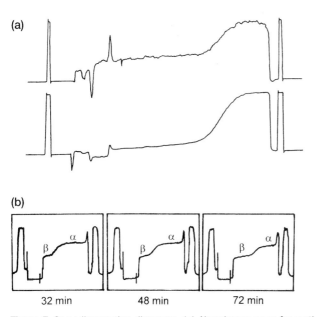

Figure 5 Co-sedimentation diagrams. (a) Absorbance scan for methylmalonyl mutase (0.7 mg/ml) and its cofactor (offset toward the top) scanned within 2 min of each other. The centre of the sedimenting boundary is virtually the same for both, and there is no significant residual absorbance left behind the boundary, suggesting that under the solvent conditions used (50 mM Tris–HCl pH 7.5 + 5mM EDTA) the cofactor ligand is bound to the protein. From ref. 82. Monochromator = 295 nm (bottom), 608 nm (top), rotor speed 44 000 r.p.m. and 20 °C. (b) Absorbance scan for the ligand constituent obtained in a sedimentation velocity experiment after centrifuging a mixture of methyl orange (490 µM) and bovine serum albumin (30 µM) for 32 min at 59 780 r.p.m. and 4 °C. Data are taken from ref. 41.

The appropriate concentration range for study depends on the strength of the interaction, which is either described by the molar association constant K or by the corresponding dissociation constant K_d. For a simple 1:1 stoichiometry, i.e. of the type A + B $\leftrightarrow$ AB, the association constant K_{AB} is related to the molar concentrations C_i of participating species by the expression:

$$K_{AB} = C_{AB}/(C_A C_B) \qquad [13]$$

where K_{AB} has units of reciprocal molarity (M^{-1}). For reactions with higher stoichiometry, e.g. A + 2B $\leftrightarrow$ AB$_2$, the stoichiometric association constant for complex formation from reactants needs to be written as:

$$K_{AB_2} = C_{AB_2}/[C_A C_B^2] \qquad [14]$$

where the units of K$_{AB2}$ are M^{-2}. Dissociation constants are just the reciprocals of these association equilibrium constants. For 1:1 stoichiometries K_d values below 1 μM tend to be classified as strong interactions, whereas those with K_d > 50 μM are often designated as weak interactions.

A complication encountered in the analysis of sedimentation velocity patterns is the non-ideality that derives from the exclusion volume and charge properties of the macromolecule and/or complex. This non-ideality, which is incorporated into the 'Gralén' parameter k_s, is described at low concentration by *Equation 4* for non-interacting globular proteins systems and by *Equation 5* for asymmetric solutes. For higher concentrations additional coefficients can be used in the expansion, as indicated in *Equations 4* and *5*. Alternatively, the concentration dependence may still be written in the form:

$$s^o{}_{20,w} = s_{20,w}(1 - gc) \qquad [15]$$

in which g now becomes the following function of c (28):

$$g(c) = \{k_s - [(cv_s)^2(2\phi_p - 1)]/\phi_p^2\}/\{k_s c - 2cv_s + 1\} \qquad [16]$$

and where v_s is the swollen specific volume (approx. 1 ml/g for globular proteins), ϕ_p a parameter known as the maximum packing fraction by volume, and k_s continues to be the limiting Gralén coefficient (in the absence of associative/dissociative phenomena).

The extraction of K_{AB} (or K_d), by means of the SA-PLOT routine is considered for the ideal and non-ideal cases in Section 5.2.

3.7 Sedimentation coefficient ratios

Another useful criterion for the extent of an interaction involving proteins with other biomolecular species is the ratio of the sedimentation coefficients of the product(s) to the reactant(s). This is particularly useful for the analysis of interactions where large irreversible complexes are formed (29). Provided that assumptions are made about the conformation(s) of reactant(s) and product(s), an estimate for the size/stoichiometry of the complex can be made on the basis of a 'Mark–Houwink–Kuhn–Sakurada' relation (30):

$$(s_{20,w})_{oligomer}/(s_{20,w})_{monomer} = (M_{oligomer}/M_{monomer})^b \qquad [17]$$

Similar relations exist for the intrinsic viscosity, translational diffusion coefficient, and radius of gyration. The magnitude of the b coefficient in *Equation 17* depends upon molecular shape: values are ~ 0.67 for spheres, ~ 0.15 for rods, and ~ 0.4–0.5 for coils. In practice the sphere value of 0.67 is usually assumed for globular proteins, together with the further assumption that the conformation of oligomer and monomer are essentially similar. These unsubstantiated assumptions mean that sedimentation velocity studies alone cannot provide unequivocal estimates of interaction stoichiometries, which therefore require confirmation by procedures such as sedimentation equilibrium.

3.8 Sedimentation velocity fingerprinting

For very large protein–biomolecular complexes the sedimentation rates are too fast for measurement even at the lowest practical operating speeds of an analytical ultracentrifuge (1000 rev/min): in such cases reaction products do not remain in solution. A technique known as sedimentation velocity fingerprinting can be used whereby the depletion of reactant concentrations is used to assess the concentration of complex(es) removed from the solution by centrifugation (31).

4 Sedimentation velocity analysis of the shape of a molecular complex

Once the sedimentation coefficient, $s^o_{20,w}$, of the product (and/or the reactants) has been established, the gross conformation or 'shape' of the reaction product can then be examined (29, 32–34), so long as the molecular mass, M, of the product is known, for example, from sedimentation equilibrium. Knowledge of $s^o_{20,w}$ and M permits the evaluation of the translational frictional ratio f/f_o, the ratio of the translational frictional coefficient of the particle to that for a spherical particle of the same mass and anhydrous volume, from the relationship:

$$f/f_o = M_A(1 - \bar{v}_A \rho_{20,w})/[N 6\pi \eta_{20,w} s^o_{20,w} \{3 M_A \bar{v}_A / 4\pi N\}^{1/3}] \qquad [18]$$

where N is Avogadro's number, $\bar{v}_A$ is the partial specific volume of the solute particle. This translational frictional ratio reflects the shape (represented by the parameter P) and state of hydration (δ) of the particle in accordance with the expression:

$$f/f_o = P(1 + \delta/\bar{v}_A \rho_{20,w})^{1/3} \qquad [19]$$

From a practical viewpoint the hydration parameter δ (sometimes denoted as w) is a very difficult parameter to measure with any precision, but can be *estimated* from the amino acid and carbohydrate content (see *Protocol 1*). Values between 0.25 and 0.5 are popularly quoted for this parameter for proteins. From *Equation 19* the shape parameter P, known either as the Perrin parameter or the frictional ratio due to shape, can be evaluated from the experimentally determined f/f_o and a selected value of δ. In practice, a range of plausible values of δ is chosen. Alternatively, δ can be eliminated by combination of f/f_o with other hydrodynamic measurements such as the intrinsic viscosity, $[\eta]$. The Gralén coefficient k_s, (35) and the ratio $k_s/[\eta]$ are also highly useful in this regard (36).

P is utilized in one of two ways:

(a) Direct evaluation of molecular shape.

(b) Selecting a plausible structure which best agrees with the data.

Some workers relate $s^o_{20,w}$ and $D^o_{20,w}$ directly with shape: we find this route can lead to confusion, especially in regard to the roles of volume and hydration: obtaining shape via the Perrin factor P is recommended.

4.1 Direct evaluation of molecular shape

The axial ratio (a/b)—the ratio of the long to small axis—of the equivalent ellipsoid of revolution (prolate or oblate) can be evaluated using the routine ELLIPS1 (32). ELLIPS1 also allows the evaluation of a/b from the complete range of other hydrodynamic measurements. An ellipsoid of revolution has the constraint of two equal axes: a prolate ellipsoid has two equal shorter axes and one longer axis, whereas an oblate ellipsoid has two equal long axes and one shorter axis. A survey of crystal structures has shown the prolate case to be the more appropriate although the distinction can be arbitrary. An alternative representation removes the requirement for two equal axes, but such action requires a more complicated approach using combination of shape functions. An easier alternative is to predict the P (and hence f/f_o, $s^o_{20,w}$) for a given structure and select the structure which best agrees with the data.

4.2 Selecting a plausible structure which best agrees with the data

For a given triaxial shape (with semi-axial dimensions a > b > c) P (and hence f/f_o, $s^o_{20,w}$) together with a comprehensive set of other hydrodynamic shape functions can be evaluated using the routine ELLIPS2 (32). The sedimentation and other hydrodynamic properties of different structures of different axial ratios (a/b, b/c) can then be compared directly. To assist this, the (a/b, b/c) ratios from a crystal structure can be first evaluated using ellipsoid fitting to crystal co-ordinates using the routine ELLIPSE (37).

Many structures however cannot be represented by ellipsoidal shapes—even general triaxial ellipsoids. The classical example is the antibody molecule. For arbitrary-shaped particles, the structure is represented by a number of spherical beads. From user specified co-ordinates the hydrodynamic properties for the composite structure can be calculated: the most advanced routine for doing this is currently SOLPRO (33, 38, 39). Unlike those for ellipsoids the hydrodynamic relations for bead constructs are not exact, but they are generally a good approximation. In practice, modelling the surface as a structure with an array of beads (called 'bead-shell' or just 'shell' modelling) appears to be the most successful, although 'filling models' where both the surface and interior structure are represented by a series of small beads can give results seriously in error:

unfortunately this means that approaches that have been presented based on representing the complete set of atoms from a crystal structure with corresponding beads should be avoided. The potential user is recommended to consult a recent work by Carrasco (40).

5 Sedimentation velocity studies of ligand binding

Having completed the general treatment of the sedimentation velocity variant of analytical ultracentrifugation for the study of equilibria, we now turn to the specific problem of quantifying an acceptor–ligand interaction by sedimentation velocity. In that regard there are two situations that can be practically considered: that in which the acceptor is macromolecular (or particulate) and the ligand is small; and that in which both reactants are macromolecular.

5.1 Interactions of a (protein) acceptor with a small ligand

Provided that the binding of ligand is without effect on the sedimentation coefficient of the acceptor ($s_{ABi} = s_A$), the free concentration of ligand in an acceptor–ligand mixture is readily determined by sedimentation velocity. In the illustrative application of *Figure 5b*, the absorption optical system has been used to monitor the sedimentation velocity behaviour of a mixture of methyl orange (B) and bovine serum albumin (A) (41). At the speed of the experiment (59 780 r.p.m.) acceptor and acceptor–ligand complexes co-migrate with the sedimentation coefficient of albumin (4.4S) but there is effectively no sedimentation of methyl orange ($s_B = 0.2S$). Consequently, the sedimentation velocity pattern reflecting the ligand constituent is biphasic, with a sedimenting boundary separating the plateau of original composition (α-phase) from a region comprising pure methyl orange (β-phase). The co-migration of A and all AB_i complexes ensures the absence of any redistribution of methyl orange as the result of migration of the acceptor constituent; and hence allows C_B^β to be identified with C_B^α, the free *molar concentration* of ligand in the mixture if the 'rectangular approximation' is made (42, 43). As noted by Steinberg and Schachman (41), this conclusion requires slight modification in sedimentation velocity because of non-compliance with assumed migration in a rectangular cell under the influence of a homogeneous field. Sedimentation in a sector-shaped cell leads to radial dilution that decreases slightly the values of C_B^β, C_B^α, $\bar{C}_B^\alpha$, and $\bar{C}_A^\alpha$ from those that would have applied to a mixture with the loaded composition. However, in view of the uncertainty surrounding the assumed identity of sedimentation coefficients for acceptor and all acceptor–ligand complexes, results are usually interpreted on the basis of the identification of C_B^β with the free ligand concentration in a mixture with the composition that was subjected to sedimentation velocity. Inasmuch as the sole objective of the ultracentrifugation is to generate an acceptor-free region for measurement of the ligand concentration, the experiments may also be performed in a preparative centrifuge (44–48).

In situations where B is a small ligand such as a metal ion or a coenzyme, the estimate of C_B^α is combined with the values of $\bar{C}_A^\alpha$ and $\bar{C}_B^\alpha$ to generate the binding function v:

$$v = (\bar{C}_B^\alpha - C_B^\alpha)/\bar{C}_A^\alpha \qquad [20]$$

the dependence of which upon C_B^α is interpreted in terms of the conventional binding equation:

$$v = pk_{AB}C_B^\alpha/(1 + k_{AB}C_B^\alpha) \qquad [21]$$

where k_{AB} denotes the intrinsic binding constant for the interaction with p equivalent and independent sites on the acceptor (49).

5.2 Interactions of an acceptor with a macromolecular ligand

For the interaction between an acceptor and a macromolecular ligand the sedimentation coefficient of the complex is likely to be greater than that of either reactant. Under those circumstances ($s_{AB} > s_A > s_B$ for a system with 1:1 complex formation) a reaction boundary and a boundary corresponding to a pure reactant are generated in a sedimentation velocity experiment (42, 50)—a feature illustrated in *Figure 6* for the electrostatic interaction between ovalbumin (A) and lysozyme (B) at neutral pH and low ionic strength (43). Schlieren patterns for the individual reactants are presented in Figures 6a and 6b, whereas *Figure 6c* refers to a mixture of lysozyme and ovalbumin in 1.5:1 molar ratio. A reaction boundary ($s = 4.2S$) and a lysozyme boundary ($s = 2.3S$) are clearly evident. However, only a single boundary is observed for a mixture with ovalbumin in molar excess (*Figure 6d*)—a reflection of incomplete resolution between a pure reactant phase (now ovalbumin) and the reaction boundary.

Even in situations where the sedimentation velocity pattern reflecting an acceptor–ligand interaction exhibits resolution of a pure reactant boundary (as in *Figure 6c*), the important point to note is that the ligand concentration in the pure-solute phase (say C_B^β) does not equal its concentration (C_B^α) in the

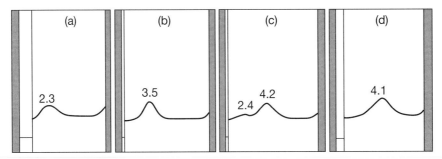

Figure 6 Schlieren patterns obtained in a study of the interaction between ovalbumin and lysozyme, pH 6.8, $I = 0.02$, by sedimentation velocity (59 780 r.p.m., 20 °C). (a) Lysozyme (0.28 mM). (b) Ovalbumin (0.14 mM). (c) Mixture of lysozyme (0.21 mM) and ovalbumin (0.13 mM). (d) Mixture of lysozyme (0.14 mM) and ovalbumin (0.16 mM). Data are taken from ref. 43.

equilibrium mixture for a system with $s_{AB} > s_A > s_B$. However, considerations of mass conservation (50–52) show that the free concentration of the other reactant, C_A^α, may be determined from the expression:

$$C_A^\alpha = [\bar{C}_A^\alpha (\bar{s}_A - s_B) - \bar{C}_B^\alpha (s' - s_B) + C_B^\beta (s' - s_B)]/(s_A - s_B) \quad [22]$$

where s_A, the average sedimentation coefficient of acceptor constituent, and s', the sedimentation coefficient of the boundary of ligand constituent within the reaction boundary, must both be taken as $s^{\alpha\beta}$, the sedimentation coefficient of the reaction boundary. A value of 3×10^4 M^{-1} for K_{AB} is obtained (43) by combining the value of C_A^α emanating from the application of *Equation 22* to the results from *Figure 6c* with the expression for the association equilibrium constant, namely:

$$K_{AB} = C_{AB}^\alpha/(C_A^\alpha C_B^\alpha) = (\bar{C}_A^\alpha - C_A^\alpha)/[C_A^\alpha (\bar{C}_B^\alpha - \bar{C}_A^\alpha + C_A^\alpha)] \quad [23]$$

In keeping with sedimentation velocity studies of acceptor interactions with small ligands, the 'rectangular approximation' is inherent in *Equation 22*, as is neglect of the composition dependence of the sedimentation coefficients of individual species (s_A, s_B). These problems also pervade the characterization of acceptor–ligand interactions by an alternative sedimentation velocity procedure —interpretation of the constituent sedimentation coefficients $\bar{s}_A$ and $\bar{s}_B$.

Inasmuch as the constituent sedimentation coefficients of the two solute components in an acceptor–ligand system undergoing 1:1 complex formation are given by:

$$\bar{s}_A = (s_A C_A^\alpha + s_{AB} C_{AB}^\alpha)/\bar{C}_A^\alpha \quad [24a]$$

$$\bar{s}_B = (s_B C_B^\alpha + s_{AB} C_{AB}^\alpha)/\bar{C}_B^\alpha \quad [24b]$$

it follows that $\bar{s}_A$ is a function of mixture composition provided that acceptor–ligand complex migrates faster than A ($s_{AB} > s_A$). For the ligand constituent the corresponding proviso that $s_{AB} > s_B$ always pertains, and hence $\bar{s}_B$ invariably shows a progressive increase for mixtures with increasing constituent concentration of one reactant but fixed constituent concentration of the other. Elimination of C_{AB}^α from Equations 24a and 24b on the grounds that $C_{AB}^\alpha = (\bar{C}_A^\alpha - C_A^\alpha) = (\bar{C}_B^\alpha - C_B^\alpha)$ for an interaction confined to 1:1 stoichiometry leads to the following expressions for the concentration of free reactant.

$$C_A^\alpha = \bar{C}_A^\alpha (s_{AB} - \bar{s}_A)/(s_{AB} - s_A) \quad [25a]$$

$$C_B^\alpha = \bar{C}_B^\alpha (s_{AB} - \bar{s}_B)/(s_{AB} - s_B) \quad [25b]$$

The value of $\bar{s}_i$ (where i = A or B) may be determined by application of the basic transport equation:

$$\bar{s}_i = -(1/2\omega^2 t) \ln [\{2\int_{r_a}^{r_p} r \bar{C}_i(r)\, dr\}/(r_p^2 \bar{C}_i^\circ) + (r_a^2/r_p^2)] \quad [26]$$

to a sedimentation velocity distribution recorded at effective time t after attainment of angular velocity ω (40). In *Equation 26* the integration covers radial distances from the air–liquid meniscus r_a to a position r_p in the α-plateau region beyond the $\alpha\beta$ reaction boundary; and $\bar{C}_i^\circ$ denotes the total concentration of

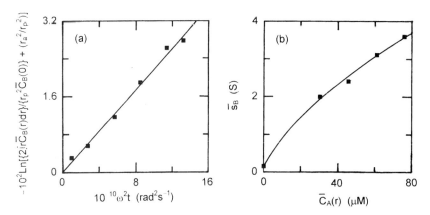

Figure 7 Use of constituent sedimentation coefficients for the characterization of an acceptor–ligand interaction. (a) Determination of $\bar{s}_B$ by the application of *Equation 7* to distributions for the ligand constituent obtained by subjecting a mixture of methyl orange (30 μM) and bovine serum albumin (45 μM) to sedimentation at 59 780 r.p.m. and 20 °C for the indicated times in a Beckman model E ultracentrifuge. (b) Dependence of $\bar{s}_B$ upon albumin concentrations in mixtures with a fixed concentration (30 μM) of methyl orange. Data are taken from ref. 41.

component i in the loaded mixture. For the application of this expression $\bar{C}_i^o$ and $\bar{C}_i(r)$ may be replaced by the corresponding optical parameters (e.g. absorbance); and the product $\omega^2 t$ is recorded as part of the printout for each recorded distribution in the XL-A and XL-I ultracentrifuges.

The measurement of $\bar{s}_B$ by means of *Equation 26* is illustrated in *Figure 7a*, which refers to a sedimentation velocity experiment conducted at 58 780 r.p.m. on a mixture of bovine serum albumin (45 μM) and methyl orange (30 μM): a value of 2.4S for $\bar{s}_B$ is obtained from the slope (41). Although the values of $\bar{s}_B$ obtained with the same methyl orange concentration and a range of albumin concentrations exhibit the predicted increase with increasing $\bar{C}_A^\alpha$ (*Figure 7b*), their quantitative interpretation by the above procedure is precluded by nonconformity with the assumed 1:1 stoichiometry of the acceptor–ligand interaction.

5.3 Sedimentation velocity studies of weak interactions

For weak interactions the negative dependence of sedimentation coefficient upon solute concentration needs to be taken into account. To that end a procedure called *SA-PLOT* has been developed around the general concentration dependence expressions (Equations 15 and 16) to allow simulation of a dependence of $\bar{s}$ (the weight-average sedimentation coefficient) upon total solute concentration that can be compared with its experimental counterpart. This procedure is designed primarily for the characterization of solute self-association, but can also be used for studies of the interaction between reactants with identical sedimentation coefficients. For a monomer–dimer system the statement of mass conservation:

$$\bar{c} = c_1 + 2c_1^2/(K_d M_1) \qquad [27]$$

is a quadratic equation with solution:

$$c_1 = K_d M_1 [-1 + \{1 + 8\bar{c}/(K_d M_1)\}^{1/2}]/4 \qquad [28]$$

which allows the monomer concentration, c_1, and dimer concentration, $c_2 = \bar{c} - c_1$, to be calculated for any specified magnitude of the dissociation constant K_d. Combination of these values of c_1 and c_2 with the magnitudes of the corresponding sedimentation coefficients (s_1 and s_2) calculated from *Equations 15* and *16* on the basis that $\bar{c}$ is the appropriate concentration then allows estimation of the dependence of $\bar{s}$ (*Equation 24*) upon $\bar{c}$. The program *SA-PLOT* utilizes Equations 15, 16, 24, and 28 to compute $\bar{s}$ as a function of $\bar{c}$ for an assigned value of the dissociation constant, which is then refined iteratively on the basis of minimizing the sums of squares of residuals between experimental data and the simulated dependencies for designated K_d values. As noted above, the *SA-PLOT* program can also to be applied to a 1:1 interaction between different proteins (53), provided that their molecular masses and hence sedimentation coefficients are within 10–15% of each other (*Figure 8*).

5.4 The shape of sedimenting boundaries for acceptor–ligand systems

Thus far we have presented characterizations of an acceptor–ligand interaction on the basis of the size of the reactant boundary (C_B^β) and the composition dependence of the magnitude of constituent sedimentation coefficients. Neither of these procedures has taken advantage of the detailed form of the sedimenting boundary system, which is undoubtedly the most striking aspect of a

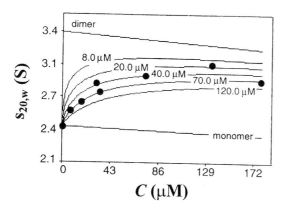

Figure 8 Concentration dependence of the sedimentation coefficient for a protein interacting with a macromolecular ligand (another protein): the cell adhesion molecule CD42 with its counter-receptor CD48. The two have similar molecular weights (~ 28 500) and the interaction can be regarded as an effective 'monomer–dimer' system. Concentration expressed in molar terms (with respect to monomer). The (weight average) sedimentation coefficient data points (•) are modelled iteratively to Equations 15, 16, 24, and 28 {with k_s (monomer) set as 5 ml/g; k_s (dimer) as 8.5 ml/g} for values of the dissociation constant K_d in the ranges 8–120 μM using the software *SA-PLOT*. From ref. 53 and reproduced courtesy of Springer–Verlag.

sedimentation velocity distribution. These boundary forms are more distinctive when plotted in derivative format (dc/dr versus r)—the distribution recorded by the schlieren optical system that has been omitted from the current generation of analytical ultracentrifuges. However, a procedure has been devised (54-56) whereby the equivalent shape of the derivative distribution is extracted from the optically recorded integral distribution (concentration or absorbance as a function of radial distance).

Results from application of the $g(s^*)$ procedure to sedimentation velocity distributions for an acceptor-ligand system are presented in *Figure 9*, which refers to the interaction between diphtheria toxin and an elicited monoclonal antibody (56). Studies at neutral pH were used to establish the forms of the normalized derivative distributions, $g(s^*)$ versus s^*, for the toxin (B) alone, the antibody (A), and the AB_2 complex (*Figure 9a*). Adjustment of a solution of AB_2 complex to pH 5 causes the complex to undergo dissociation in a manner such that the distribution remains essentially unimodal despite the coexistence of species with molecular masses of 266, 150, and 58 kDa (*Figure 9b*). Such behaviour is typical of a system in rapid association equilibrium, for which the major indicator of dissociation is the observation that dilution leads to a progressive decrease in the value of s^* at the peak of the distribution. Clearly, there is far more potential information to be gained from the shapes of these patterns than simply the value of a constituent sedimentation coefficient, $\bar{s}$ (or $\bar{s}^*$).

What is really required is an analytical solution to the differential equation describing mass migration in a sedimentation velocity experiment—the Lamm equation—which for a single non-interacting solute is given by *Equation 8*. However, that problem is seemingly intractable. Our understanding of the shapes of sedimentation velocity patterns has therefore stemmed from the pioneering studies of Gilbert (42, 57), who established the forms of such distributions by obtaining analytical solutions to the differential equations describing diffusion-

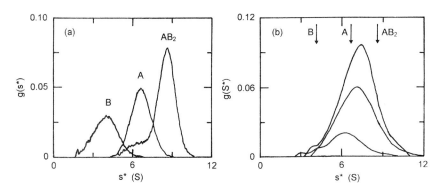

Figure 9 Use of the $g(s^*)$ analysis (13) to deduce the equivalent of schlieren patterns from integral sedimentation velocity distributions for a system comprising the interaction of diphtheria toxin (B) with biospecific monoclonal antibody (A). (a) $g(s^*)$-s^* distributions for the two separate reactants and for the stable AB_2 complex at neutral pH. (b) Corresponding patterns deduced from integral distributions for the indicated concentrations of complex at pH 5.0. Data are taken from ref. 56.

free migration. Despite the passage of nearly four decades, those publications and another that tackled the same problem in a different manner (50, 58) are pivotal to our understanding of the effects of chemical re-equilibration in sedimentation velocity experiments. Indeed, use has already been made of those findings to characterize systems with $s_{AB} = s_A > s_B$ (*Figure 6*) and $s_{AB} > s_A > s_B$ (*Figure 9*).

From the viewpoint of comparing experimental patterns with such predicted behaviour, the absence of diffusional effects in the latter has been a large impediment to the exercise. *Figures 10a* and *10b* depict the theoretical diffusion-free behaviour that led to the interpretation of sedimentation velocity patterns for the ovalbumin–lysozyme system (*Figures 6c* and *6d*). In that regard the failure to observe an ovalbumin (A) boundary (*Figure 6d*) under conditions comparable with those pertaining in *Figure 11b* has been explained on the grounds that diffusional spreading would have disguised the predicted resolution. Although such rationalization is certainly reasonable, the inference would obviously benefit from the generation of a predicted distribution that also takes into account the effects of diffusional spreading.

Boundary spreading due to the effects of diffusion is now usually incorporated into theoretical sedimentation velocity distributions by solving numerically the Lamm equation by the finite element treatment of Claverie (59–61). To date the major use of this approach has been to accommodate the effects of concentration-

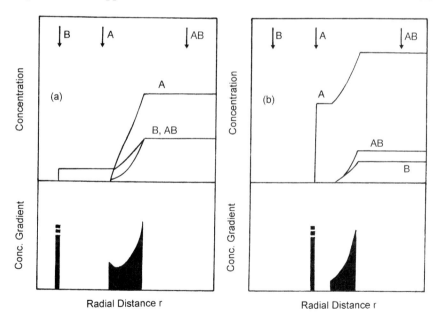

Figure 10 Illustrative diffusion-free sedimentation velocity patterns for acceptor–ligand interactions with $s_{AB} > s_A > s_B$. (a) Sedimentation velocity distributions for a mixture of acceptor and ligand with ligand (B) in molar excess. (b) Corresponding distribution for a mixture with acceptor (A) in molar excess. In each case the upper pattern is the integral whereas the lower represents the derivative (schlieren) pattern. Details of the manner in which such patterns are deduced are to be found in refs 42, 50, and 57.

dependence of s and D for a single, non-interacting solute on migration and boundary spreading. Use of the Claverie method to obtain the best-fit description of the migration and boundary spreading in terms of *Equation 4* and the corresponding equation for the translational diffusion coefficient:

$$(D_A)_{20,w} = (D_A^\circ)_{20,w}(1 + k_D c + ..) \qquad [29]$$

leads to unique identification of the molar mass by combining the estimates of $(s_A^\circ)_{20,w}$ and $(D_A^\circ)_{20,w}$ (62–65).

Application of the technique to a system undergoing chemical re-equilibration entails alternating rounds of simulated transport and chemical reaction—the procedure introduced by Cann and Goad (66) to modify the values of the sedimentation and diffusion coefficients (now s_i and D_i) for solution of the Lamm equation by the finite difference method. Details of the finite element and finite difference approaches are reviewed by Cox and Dale (67), who discuss the potential of the Claverie method for simulating the sedimentation velocity behaviour of chemically reacting systems involving solute self-association as well as interactions between dissimilar reactants. This approach is now being actively pursued in the realm of solute self-association (62, 63) and also interactions between two solute components (68, 69).

Despite its sophistication and ability to generate sedimentation velocity patterns with a greater sense of experimental realism, this numerical solution of the Lamm equation is not necessarily providing an accurate description of the sedimentation behaviour of an interacting system. A major limitation is likely to be inadequacy of the expressions (the counterparts of *Equations 4* and *29*) invoked to describe the composition dependence of s and D for the individual species. In that regard the necessity to assign magnitudes to sedimentation coefficients (s_i°) for any postulated complex species AB_i has already been addressed in discussing the use of constituent sedimentation coefficients for characterizing interactions. There is also a problem with specifying the forms of the composition dependence of s and D arising from non-chemical interactions between species, there being no theoretical justification for the commonly used substitution of total solute concentration $\bar{c}$ for c in *Equations 4* and *29*. Furthermore, in view of the number of parameters requiring evaluation by curve-fitting, the method is unlikely to become a major contender for deducing the stoichiometry and strength of acceptor-ligand interactions. Nevertheless, it has considerable potential for testing further the adequacy of a quantitative description of an acceptor–ligand interaction that has been obtained by other means.

6 The study of ligand-mediated conformational changes

Elucidation of the mechanism responsible for the allosteric behaviour of enzymes has inevitably posed a problem because of the need to distinguish between models based on pre-existence (70) and ligand-induction (71) of the enzyme isomerization. Sedimentation velocity provides a powerful means of detecting the change in enzyme shape; and, in favourable circumstances, the means of distinguishing between pre-existing and ligand-induced isomerization of the acceptor.

Difference sedimentation velocity (72, 73) was introduced over 30 years ago as a means of quantifying ligand-mediated conformational changes in enzymes in terms of differences in hydrodynamic volume (74). Such changes were quantified initially on the basis of the difference between values of the sedimentation coefficients obtained from simultaneous velocity runs on enzyme solutions with and without ligand. However, more recent studies (75–77) have employed the expression (75):

$$d(\ln r_- - \ln r_+)/dt = \omega^2(s_- - s_+) \qquad [30]$$

where r_- and r_+ are the respective radial positions of the boundaries in the ligand-free and ligand-containing solutions after centrifugation at angular velocity ω for time t. Provided that the optical records for both solutions are recorded simultaneously, the difference in sedimentation coefficients is obtained from the slope of the dependence of $(\ln r_- - \ln r_+)$ upon t.

Because the design of the Beckman XL-A and XL-I ultracentrifuges precludes the simultaneous recording of solute distributions in two cells, the two distributions being compared must be recorded sequentially. Provided that the time increment between the recording of the distributions in the two cells is constant, *Equation 30* with t taken as the time for the first of the paired distributions, continues to provide an exact description of the difference in sedimentation coefficients. Although the fluctuation of the time increment by 3–4% about a mean in the XL-A ultracentrifuge is at variance with this proviso, the random error associated with boundary location is likely to render insignificant the relatively minor departure from the predictions of *Equation 28*; and accordingly difference sedimentation velocity studies can be pursued with confidence in the current (as well as older) generation of analytical ultracentrifuges (78). We illustrate the potential of difference sedimentation velocity for the detection and quantification of the small changes in sedimentation coefficient of rabbit muscle pyruvate kinase in the presence of phenylalanine, an allosteric inhibitor of the enzyme.

Determination of sedimentation coefficients from the two separate time dependencies of the logarithm of radial distance migrated is presented in *Figure 11a*, which signifies a slightly faster migration rate for enzyme alone than for enzyme in the presence of a saturating concentration (5 mM) of phenylalanine. Although the independent estimates of 9.5 ($\pm$ 0.3) and 9.2 ($\pm$ 0.2)S for pyruvate kinase in the absence and presence of phenylalanine indicate a probable difference of 0.3S between the sedimentation coefficients of enzyme and enzyme–inhibitor complex, the result, 0.3 ($\pm$ 0.5)S is clearly equivocal. On the other hand, the difference plot of results according to *Equation 30* is far more definitive in that regard (*Figure 11b*) inasmuch as linear regression analysis yields a slope, $\Delta s = (s_- - s_+)$, of 0.31 ($\pm$ 0.08)S (78). Earlier results (77) for the dependence of the sedimentation coefficient difference upon phenylalanine concentration are

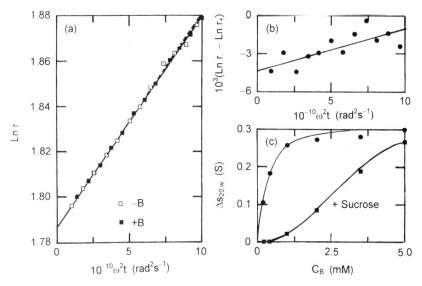

Figure 11 Studies of ligand-mediated conformational changes in rabbit muscle pyruvate kinase (pH 7.5, $I = 0.13$) by difference sedimentation velocity. (a) Separate measurement of the sedimentation coefficients of enzyme alone and in the presence of phenylalanine (5 mM). (b) Direct comparison of the two sedimentation coefficients by difference sedimentation velocity, the results being plotted according to *Equation 30*. (c) Dependence of the difference in sedimentation coefficient upon phenylalanine concentration, together with the effect of molecular crowding by sucrose (0.1 M) on that difference. Data in (a) and (b) are taken from ref. 78, and those in (c) from ref. 77.

summarized (●) in *Figure 11c*. A corresponding comparison of sedimentation coefficients for pyruvate kinase in the absence and presence of phospho*enol*pyruvate (1 mM) yielded a value of −0.03 (± 0.01)S for Δs, which signifies the likelihood that the sedimentation coefficient of 9.5S for enzyme alone is the weight-average for an equilibrium mixture of species with sedimentation coefficients of 9.47 and 9.81S. In that regard the consequent isomerization constant of 0.09 so determined matches the value deduced (79) by analysis of enzyme kinetic data in terms of the Monod model.

The question of the pre-existence or ligand-induction of the conformational change in the enzyme giving rise to the sedimentation coefficient difference can be addressed further by taking advantage of thermodynamic non-ideality arising from the crowding effect of a high concentration of an inert co-solute. Entropic considerations dictate that a crowded environment should displace any enzyme isomeric equilibrium in favour of the smaller isomer—a phenomenon illustrated in *Figure 11c* by the diminished magnitudes of Δs observed in the presence of 0.1 M sucrose (■).

To test whether the difference sedimentation velocity result obtained with phospho*enol*pyruvate reflected perturbation of a pre-existing isomerization in favour of the smaller enzyme state, the experiment was repeated in buffer supplemented with 0.1 M sucrose (77). The lack of an effect of phosph*enol*pyruvate on this occasion, $\Delta s_{20,w} = -0.003\ (\pm\ 0.005)$S, indicates that a high concentration of inert co-solute can also bring about the change in sedimentation coefficient effected by substrate. Such displacement of an isomerization equilibrium in the absence of substrate (ligand) establishes its pre-existence; and hence justifies consideration of the rabbit muscle pyruvate kinase system in terms of the Monod model of allostery.

This experimental illustration of the use of an inert co-solute for detecting protein isomerizations demonstrates the potential of thermodynamic non-ideality as a means of probing such phenomena. Indeed, the above combination of difference sedimentation velocity and molecular crowding effects has been used subsequently to establish that the conformational change undergone by yeast hexokinase as the result of glucose binding also reflects preferential interaction of substrate with an equilibrium mixture of isomeric enzyme states (78). Determining the nature of an isomerization (pre-existing or ligand-induced) had previously been a seemingly intractable problem to which there was no unequivocal solution: but now there is one.

7 Concluding remarks

Sedimentation velocity is frequently the method by which a reversible macromolecular interaction is detected during routine monitoring of the purification and properties of a protein or enzyme. The existence of solute self-association or reversible interaction between dissimilar reactants gives rise to distinctive sedimentation velocity behaviour, which may be used not only as a diagnostic of species interconversion but also as a means of obtaining a preliminary characterization of the interaction. Indeed, sedimentation velocity has proven the method of choice for examining the effects of small ligands on the interconversion between the two isomeric states of allosteric enzymes.

Absolute characterization of the equilibrium constant and reaction stoichiometry for an interaction involving a change in molecular mass is precluded by the dependence of the sedimentation coefficient upon the shape as well as size of the resulting complex species—a situation that necessitates resort to a model of any putative complex species in order to specify the magnitude of its sedimentation coefficient. However, the preliminary characterization afforded by the analysis of sedimentation velocity behaviour can be used to advantage in the design of subsequent sedimentation equilibrium studies. The latter have the potential to afford a more definitive characterization of the interaction because the molecular mass of any postulated complex species may be assigned unequivocally from those of the reactants and the specified stoichiometry. Such characterization of macromolecular interactions by sedimentation equilibrium is discussed in the next chapter.

References

1. Gerhart, J. C. and Schachman, H. K. (1968). *Biochemistry*, **7**, 538.
2. Harris, S. J. and Winzor, D. J. (1988). *Arch. Biochem. Biophys.*, **265**, 458.

3. Fujita, H. (1956). *J. Chem. Phys.*, **24**, 1084.
4. Baldwin, R. L. (1957). *Biochem. J.*, **65**, 503.
5. Creeth, J. M. and Winzor, D. J. (1962). *Biochem. J.*, **83**, 566.
6. Van Holde, K. E. (1960). *J. Phys. Chem.*, **64**, 1582.
7. Inkerman, P. A., Winzor, D. J., and Zerner, B. (1975). *Can. J. Biochem.*, **53**, 547.
8. Claverie, J.-M., Dreux, H., and Cohen, R. (1975). *Biopolymers*, **14**, 1685.
9. Cohen, R. and Claverie, J.-M. (1975). *Biopolymers*, **14**, 1701.
10. Claverie, J.-M. (1976). *Biopolymers*, **15**, 843.
11. Behlke, J. and Ristau, O. (1997). *Biophys. J.*, **72**, 428.
12. Demeler, B. and Shaker, H. (1997). *Biophys. J.*, **74**, 444.
13. Schuck, P. and Millar, D. B. (1998). *Anal. Biochem.*, **259**, 48.
14. Schuck, P. (1998). *Biophys. J.*, **75**, 1503.
15. See the Web site: http://www.nottingham.ac.uk/ncmh/ and Helmut Coelfen<coelfen@mpikg-teltow.mpg.de>
16. Philo, J. (1994). *Biophys. J.* **72**, 435.
17. Lamm, O. (1929). *Z. Physik. Chem. (Leipzig)*, **A143**, 177.
18. Stafford, W. F. (1992). *Anal. Biochem.*, **203**, 295.
19. Van Holde, K. E. and Weischet, W. O. (1978). *Biopolymers*, **17**, 1387. See also the Web site: http://www.biochem.uthscsa.edu/UltraScan
20. See the Web site: ftp://alpha.bbri.org/rasmb/spin/ms_dos/sednterp.
21. Kratky, O., Leopold, H., and Stabinger, H. (1973). In *Methods in Enzymology*, Vol. 27, p. 98.
22. Schachman, H. K. (1959). *Ultracentrifugation in biochemistry*. Academic Press, New York.
23. Lloyd, P. H. (1974). *Optical methods in ultracentrifugation, electrophoresis and diffusion with a guide to the interpretation of optical records.* Oxford Univ. Press, Fairlawn, NJ.
24. Johann, C., Thiessen, A., Deacon, M. P. and Harding, S. E. (1999). *Refractive increment data book*. Nottingham Univ. Press, Nottingham, UK.
25. See Ertughrul, O. W. D., Errington, N., Raza, S., Sutcliffe, M. J., Rowe, A. J., and Scrutton, N. S. (19XX). *Protein Eng.*, **11**, 447
26. Cherwell Scientific Limited, Oxford, UK.
27. Johnston, J. P. and Ogston, A. G. (1946). *Trans. Faraday Soc.*, **42**, 789.
28. Rowe, A. J., Wynne Jones, S., Thomas, D. G., and Harding, S. E. (1992). In *Analytical ultracentrifugation in biochemistry and polymer science* (ed. S. E. Harding, A. J. Rowe, and S. E. Horton), pp. 49–62. Royal Society of Chemistry, Cambridge, UK.
29. Harding, S. E. (1995). *Biophys. Chem.*, **55**, 69.
30. Harding, S. E., Davis, S. S., Deacon, M. P., and Fiebrig, I. (1998). *Biotechnol. Genet. Eng. Rev.*, **16**, 41.
31. Deacon, M. P., Davis, S. S., White, R. J., Nordman, H., Carlstedt, I., Errington, N., *et al.* (1999). *Carbohydr. Polym.*, **38**, 235.
32. Harding, S. E., Horton, J. C., and Colfen, H. (1997). *Eur. Biophys. J.*, **25**, 347.
33. Garcia de la Torre, J., Carrasco, B., and Harding, S. E. (1997). *Eur. Biophys. J.*, **25**, 361.
34. Pavlov, G. M., Rowe, A. J., and Harding, S. E. (1997). *Trends Anal. Chem.*, **16**, 401.
35. Gralén, N. (1944). PhD Dissertation, University of Uppsala.
36. Creeth, J. M. and Knight, C. G. (1965). *Biochim. Biophys. Acta*, **102**, 549.
37. Taylor, W. R., Thornton, J. M., and Turnell, R. J. (1980). *J. Mol. Graph.*, **1**, 30.
38. Garcia de la Torre, J., Harding, S. E., and Carrasco, B. (1999). *Eur. Biophys. J.*, **28**, 119.
39. Garcia de la Torre, J., Harding, S. E., and Carrasco, B. (1998). *Biochem. Soc. Trans.*, **26**, 716.
40. Carrasco, B. (1998). PhD Thesis, Universidad de Murcia.
41. Steinberg, I. Z. and Schachman, H. K. (1966). *Biochemistry*, **5**, 3728.

42. Gilbert, G. A. and Jenkins, R. C. L. (1959). *Proc. R. Soc. London, Ser. A*, **253**, 420.
43. Nichol, L. W. and Winzor, D. J. (1964). *J. Phys. Chem.*, **68**, 2455.
44. Chanutin, A., Ludewig, S., and Masket, A. V. (1942). *J. Biol. Chem.*, **143**, 737.
45. Velick, S. F., Hayes, J. E., Jr., and Harting, J. (1953). *J. Biol. Chem.*, **203**, 527.
46. Velick, S. F. and Hayes, J. E., Jr. (1953). *J. Biol. Chem.*, **203**, 545.
47. Arnold, H. and Pette, D. (1968). *Eur. J. Biochem.*, **6**, 163.
48. Clarke, F. M. and Masters, C. J. (1972). *Arch. Biochem. Biophys.*, **153**, 258.
49. Klotz, I. M. (1946). *Arch. Biochem.*, **9**, 109.
50. Nichol, L. W. and Ogston, A. G. (1965). *Proc. R. Soc. London, Ser. B*, **163**, 343.
51. Nichol, L. W. and Ogston, A. G. (1965). *J. Phys. Chem.*, **69**, 1754.
52. Longsworth, L. G. (1959). In *Electrophoresis, theory, methods, and applications* (ed. M. Bier), p. 91. Academic Press, New York.
53. Silkowski, H., Davis, S. J., Barclay, A. N., Rowe, A. J., Harding, S. E., and Byron, O. (1997). *Eur. Biophys. J.*, **25**, 455.
54. Stafford, W. F. (1992). *Anal. Biochem.*, **203**, 295.
55. Stafford, W. F. (1994). In *Methods in Enzymology*, Vol. 240, p. 478.
56. Raso, V., Brown, M., McGrath, J., Liu, S., and Stafford, W. F. (1997). *J. Biol. Chem.*, **272**, 27618.
57. Gilbert, G. A. (1959). *Proc. R. Soc. London, Ser. A*, **250**, 377.
58. Nichol, L. W. and Winzor, D. J. (1972). *Migration of interacting systems*. Clarendon Press, Oxford.
59. Claverie, J.-M., Dreux, H., and Cohen, R. (1975). *Biopolymers*, **14**, 1685.
60. Cohen, R. and Claverie, J.-M. (1975). *Biopolymers*, **14**, 1701.
61. Claverie, J.-M. (1976). *Biopolymers*, **15**, 843.
62. Behlke, J. and Ristau, O. (1997). *Biophys. J.*, **72**, 428.
63. Demeler, B. and Shaker, H. (1997). *Biophys. J.*, **74**, 444.
64. Schuck, P. and Millar, D. B. (1998). *Anal. Biochem.*, **259**, 48.
65. Schuck, P. (1998). *Biophys. J.*, **75**, 1503.
66. Cann, J. R. (1970). *Interacting macromolecules*. Academic Press, New York.
67. Cox, D. J. and Dale, R. S. (1981). In *Protein-protein interactions* (ed. C. Frieden and L. W. Nichol), p. 173. Wiley, New York.
68. Behlke, J., Ristau, O., and Schonfeld, H. J. (1997). *Biochemistry*, **36**, 5149.
69. Behlke, J. and Ristau, O. (1997). *Eur. Biophys. J.*, **25**, 325.
70. Monod, J., Wyman, J., and Changeux, J.-P. (1965). *J. Mol. Biol.*, **12**, 88.
71. Koshland, D. E., Jr., Némethy, G., and Filmer, D. (1966). *Biochemistry*, **5**, 365.
72. Gerhart, J. C. and Schachman, H. K. (1968). *Biochemistry*, **7**, 538.
73. Schumaker, V. N. and Adams, P. (1968). *Biochemistry*, **7**, 3422.
74. Schumaker, V. N. (1968). *Biochemistry*, **7**, 3427.
75. Howlett, G. J. and Schachman, H. K. (1977). *Biochemistry*, **16**, 5077.
76. Oberfelder, R. W., Barisas, B. G., and Lee, J. C. (1984). *Biochemistry*, **23**, 3822.
77. Harris, S. J. and Winzor, D. J. (1988). *Arch. Biochem. Biophys.*, **265**, 458.
78. Jacobsen, M. P. and Winzor, D. J. (1997). *Prog. Colloid Polym. Sci.*, **107**, 82.
79. Oberfelder, R. W., Lee, L. L.-Y., and Lee, J. C. (1984). *Biochemistry*, **23**, 3813.
80. Ralston, G. (1993). *Introduction to analytical ultracentrifugation*. Beckman Instruments Inc., Fullerton, California, USA.
81. Morgan, P. J., Harding, S. E., Plyte, S. E., and Kneale, G. G. (1989). *Biochem. Soc. Trans.*, **17**, 234.
82. Marsh, E. N. and Harding, S. E. (1993). *Biochem. J.*, **290**, 551.

Chapter 5
Sedimentation equilibrium in the analytical ultracentrifuge

Donald J. Winzor
Centre for Protein Structure, Function and Engineering, Department of Biochemistry, University of Queensland, Brisbane, Queensland 4072, Australia.

Stephen E. Harding
NCMH Physical Biochemistry Laboratory, University of Nottingham, School of Biosciences, Sutton Bonington, Leicestershire LE12 5RD, UK.

1 Introduction

For many years analytical ultracentrifugation was the major source of information on the heterogeneity and molecular size of macromolecules. In the field of protein chemistry the question of solute heterogeneity is now usually addressed by gel electrophoretic and gel chromatographic techniques, and the molecular weight is either calculated from the amino acid sequence or obtained by mass spectrometry. Because such molecular weight values refer only to the covalently-linked polypeptide chain(s), they provide no information about the macromolecular state of the functional protein or enzyme. In its simplest application molecular weight measurement by analytical ultracentrifugation is therefore used to characterize quaternary structure, which affords an example of a self-association equilibrium that has gone to completion.

For many proteins, however, the monomeric and polymeric forms coexist in association equilibrium, the relative proportions of the two macromolecular states varying with total solute concentration in accordance with *Le Chatelier's* principle: the polymeric state is favoured by an increase in concentration, whereas dilution favours the monomeric state. Analytical ultracentrifugation has great potential for characterizing the self-association equilibrium by virtue of these concentration-dependent changes in the average macromolecular state of the solute. The requirement that a rapidly re-equilibrating solute system be characterized without perturbing the equilibrium state is readily accommodated by either of the two commonly used techniques in analytical ultracentrifugation —sedimentation velocity and sedimentation equilibrium. In the former the ultracentrifuge is operated at a sufficiently high angular velocity for the centrifugal force on a solute molecule to dominate its migration. In sedimentation equilibrium the instrument is operated at a much lower angular velocity to allow a balance to be achieved between the radially-outward flow of solute and the back-diffusional flow in response to the concentration gradient being developed by the centrifugal force.

Despite the greater biological prevalence of interactions between dissimilar macromolecular reactants, protein self-association has been the predominant phenomenon studied by analytical ultracentrifugation. However, the introduction of a new generation of analytical ultracentrifuges has also kindled interest in use of the technique for characterizing interactions between dissimilar reactants. The main emphasis in current ultracentrifuge studies is thus the study of non-covalent macromolecular association equilibria: protein–protein interactions such as those involved in enzyme self-association or the binding of an antibody to its eliciting protein antigen; protein–nucleic acid interactions such as those associated with regulation of the transcription and translation of genetic information; and protein–carbohydrate interactions such as those between a lectin and the sugar moiety of a glycoprotein.

2 Experimental aspects of sedimentation equilibrium

As noted above, the sedimentation equilibrium variant of analytical ultracentrifugation entails operation of the instrument at a relatively low angular velocity that allows the centrifugally driven flow of solute to be matched by the diffusion-driven counterflow in response to the gradient in solute concentration being generated by the sedimentation process. From the viewpoint of molecular weight determination, the magnitude of the sedimentation coefficient alone does not suffice because of its dependence upon shape as well as size of the solute. In the limit of zero solute concentration the sedimentation coefficient of a solute, s_A°, is related to molecular parameters by the expression:

$$s_A^\circ = M_A(1 - \bar{v}_A \rho_s)/(Nf_A) \qquad [1]$$

where f_A denotes the shape-dependent translational frictional coefficient of the solute with molecular weight M_A and partial specific volume $\bar{v}_A$: ρ_s is the solvent density and N is Avogadro's number. Because the corresponding diffusion coefficient, D_A°, is governed by the magnitude of the translational frictional coefficient according to the expression:

$$D_A^\circ = RT/(Nf_A) \qquad [2]$$

where R and T refer to the universal gas constant and absolute temperature respectively, the influence of f_A on the separate magnitudes of the sedimentation and diffusion coefficients disappears from their ratio: specifically,

$$s_A^\circ/D_A^\circ = M_A(1 - \bar{v}_A \rho_s)/(RT) \qquad [3]$$

Inasmuch as the solute distribution at sedimentation equilibrium is governed by this ratio, the parameter to emerge from analysis of such distributions is the buoyant molecular weight, $M_A(1 - \bar{v}_A\rho_s)$.

2.1 Procedural details of a sedimentation equilibrium experiment

Sedimentation equilibrium experiments are conducted in a double-sector cell. One sector contains the macromolecular solution and the other the appropriate solvent. Because proteins, nucleic acids, and many polysaccharides bear net charge, the reference sector needs to contain buffer with which the macro-ion is in dialysis equilibrium so that the chemical potential of the macromolecule (the driving force of diffusion) is defined under conditions of constant chemical potential of solvent (1), which then comprises all diffusible components—buffer species, supporting electrolyte, small ligands, etc., as well as water. This dialysis step leads to a situation wherein the concentration of counterions in the macromolecular solution exceeds their concentration in the reference sector by the product $|Z_A|C_A/2$, where C_A is the molar concentration of solute with net charge Z_A, whereas the non-counterion concentration in the solution is lower by the same factor. Only for an uncharged macromolecule are the concentrations of diffusible ions identical in the solution and solvent sectors.

Various procedures may be used to effect the required distribution of dialysable ions. Of those, the classical procedure of exhaustive dialysis may be avoided by subjecting the macromolecular solution to zonal gel chromatography on a column pre-equilibrated with the buffer to be used in the solvent sector. Alternatively, the use of centrifugal ultrafiltration assemblies can achieve the same result. Although dialysis is precluded for a small solute, the condition of dialysis equilibrium can still be achieved by adjusting the composition of solvent in the reference sector to meet the above distribution requirements—a procedure used to investigate the micellization of chlorpromazine by sedimentation equilibrium (2).

A second consideration in the design of an experiment is the length of the solution column to be subjected to sedimentation equilibrium. Inasmuch as the time to attain equilibrium varies inversely with the square of the column length (3), columns larger than 3 mm are rarely used; and very short columns (< 1 mm) can be used to accelerate the attainment of sedimentation equilibrium (4). Whereas 16–36 hours may be required to achieve the required time-independence of solute distribution in 2–3 mm columns, effective sedimentation equilibrium can be attained (admittedly with decreased accuracy) in less than an hour by decreasing the column length to below 1 mm (4). The other means of decreasing the duration of an experiment is by resorting to initial overspeeding protocols (5, 6). Inasmuch as equilibrium is, by definition, only reached in the limit of infinite time, an experiment is deemed to have attained the condition of effective sedimentation equilibrium when distributions recorded several hours apart become time-invariant.

Protocol 1

Sedimentation equilibrium: basic operation

Equipment and reagents
- Ultracentrifuge
- Optical system
- Protein
- Buffer

Method

1. Concentration requirements of the protein: see Chapter 4, *Protocol 1*.

2. Choice of optical system: as Chapter 4, *Protocol 1*. However, with interference optics, the lower limit is ~ 0.5 mg/ml (as opposed to ~ 0.1 mg/ml for simple boundary identification with sedimentation velocity) with the maximum path length cell (12 mm) available in the XL-I centrifuge. (A 2.5-fold lower concentration is permissible with the 30 mm path length cells that can be used in the Model E.) Similarly with UV absorption optics a slightly higher loading concentration (compared with sedimentation velocity) of ~ 0.3 absorbance units should be used. For higher concentrations (> 5 mg/ml for interference optics or > 1.4 absorbance units) use shorter path length cells.

3. Choice of appropriate buffer/solvent. As Chapter 4, *Protocol 1*. For charged macromolecular systems, dialyse solutions against the reference buffer before analysis. Alternatively, employ gel chromatography to achieve dialysis equilibrium (see above).

4. Loading the solutions into the cell. As with sedimentation velocity, double sector cells need to be used. However, solution columns should be shorter to keep the time to reach equilibrium down to acceptable levels. For a 12 mm optical path length cell, a loading volume of 0.1 ml will give a ~ 0.25 cm solution column. Equilibrium is normally attained after ~ 12 h. In the ultra-short column method (4), solution columns of as little as 0.7 mm (~ 0.02 ml) can attain equilibrium within a few hours. The user may also wish to use a multichannel centrepiece in the cell(s) (see Chapter 4, *Figure 1*). These permit three or more solution/solvent pairs. Although they are generally unsuitable for sedimentation velocity work (because of an upper speed limit of ~ 40 000 r.p.m.) they are generally ideal for sedimentation equilibrium. However, because of the non-sector shape of the channels, the use of an inert fluorocarbon oil to give a 'false bottom' in each channel is recommended, although the 'intertness' should be checked beforehand.

5. Choose the appropriate temperature: as in Chapter 4, *Protocol 1*.

6. Speeds are considerably lower than speeds used for velocity sedimentation, by a factor of 2.5–3, unless the 'meniscus depletion' method is being used (about twofold lower compared with sedimentation velocity). Thus for ovalbumin ($M \sim 45\,000$) a regular low speed equilibrium experiment would be conducted at 15 000–20 000 r.p.m., whereas a meniscus depletion experiment would be run at 25 000–30 000

Protocol 1 continued

r.p.m. If meniscus depletion is being attempted, familiarize yourself carefully with its limitations (17, 18, 61). If the regular low- or intermediate speed method is being used, the meniscus concentration needs to be found either by simple extrapolation for UV absorbance optics, or by more sophisticated procedures for interference optics (see later). For schlieren optics (yielding $M_{z,app}$), no such determination is required. The larger the redistribution of solute concentration in the centrifuge cell, the greater is the accuracy of the result. However, attention should be paid to possible loss of optical registration at or near the cell base if too high a speed is chosen. Where possible, two or three equilibrium speeds should be chosen, although this can extend the length of an experiment to several days: care needs to be taken over protein stability.

7. Check for equilibrium by comparing scans/traces recorded several hours apart. Then perform a baseline determination (for UV absorption) by overspeeding (caution required with multichannel cells) and sedimenting all macromolecular solute before recording the residual absorbance. Where this is not possible—namely for smaller proteins (M below about 10 000) where the equilibrium speed is going to be 40 000 r.p.m. or higher—make sure that you have dialysed carefully beforehand and that dialysate is in the reference channel(s). For interference optics a baseline is not necessary, but it may be important to perform a 'blank' correction to correct for window distortion: this is achieved by either:

   (a) Stopping the run, agitating the cell to uniformly redistribute the solute, and returning to the equilibrium speed used before recording a scan immediately.

   (b) Stopping the run, removing the solution and reference solvent (without dismantling or loosening the torque on the cell buttress ring), flushing with water, drying with a current of air, refilling with water in both reference and solution sectors (to the same level as the equilibrium experiment), and then returning to the equilibrium speed used for an immediate scan. These methods are discussed in ref. 62.

8. Evaluate the partial specific volume of the protein (and any macromolecular ligand) using the programme *SEDNTERP* or by densimetry (see Chapter 4, *Protocol 1*).

9. Choose the appropriate software (*Protocol 2*) for analysis.

2.2 Extraction of the molecular weight of a single solute

A sedimentation equilibrium distribution comprises a relatively featureless, monotonic increase in solute concentration, c_A, with radial distance from the air–liquid meniscus (r_a) to the bottom (r_b) of the cell (*Figure 1*). Depending upon whether the absorbance or interference optical system of the Beckman XL-I ultracentrifuge is used to record the distribution, the ordinate is expressed as an absorbance at a given wavelength, $A_\lambda(r)$, or a Rayleigh fringe number, $J(r)$, both being directly proportional to the corresponding solute concentration, $c_A(r)$.

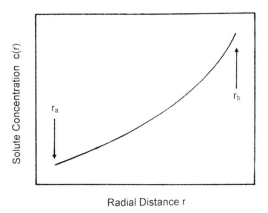

Figure 1 Schematic representation of the concentration distribution in a sedimentation equilibrium experiment on a solution with radial extremities at r_a and r_b.

Sedimentation equilibrium was initially considered in terms of the balance between the sedimentation and diffusional migration processes (7); but it was then realized that the results from such experiments were amenable to rigorous thermodynamic analysis (8, 9). Consequently, even though the experimental record is in terms of solute concentration as a function of radial distance, the distribution of a single solute at sedimentation equilibrium is defined in terms of thermodynamic activity, z_A, and the relationship (10, 11):

$$z_A(r) = z_A(r_F)\exp[M_A(1 - \bar{v}_A\rho_s)\,\omega^2(r^2 - r_F^2)/(2RT)] \qquad [4]$$

In this expression the thermodynamic activity at any given radial distance r is related to its activity at a chosen reference radial distance r_F by an exponential term involving the buoyant molecular weight, the square of the angular velocity (ω) and the difference between the squares of the two radial distances: $M_A(1 - \bar{v}_A\rho_s)\,\omega^2/(2RT)$ is a combination of parameters termed the reduced molecular weight. Such provision of information on the thermodynamic activity of solute is extremely important from the viewpoint of incorporating rigorous allowance for effects of thermodynamic non-ideality into the analysis of sedimentation equilibrium distributions (12–15). However, most studies are performed under conditions approaching thermodynamic ideality; and we therefore simplify presentation of the analysis by considering that the weight-concentrations of solute, $c_A(r)$ and $c_A(r_F)$, may be substituted for $z_A(r)$ and $z_A(r_F)$ respectively in the above expression. From the logarithmic form of *Equation 4* written in those terms it is evident that:

$$M_A(1 - \bar{v}_A\rho_s) = (2RT/\omega^2)\,d[\ln c_A(r)]/dr^2 \qquad [5]$$

which allows the buoyant molecular weight to be determined from the slope of the dependence of the natural logarithm of the concentration upon the square of radial distance. Examples of molecular weight measurement by direct analysis of a sedimentation equilibrium distribution (*Equation 4*) and by means of the

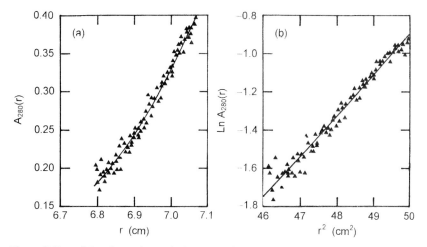

Figure 2 Use of the absorption optical system of the Beckman XL-I ultracentrifuge to determine the buoyant molecular weight of ovalbumin (pH 4.59, $I = 0.16$) by sedimentation equilibrium at 9000 r.p.m. and 20 °C. (a) Absorbance distribution at 280 nm and the best-fit description, $M_A(1 - \bar{v}_A\rho_s) = (11.6 \pm 0.4)$, in terms of *Equation 4* with $r_F = 7.000$ cm. (b) Corresponding analysis of the same distribution according to *Equation 5*, together with the best-fit relationship, $M_A(1 - v_{A\rho s}) = (11.6 \pm 0.4)$, obtained by linear regression analysis.

logarithmic transform (*Equation 5*) are presented in *Figures 2a* and *2b* respectively, which refer to results obtained with the absorption optical system.

Direct application of *Equations 4* and *5* to sedimentation equilibrium distributions recorded by means of the Rayleigh interference optical system is precluded by the fact that the ordinate of the distribution is recorded refractometrically in terms of the difference between the solute concentration at radial distance r and that at the air–liquid meniscus, r_a. This concentration difference is expressed in terms of Rayleigh fringes $j(r)$, where 3.33 fringes corresponds to a concentration difference of 1 mg/ml for proteins (16). $J(r)$, the number of fringes corresponding to the solute concentration $c_A(r)$ must therefore be obtained from the expression:

$$J(r) = J(r_a) + j(r) \qquad [6]$$

which clearly requires knowledge of $J(r_a)$, the refractive index counterpart of $c_A(r_a)$, the solute concentration at the air–liquid meniscus.

An elegant means of overcoming the need to measure $J(r_a)$ is the selection of an angular velocity that ensures a value of essentially zero for the solute concentration at the meniscus (*Figure 3a*), whereupon $J(r) \approx j(r)$. Such practice forms the basis of the very popular high-speed or meniscus-depletion variant of sedimentation equilibrium (17). However, this obvious advantage of the high-speed technique is accomplished at the expense of the range of r over which the radial dependence of $J(r)$ can be used; and hence of the inherent accuracy of the molecular weight measurement. Greater accuracy is attainable from distributions obtained by the low-speed sedimentation equilibrium technique (3) because

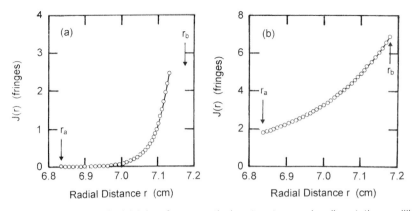

Figure 3 Use of the Rayleigh interference optical system to record sedimentation equilibrium distributions for α-chymotrypsin. (a) Distribution obtained by subjecting an enzyme solution (pH 4.1, I 0.08) to centrifugation at 34 000 r.p.m. to achieve solute depletion at the meniscus. (b) Distribution from a sample of α-chymotrypsin (pH 3.9, I 0.20) spun at 14 000 r.p.m.

of the ability to analyse the entire distribution (*Figure 3b*). However, its use is conditional upon the measurement of $J(r_a)$ to allow the calculation of $J(r)$ from $j(r)$ via *Equation 6*.

Various methods are available for the determination of $J(r_a)$. In the 'intercept over slope' procedure (18) which has been incorporated in the program 'MSTARI' for molecular weight analysis from interference optical records (19) the fundamental differential equation of sedimentation equilibrium is manipulated to yield the expression:

$$2RT\, j(r)/[M^*(r)\,(1 - \bar{v}_A\rho_s)\,\omega^2] = J(r_a)\,(r^2 - r_a^2) + 2\int_{r_a}^{r} [r\, j(r)]\,dr \qquad [7]$$

where $M^*(r)$ is an operational point-average molecular weight at radial distance r: {under thermodynamically ideal conditions for a single solute, $M^*(r) = M_A$ for all r: other useful identities for heterogeneous and non-ideal systems are that $M^*(r_a) = M_{w,app}(r_a)$ and $M^*(r_b) = M_{w,app}$ for the whole distribution (i.e. from meniscus to cell base) in the ultracentrifuge cell (18)}. Rearrangement of *Equation 7* to the form:

$$j(r)/(r^2 - r_a^2) = J(r_a)M^*(r)\,(1 - \bar{v}_A\rho_s)\,\omega^2/(2RT)$$
$$+ [M^*(1 - \bar{v}_A\rho_s)\,\omega^2/\{2RT(r^2 - r_a^2)\}] \int_{r_a}^{r} [r\, j(r)]\,dr \qquad [8]$$

shows that the dependence of $j(r)/(r^2 - r_a^2)$ upon $[\int\{r\, j(r)\}dr]/(r^2 - r_a^2)$ has an ordinate intercept of $J(r_a)M^*(r_a)\,(1 - \bar{v}_A\rho_s)\,\omega^2/(2RT)$ and a limiting slope, as $r \to r_a$, of $M^*(r_a)\,(1 - \bar{v}_A\rho_s)\,\omega^2/(RT)$. The ratio of the ordinate intercept to the limiting slope eliminates $M^*(r_a)$ and is just $J(r_a)/2$. The practical disadvantage of this procedure is that the *limiting slope* is required near the meniscus: data near the meniscus can be notoriously noisy because of the small fringe increments: this problem can be partly obviated by multiple sampling of data sets for a range of radial positions and extrapolating estimated $J(r_a)$ values to $r \to r_a$ (19). Of course once $J(r_a)$ has been found in this way, an estimate for $M^*(r_a) = M_{w,app}(r_a)$ can be obtained from the slope.

Provided that the entire sedimentation equilibrium distribution is resolved, the problems of extrapolation procedure to obtain $J_a(r)$ can be obviated by performing the integration in *Equations 7* and *8* over the entire range of the distribution, r_a to r_b (the cell base), and invoking the mass conservation requirement that:

$$J_o(r_b^2 - r_a^2) = 2\int_{r_a}^{r_b} [r\{J(r_a) + j(r)\}]dr = \int_{r_a}^{r_b} [J(r_a) + j(r)]dr^2 \quad [9]$$

where J_o is the number of Rayleigh fringes observed by forming a boundary between solvent and the loaded solute solution in a separate experiment with a synthetic boundary cell. It then follows that:

$$J(r_a) = [J_o(r_b^2 - r_a^2) - \int_{r_a}^{r_b} j(r)dr^2]/(r_b^2 - r_a^2) \quad [10]$$

The integration required for the application of this procedure is incorporated into the *MSTAR* program (19) referred to in *Protocol 2*, which allows for revision of the $J(r_a)$ value on the basis of the J_o value thereby deduced (20). It should be noted that mass conservation arguments may also be used to yield an equivalent expression for $J(r_a)$, namely (21):

$$J(r_a) = J_o - [j(r_b)r_b^2 - \int_{r_a}^{r_b} r^2 dj]/(r_b^2 - r_a^2) \quad [11]$$

An alternative procedure for determining a model-independent value of $J_a(r)$ entails location of the hinge-point r_h, the radial distance where the solute concentration may be identified with the loading concentration, i.e. the point where $J(r) = J_o$. In current-generation ultracentrifuges the adoption of this procedure relies upon concurrent use of the absorption optical system to record not only the equilibrium distribution but also the distribution immediately after attainment of rotor speed, when $c_A(r)$ is the loading concentration for all r (22). After ascertaining r_h from the intersection point of the two absorbance distributions, the magnitude of $J(r_a)$ is obtained as $J_o - j(r_h)$, whereupon the absolute fringe distribution, $J(r)$ versus r, is again determined from *Equation 6*.

Protocol 2

Sedimentation equilibrium: software

1   Choose first of all a molecular weight program that does not assume a model (monomer, ideal, associating etc.). One such programme is *MSTAR* (19) with two versions: *MSTARA* for UV absorption optics, *MSTARI* for interference optics (a new version *MSTARS* is being constructed for schlieren optics). *MSTAR* works out, amongst other things (i) the (apparent weight) average molecular weight $M_{w,app}$ over the whole distribution of solute in the centrifuge cell (from meniscus r_a to base r_b) and (ii) the (apparent) point weight-average molecular weight $M_{w,app}(r)$ as a function of radial position, r (or the equivalent local concentration $c(r)$. (N.B. If the ultrashort-column

Protocol 2 continued

technique is used it will be difficult to obtain reliable point average data.) The difference between the value of $M_{w(app)}$, recorded over a range of loading concentration and the monomer molecular weight M_1 (from sequence or MALDI mass spectrometry), should give an idea of the presence of protein–macromolecule interactions, and the stoichiometry. Any increase in $M_{w,app}(r)$ vs $c(r)$ should give an additional indication of protein–macromolecular ligand interaction phenomena. $M_{z,app}(r)$ and $M_{z,app}(r)$ data can also be obtained from absorption or interference records (consult the help file for instructions), but reliable data of this sort may not be achievable with UV absorption and is difficult with interference optics because it requires a double mathematical differentiation of the raw concentration data—and each successive differentiation amplifies noise): if $M_{z,app}$ data is required, it is better to use schlieren optics and to consult an advanced user. (The purpose of determining both M_z and M_w is that the ratio M_z/M_w is a useful indicator of heterogeneity.)

2   The routines described in step 1 are based on the extraction of apparent molecular weight (i.e. not corrected for non-ideality). For non-interacting systems non-ideality is either negligible at low concentration (for proteins, usually < 0.5 mg/ml) or is eliminated by extrapolation of $M_{w,app}$ or $M_{w,app}(r)$ to zero concentration. However, care has to be exercised with interacting systems because the extrapolation to zero concentration favours the dissociated state (depending on the reaction strength).

3   If a protein–macromolecular interaction has been identified (using e.g. step 1) then use a program specifically designed for the analysis of interacting systems: for (i) checking the stoichiometry of an interaction, (ii) estimating interaction constants X_i (in g/l) or K_i (in l/mol). Three examples are *ASSOC4* provided by Beckman (64), *NONLIN* (65), and *PSI* (66). *ASSOC4* and *NONLIN* are based on direct analysis of the $c(r)$ vs r data (essentially a truncated form of *Equation 14*). *NONLIN* also permits 'global analysis' in the sense of incorporating multiple sets of data recorded at *different speeds* or *different loading concentrations* in the analysis. Both the Beckman software and *NONLIN* permit the second thermodynamic virial coefficient to be entered if known: this can be predicted from knowledge of the triaxial molecular dimensions, hydration and charge properties (67). However, floating this parameter as a variable to extract association constants is not recommended (see Sections 3.2 and 4.1.2). *PSI* (based on *Equation 24*) allows for the use of arbitrary reference points as opposed to the meniscus (where the data is the least reliable), and permits 'global' analysis in the sense of different species recorded using different optical systems. It is essentially a model independent version of an earlier programme called *OMEGA* (19) based on the omega function (*Equations 15* and *16*). PSI also facilitates incorporation of non-ideality and, conversely for single solute systems, provides a means for estimating the second thermodynamic virial coefficient.

2.3 Extraction of point average molecular weights for interacting systems

In a sedimentation equilibrium study of the simplest acceptor–ligand system (A + B ⟷ C) the solute distribution of each species is described by *Equation 4* written for that specific species, together with the additional restriction that the thermodynamic activities (concentrations for an ideal system) of the two reactants and complex must comply with the law of mass action for the equilibrium reaction (23, 24). From the viewpoint of molecular weight determination, any single estimate must clearly be an average value that takes into account the proportion as well as the molecular weight of each species. Furthermore, the relative composition varies with radial distance (*Figure 4a*) and hence the average molecular weight also reflects that variation, which is manifested (*Figure 4b*) as curvilinearity in the plot of total concentration $c(r)$ (—) according to *Equation 5*. The traditional molecular weight approach to ultracentrifugal analysis of such a system thus requires delineation of the radial dependence of the weight-average molecular weight, $\bar{M}_w(r)$, which is governed by the expressions:

$$\bar{M}_w(r) = [c_A(r)M_A + c_B(r)M_B + c_C(r)M_C]/\bar{c}(r) \qquad [12a]$$

$$\bar{c}(r) = c_A(r) + c_B(r) + c_C(r) \qquad [12b]$$

$$c_C(r) = X_{AB}c_A(r)c_B(r) \qquad [12c]$$

where X_{AB}, the association equilibrium constant with species concentrations (c_i) expressed in g/litre, is related to its more traditional molar counterpart, K_{AB}, by:

$$X = K_{AB}M_C/(M_AM_B) \qquad [12d]$$

Larger changes in average molecular weight are observed when the optical system allows delineation of a constituent distribution, say $\bar{c}_B(r) = c_B(r) + (M_B/M_C)c_C(r)$, rather than (or as well as) the distribution in terms of total concentration, $\bar{c}(r)$. This gives rise to a more curvilinear plot for the dependence of ln $[\bar{c}_B(r)]$ upon square of the radial distance (– – –, *Figure 4b*); and reflects the fact that the constituent weight-average molecular weight, $\bar{M}_B(r)$ is given by the relationship:

$$\bar{M}_B(r) = [c_B(r) + c_C(r)]M_B/\bar{c}_B(r) \qquad [12e]$$

Methods have certainly been developed (18, 25–28) for determining the required point-average molecular weights $\bar{M}_w(r)$—a task that amounts to defining the slope of the tangent to the curvilinear plot in *Figure 4b* for each value of r. Furthermore, the analysis of the consequent [$\bar{M}_w(r)$, $\bar{c}(r)$] or [$\bar{M}_B(r)$, $\bar{c}_B(r)$] data has also been the subject of theoretical deliberations (23, 24), but the method does not seem to have been put into practice. It transpires that direct analysis of the sedimentation equilibrium distribution affords a simpler approach which avoids the undesirable magnification of experimental error associated with differentiation to obtain d[ln $\bar{c}(r)$]/dr^2 or d[ln $\bar{c}_B(r)$]/dr^2 at each radial distance.

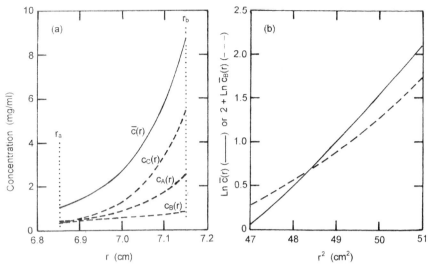

Figure 4 Characterization of an acceptor–ligand interaction by molecular weight analysis of a sedimentation equilibrium distribution. (a) Concentration distributions from a simulated sedimentation equilibrium experiment (12 000 r.p.m., 20 °C) based on *Equation 4* and the following set of parameters for a 1:1 interaction between acceptor and ligand with respective molecular weights of 60 000 and 20 000 (buoyant molecular weights of 15.6 and 5.2): r_a = 6.85 cm, r_b = 7.15 cm, r_F = 7.00 cm; $C_A(r_a)$ = 5 μM, $C_B(r_a)$ = 20 μM, K_{AB} = 40 000 M^{-1}. Broken lines signify the simulated distributions for the indicated species, whereas the solid line is the distribution in terms of total solute concentration $\bar{c}(r)$. (b) Analysis of the distributions in terms of total solute concentration $\bar{c}(r)$ and ligand constituent concentration $\bar{c}_B(r)$ according to *Equation 5*.

2.4 Direct curve-fitting of concentration distributions

For a reversible interaction involving complex formation between a multivalent acceptor A and a univalent ligand B there are only two independent sedimentation equilibrium distributions to consider. That for the A component, which includes acceptor contributing to the various complexes AB_i as well as free reactant, is given by:

$$\bar{C}_A(r) = C_A(r) + K_{AB}C_A(r)C_B(r) + K_{AB_2}C_A(r)[C_B(r)]^2 + \ldots \quad [13a]$$

whereas the sedimentation equilibrium distribution for the ligand component is described by:

$$\bar{C}_B(r) = C_B(r) + K_{AB}C_A(r)C_B(r) + 2K_{AB_2}C_A(r)[C_B(r)]^2 + \ldots \quad [13b]$$

K_{AB} and K_{AB2} are binding constants that describe the combined concentrations of all complexes with a given stoichiometry. Molar concentrations (C_i) have been used to allow expression of the proportions of A and B in the various complexes as simple integers. The condition of sedimentation equilibrium, *Equation 4* with $C_i(r)$ substituted for $z_A(r)$ in an ideal system, is now introduced to allow *Equations 13a* and *13b* to be written in the form:

$$\bar{C}_A(r) = C_A(r_F) \psi_A(r) + K_{AB}C_A(r_F)C_B(r_F) \psi_A(r) \psi_B(r)$$
$$+ K_{AB_2}C_A(r_F)[C_B(r_F)]^2\psi_A(r)[\psi_B(r)]^2 + \ldots \quad [14a]$$

$$\bar{C}_B(r) = C_B(r_F) \psi_B(r) + K_{AB}C_A(r_F)C_B(r_F) \psi_A(r) \psi_B(r)$$
$$+ 2K_{AB_2}C_A(r_F)[C_B(r_F)]^2\psi_A(r)[\psi_B(r)]^2 + \ldots \quad [14b]$$

$$\psi_i(r) = \exp[M_A(1 - \bar{v}_{ips}) \omega^2(r^2 - r_F^2)/(2RT)]$$
$$i = A, B \quad [14c]$$

On noting that $\psi_A(r) = [\psi_B(r)]^u$ where $u = [M_A(1 - \bar{v}_A\rho_s)]/[M_B(1 - \bar{v}_B\rho_s)]$, the right-hand sides of *Equations 14a* and *14b* are discrete multinomials in $\psi_B(r)$, with the coefficients of the series defined in terms of the constant parameters $c_A(r_F)$, $c_B(r_F)$, K_{AB}, K_{AB2}, etc. Furthermore, determination of the buoyant molecular weight, $M_i(1 - \bar{v}_i\rho_s)$, of the two reactants from separate sedimentation equilibrium experiments on each reactant in the absence of the other allows $\psi_B(r)$ to be regarded as the independent variable (22, 29). The extent to which advantage may be taken of direct curve-fitting to *Equation 14* for evaluation of the equilibrium constant(s) and the reference radial concentrations of the two free reactants clearly depends upon the nature and number of sedimentation equilibrium distributions available for analysis.

Inasmuch as the sedimentation equilibrium distributions are recorded optically in the analytical ultracentrifuge, there are several situations that may be encountered. The maximal potential for quantitative analysis of a sedimentation equilibrium experiment pertains when the optical system allows access to the radial distributions of the separate concentrations of acceptor and ligand constituents, $\bar{C}_A(r)$ and $\bar{C}_B(r)$, whereupon the objective of direct curve-fitting is the best description of both distributions in terms of a single set of parameters (30–33). To date this approach has been restricted to the analysis of sedimentation equilibrium distributions recorded at different wavelengths to resolve the separate distributions for the A and B constituents—an approach that relies upon additivity of the absorbances of reactants and complex(es) at each given wavelength. It is therefore necessary to demonstrate the validity of this assumption/approximation by recording spectra of mixtures with a range of acceptor/ligand ratios.

Because the Beckman XL-A ultracentrifuge is only equipped with the absorption optical system, the study of interactions such as those between proteins and polysaccharides is disadvantaged in that the only recorded sedimentation equilibrium distribution is in terms of the protein constituent. This situation needs to be accommodated despite a relative lack of information on which to base the quantitative analysis. In similar vein there is a need to consider the circumstance in which the only available sedimentation equilibrium distribution is related to the combined constituent concentrations of acceptor and ligand, $\bar{C}_A(r) + \bar{C}_B(r)$. This situation is the norm in sedimentation equilibrium studies of interactions by means of the interference optical system, but has also been encountered in an XL-A investigation of an interaction between two flavoproteins (34).

The latter two situations are clearly suboptimal from the viewpoint of interpreting quantitatively a sedimentation equilibrium experiment by direct curve-fitting of the single distribution that is available. Alternative procedures based on the integrated form of the sedimentation equilibrium expression for a single solute (*Equation 4*) have been devised which illustrate the feasibility of a quantitative analysis even under these adverse circumstances (22, 34).

2.5 Omega and psi analyses of sedimentation equilibrium distributions

A major breakthrough in the interpretation of sedimentation equilibrium distributions for interacting systems was the decision to abandon the use of molecular weight analysis in favour of evaluation of the thermodynamic activity of the smallest species contributing to a sedimentation equilibrium distribution (11, 35, 36). For this purpose the omega function for the smaller (ligand) reactant, $\Omega_B(r)$, was obtained from the Rayleigh sedimentation equilibrium distribution by means of the relationship:

$$\Omega_B(r) = [\bar{c}(r)/\bar{c}(r_F)]\exp[M_B(1 - \bar{v}_B\rho_s)\omega^2(r_F^2 - r^2)/(2RT)] \quad [15]$$

where $\bar{c}(r)$ and $\bar{c}(r_F)$ denote the total solute concentrations at the respective radial and reference radial positions. On the grounds that:

$$\lim_{\bar{c}(r) \to 0} \Omega_B(r) = z_B(r_F)M_B/\bar{c}(r_F) \quad [16]$$

the thermodynamic activity of free ligand at the reference radial position, $z_B(r_F)$, was then obtained from the ordinate intercept of the dependence of $\Omega_B(r)$ upon $\bar{c}(r)$; and the thermodynamic activity of free ligand throughout the distribution determined by applying *Equation 4* (11). Subject to the validity of assumed thermodynamic ideality, each experimentally determined value of $c_B(r)$ could be subtracted from $\bar{c}_B(r)$ to yield a revised concentration distribution, $\bar{c}_A^*(r)$ versus r, with the acceptor the smallest contributor. Repetition of the above steps with the omega function defined as $\Omega_A(r)$ could then be used to yield $c_A(r_F)$ and hence $c_A(r)$ throughout the distribution. As noted at that early stage (35, 36), successive application of the omega procedure to the residual distributions has the potential to define the concentrations of all reactant and product species contributing to the original dependence of $\bar{c}(r)$ upon radial distance; and hence to define the equilibrium constant(s) describing complex formation between acceptor and ligand.

In the event that separate concentration distributions are available for the acceptor and ligand constituents, the same approach can be applied to each of the distributions on the basis of a redefined definition of $\Omega_i(r)$, namely:

$$\Omega_i(r) = [\bar{C}_i(r)/\bar{C}_i(r_F)]\exp[M_i(1 - \bar{v}_i\rho_s)\omega^2(r_F^2 - r^2)/(2RT)] \quad [17]$$

to obtain the free acceptor and free ligand activity distributions from the respective sedimentation equilibrium distributions in terms of $\bar{C}_A(r)$ and $\bar{C}_B(r)$. For the ideal case subtraction of each $\bar{C}_i(r)$ from $C_i(r)$ allows access to $C_{AB}(r)$ from the next round of omega analysis on the residual distributions for acceptor and ligand constituents. Equilibrium constants evaluated under such circumstances should exhibit less experimental error because of the greater amount of information upon which the analysis is based.

A drawback of the omega analysis is the extent of reliance placed upon the accuracy of a curvilinear extrapolation to obtain $C_i(r_F)$ from the ordinate intercept of the dependence of $\Omega_i(r)$ upon $\bar{C}_i(r)$ (Figure 5a). Furthermore, the method is open to criticism on the statistical grounds that there is no independent variable. These deficiencies are readily overcome by resorting to the psi function (13–15, 22), which has already been defined in Equation 14c. As well as providing the basis for the model-dependent curve-fitting procedure described above, Equations 14a and 14b may be used to evaluate the free concentrations of acceptor and ligand at the reference radial position independently of any model of the interaction. Specifically, division of Equation 14b by $\psi_B(r)$ yields the expression:

$$\bar{C}_B(r)/\psi_B(r) = C_B(r_F) + K_{AB}C_A(r_F)C_B(r_F)\psi_A(r) + \ldots \quad [18a]$$

which signifies that the concentration of free ligand at the reference radial position may be obtained as the ordinate intercept of the dependence of $\bar{C}_B(r)/\psi_B(r)$ upon $\psi_A(r)$. This dependence is linear if complex formation is restricted to 1:1 stoichiometry and acceptably curvilinear if higher complexes are formed (Figure 5b). Corresponding analysis of the acceptor constituent concentration distribution in terms of the expression:

$$\bar{C}_A(r)/\psi_A(r) = C_A(r_F) + K_{AB}C_A(r_F)C_B(r_F)\psi_B(r) + \ldots \quad [18b]$$

has the potential to yield $C_A(r_F)$ (Figure 5c) and hence $C_A(r)$ throughout the distribution. K_{AB}, K_{AB_2}, etc., may therefore be evaluated by curve-fitting the [$\bar{C}_A(r)$,

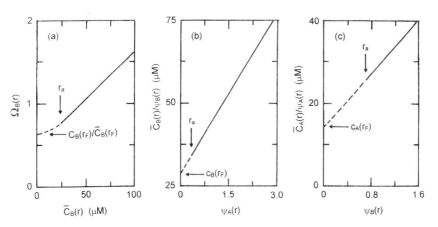

Figure 5 Comparison of the omega and psi procedures for analysis of the sedimentation equilibrium distributions simulated in Figure 3. (a) Evaluation of $C_B(r_F)/\bar{C}_B(r_F)$ as the ordinate intercept of the plot of $\bar{C}_B(r)$ data according to Equation 17 with $r_F = 7.00$ cm. (b) Corresponding psi analysis of the same data according to Equation 18a to obtain $C_B(r_F)$ as the ordinate intercept. (c) PSI analysis (Equation 18b) of the distribution for acceptor constituent to obtain $C_A(r_F)$.

$\bar{C}_B(r)$, $C_A(r)$, $C_B(r)$] data set in terms of *Equations 14a* and *14b*. Alternatively, the combination of $C_B(r)$ from analysis of the ligand constituent concentration distribution with the corresponding values of $\bar{C}_B(r)$ and $\bar{C}_A(r)$ allows analysis of the results in conventional binding fashion, namely:

$$\nu(r) = [\bar{C}_B(r) - C_B(r)]/\bar{C}_A(r) \qquad [19a]$$

$$= \{K_{AB}C_B(r) + 2K_{AB2}[C_B(r)]^2 + ..\}/\{1 + K_{AB}C_B(r) + K_{AB2}[C_B(r)]^2 + ...\} \qquad [19b]$$

Such analysis has been used to characterize an electrostatic interaction between ovalbumin and cytochrome *c* at low ionic strength and neutral pH (21).

Although lacking the sophistication of the direct curve-fitting procedure, the psi analysis does have some advantages from the experimental viewpoint.

(a) In keeping with its predecessor, the omega analysis, the method based on the psi function extracts experimental estimates of the concentration distributions for the free reactants independently of any model of the acceptor–ligand interaction.

(b) $\psi_i(r)$ is a transformed but acceptable independent variable because of the essential absence of uncertainty in the measurement of radial distance.

(c) If desired, complete separation of the dependent ($\bar{C}_B(r)$) from the independent ($\psi_B(r)$) variable can be effected by determining $C_B(r_F)$ as the limiting slope of the dependence of $\bar{C}_B(r)$ upon $\psi_B(r)$ (*Equation 14b*) rather than as the ordinate intercept of the dependence of $\bar{C}_B(r)/\psi_B(r)$ upon $\psi_A(r)$ (*Equation 18a*).

A description of the execution of this procedure is given in *Protocol 3*.

Protocol 3

Characterization of a macromolecular interaction between dissimilar reactants by sedimentation equilibrium

Equipment and reagents

- See *Protocol 1*

Method

1. Evaluate the buoyant molecular mass, $M_1(1 - \bar{v}\rho_s)$, of each reactant (A, B) by sedimentation equilibrium of the individual reactants in the buffer to be used for mixtures thereof (see *Protocol 1*).

2. On the basis of a rough estimate of the binding constant for the acceptor–ligand interaction, prepare a mixture with composition [$\bar{C}_A$, $\bar{C}_B$] that should give rise to an equilibrium mixture with significant concentrations of complex and reactant species.

3. Select a rotor speed for the equilibrium run from the weight-average molecular mass predicted (*Equation 12*) for such an equilibrium mixture.

4. Subject the reaction mixture (150 μl) to centrifugation at the designated rotor

Protocol 3 continued

speed for 16-28 hours, by which time the solute distribution(s) recorded at 4 hourly intervals should be superimposable. Where possible, use a combination of optical records of the distribution that allows delineation of $\bar{C}_A(r)$ and $\bar{C}_B(r)$.

5. Evaluate $C_B(r)$, the concentration of smaller reactant (ligand) at a selected reference radial position (e.g. column midpoint) from the ordinate intercept of the dependence of $\bar{C}_B(r)/\psi_B(r)$ upon $\psi_A(r)$ (*Equation 18*).

6. Determine $C_B(r) = C_B(r_F)\psi_B(r)$ throughout the entire distribution.

7. Calculate the binding function, $v(r)$, from $\bar{C}_A(r)$, $\bar{C}_B(r)$, and $C_B(r)$ for the evaluation of the binding constant(s) via *Equation 19*.

8. In the event that the only solute distribution available is in terms of total solute concentration, $[\bar{C}_A(r) + \bar{C}_B(r)]$, substitute this parameter for $\bar{C}_B(r)$ in step 5 to obtain $C_B(r_F)$ and hence $C_B(r)$ from step 6.

9. Subtract $C_B(r)$ from $[\bar{C}_A(r) + \bar{C}_B(r)]$ to obtain a revised total solute distribution in which A is now the smallest species.

10. Obtain $C_A(r_F)$ from the ordinate intercept of the dependence of $[\bar{C}_A(r) + \bar{C}_B(r) - C_B(r)]/\psi_A(r)$ upon $\psi_{AB}(r)$; and hence $C_A(r) = C_A(r_F)/\psi_A(r)$ throughout the distribution.

11. Refer to refs 22 and 34 for the evaluation of equilibrium constants (K_{AB}, K_{AB_2}, etc.) from the $[\bar{C}_A(r), \bar{C}_B(r), C_A(r), C_B(r)]$ data set.

3 Sedimentation equilibrium studies of ligand binding

A major limitation of *sedimentation velocity* procedures (as described in the previous chapter) for the characterization of acceptor–ligand interactions is the need to adopt a hydrodynamic model of any postulated complex in order to assign a magnitude to s_{AB}. Inasmuch as the molecular weight of the same complex can be assigned unambiguously on the basis of M_A, M_B and the postulated stoichiometry (i), the analysis of sedimentation equilibrium distributions has obvious advantages over sedimentation velocity methods for characterizing acceptor–ligand interactions. Apart from an isolated study of the ovalbumin–lysozyme interaction 20 years ago by means of the omega analysis (36), the use of sedimentation equilibrium for the characterization of acceptor–ligand interactions has been restricted to post-1990. By then the advantages of direct analysis of concentration (absorbance) distributions had been realized; and accordingly the use of average molecular weights for such characterization has largely been by-passed.

3.1 Evaluation of the concentration distributions of individual species

To date there have been only three studies in which the characterization of an acceptor–ligand interaction has been based on the extraction of model-independent concentration distributions for the two free reactants from sedimentation equilibrium experiments on mixtures of the two reactants. Because the inaugural experimental study (36) employed the Rayleigh optical system, it inevitably encountered the least desirable situation where the only information available is in terms of the total solute concentration distribution. A similar situation was encountered in a recent study of the interaction between an electron transferring flavoprotein and trimethylamine dehydrogenase by means of the XL-A absorption optical system (34), both reactants being flavoproteins with comparable spectral characteristics throughout the entire range of accessible ultra-violet and visible wavelengths. In the other study (22) the two reactants, ovalbumin and cytochrome c, were chosen deliberately to illustrate the optimal situation with access to the separate constituent concentration distributions, $\bar{C}_A(r)$ and $\bar{C}_B(r)$.

3.1.1 Analysis of the total concentration distribution

In the characterization of the ovalbumin–lysozyme interaction from Rayleigh optical records of the total concentration distribution (36), the omega analysis (*Equation 15*) was first used to evaluate $c_B(r_F)$ (*Figure 6a*) and hence the concentration distribution of free lysozyme throughout the distribution via *Equation 4*. Subtraction of $c_B(r)$ from $\bar{c}_B(r)$ then led to a revised total distribution, $\bar{c}^*(r)$ versus r, with ovalbumin as the smallest solute species. Repetition of the omega analysis on the revised distribution (*Figure 6b*) then yielded $c_A(r_F)$ and hence $c_A(r)$ throughout the distribution.

A comparable procedure was adopted in a sedimentation equilibrium study (34) of the interaction between an electron transferring flavoprotein (B) and tri-

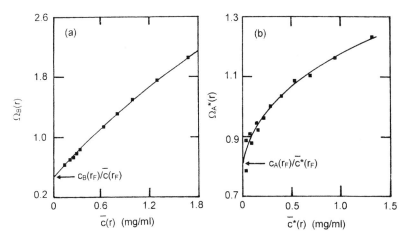

Figure 6 Determination of the free concentrations of lysozyme (B) and ovalbumin (A) by omega analysis of a Rayleigh sedimentation equilibrium distribution for a mixture of the two reactants. (a) Dependence of $\Omega_B(r)$ (*Equation 15*) upon total concentration $\bar{c}(r)$ to obtain $c_B(r_F)$ in a mixture with a total concentration, $\bar{c}(r_F)$, of 0.52 mg/ml. (b) Corresponding dependence of $\Omega_A^*(r)$ residual total concentration $c^*(r)$ to obtain $c_A(r_F)$ at the same reference radial position. Data are taken from ref. 36.

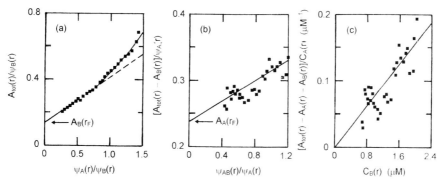

Figure 7 Characterization of the interaction between an electron transferring flavoprotein (B) and trimethylamine dehydrogenase (A) by psi analysis of a sedimentation equilibrium distribution (A_{280}) for a mixture of the two reactants. (a) Determination of $A_B(r_F)$ as the ordinate intercept of the dependence of $A_{tot}(r)/\psi_B(r)$ upon $\psi_A(r)/\psi_B(r)$ for a mixture with $A_{tot}(r_F) = 0.45$. (b) Corresponding plot of the residual total absorbance distribution to obtain $A_A(r_F)$ as the ordinate intercept. (c) Evaluation of the equilibrium constant K_{AB}. Data are taken from ref. 34.

methylamine dehydrogenase (A) except that psi rather than the omega function was used to evaluate absorbances corresponding to $c_B(r_F)$ and $c_A(r_F)$ (*Figures 7a* and *7b*). The equilibrium constant, K_{AB}, was then obtained by plotting the residual absorbance at 280 nm divided by the free concentration of dehydrogenase as a function of free electron transferring flavoprotein (*Figure 7c*). Because trimethylamine dehydrogenase (a dimer) possesses two sites for electron transferring flavoprotein, the slope of *Figure 7c* (K_{AB} multiplied by the molar absorption coefficient of complex) almost certainly defines $2k_{AB}$, where k_{AB} is the intrinsic binding constant (37). In that regard the essentially linear form of *Figure 7b* signifies the contribution of essentially a single species to the residual absorbance (see *Figure 5*), whereupon the contribution of AB_2 must be negligible—an inference supported by the essentially linear nature of *Figure 7c*.

3.1.2 Analysis of separate constituent concentration distributions

As noted above, the only other attempt to extract a reactant concentration distribution from sedimentation equilibrium results for an acceptor–ligand system (22) involved a study of the electrostatic interaction between cytochrome *c* (B) and ovalbumin (A) at low ionic strength (pH 6.3, $I = 0.03$). *Figure 8a* presents sedimentation equilibrium distributions recorded at 410 and 280 nm for a mixture of ovalbumin and cytochrome *c* that had been centrifuged to equilibrium at 15 000 r.p.m. and 20 °C. Because the cytochrome *c* constituent is the only contributor to the distribution recorded at 410 nm, the radial dependence of $\bar{c}_B(r)$ is readily obtained as $A_{410}(r)$ divided by the molar absorption coefficient of cytochrome *c* at that wavelength (*Figure 8b*). Knowledge of the relative magnitudes of the molar absorption coefficients of cytochrome *c* at 280 and 410 nm then allows calculation of the contribution of $\bar{c}_B(r)$ to $A_{280}(r)$, whereupon $\bar{c}_A(r)$ is obtained from the residual absorbance at each radial distance (*Figure 8b*).

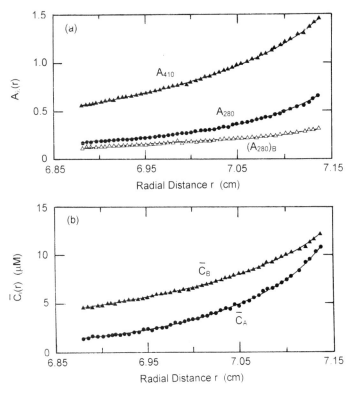

Figure 8 Sedimentation equilibrium distributions at 15 000 r.p.m. and 20 °C for a mixture of ovalbumin (A) and cytochrome c (B). (a) Distributions in terms of absorbances at 410 and 280 nm, together with the estimated contribution of the cytochrome c constituent to the absorbance at 280 nm. (b) Constituent concentration distributions for the two components. Data are taken from ref. 22.

Analysis of the $\bar{C}_B(r)$ distribution in terms of *Equation 18a* is presented in *Figure 9a*, where the essentially linear dependence of $\bar{C}_B(r)/\psi_B(r)$ upon $\psi_A(r)$ indicates the dominance of 1:1 complex formation between cytochrome c and ovalbumin under these conditions. Substitution of the value of $C_B(r_F)$ obtained from the ordinate intercept into *Equation 4* again allows calculation of $C_B(r)$ throughout the distribution, and hence of the binding function $v(r)$ via *Equation 19*. Results from a series of sedimentation equilibrium experiments on acceptor-ligand mixtures are presented as a binding curve in *Figure 9b*, which signifies a binding constant K_{AB} (*Equation 19b*) of (60 000 ± 2000) M^{-1} for 1:1 complex formation between ovalbumin and cytochrome c (22).

3.2 Direct modelling of sedimentation equilibrium distributions

The majority of studies of acceptor-ligand interactions by sedimentation equilibrium have employed direct curve-fitting of absorbance distributions at two or more wavelengths in order to determine the binding constant for 1:1 inter-

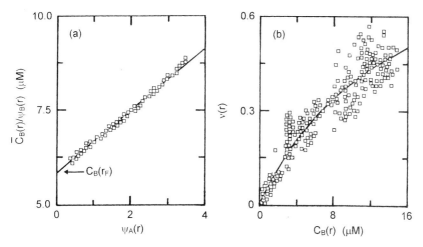

Figure 9 Characterization of the interaction between ovalbumin (A) and cytochrome c (B), pH 6.3, I = 0.03, by sedimentation equilibrium. (a) Evaluation of $C_B(r_F)$ at the reference radial position (7.000 cm) by psi analysis (*Equation 18a*). (b) Binding curve, together with the best-fit description (p = 1, K_{AB} = 60 000 M^{-1}) in terms of *Equation 19b*. Data are taken from ref. 22.

action (30–33) and for interactions with greater stoichiometry (38). Software programs are supplied with the Beckman instrument to facilitate such endeavours, which amount to curve-fitting of the distributions to *Equations 14a* and *14b* (truncated at the second term on the right-hand side for 1:1 complex formation) or their equivalents in terms of absorbances and molar absorption coefficients. The steps involved in the characterization of an acceptor–ligand interaction by direct curve-fitting are outlined in a recent discussion (39) of the characterization of heterogeneous associations. Furthermore, that investigation (39) extends the analysis by incorporating into *Equations 14a* and *14b* the capacity to make rigorous allowance for the effects of thermodynamic non-ideality on the statistical-mechanical basis of excluded volume.

Although more elegant and potentially far more accurate than the procedures described in the previous two sections, the direct curve-fitting procedures are, of course, model-dependent from the outset. Such model-dependence poses no great problem when the experimenter is certain of the reaction stoichiometry, but it is rather disconcerting to find $C_A(r_F)$ and $C_B(r_F)$ undergoing variation as the result of including extra terms in *Equations 14a* and *14b* to encompass a range of possible reaction stoichiometries. Despite the greater inaccuracy of the procedures illustrated in *Figures 6–9* because of promulgated errors in $C_A(r)$ (*Figures 6* and *7*) or $\bar{C}_B(r) - C_B(r)$ (*Figure 9*) as the result of uncertainty inherent in the estimates of $C_B(r)$, those methods do have the advantage of being model-independent until the final step. They may therefore be preferable for delineating the model of the interaction, which is based on fitting the experimental data $[\bar{C}_A(r), \bar{C}_B(r)]$ to specified expressions (*Equations 14a* and *14b*) in the two model-independent variables $C_A(r)$ and $C_B(r)$. Subsequent refinement of the characterization in terms of affinity (K_{AB}, etc.) may then well ensue from direct curve-fitting of the results in terms of the model emanating from that first analysis.

4 Ligand perturbation of acceptor self-association

The major emphasis in ultracentrifugal studies of interacting systems has undoubtedly been the characterization of protein self-association (11, 13, 27, 40–48). In the present context advantage has been taken of that experience to delineate binding parameters for the interactions of small ligands with the various oligomeric states of an acceptor (49–56). Introduction of the concepts involved in these preferential binding studies is predicated upon an understanding of the characterization of acceptor self-association by sedimentation equilibrium—a topic that is therefore considered first.

4.1 Characterization of acceptor self-association by sedimentation equilibrium

Because studies of solute self-association heralded the introduction of sedimentation equilibrium for the characterization of macromolecular interactions, the field was developed initially in the context of molecular weight measurement. We therefore begin with a discussion of that approach, which has, however, been superseded by direct analysis of the sedimentation equilibrium distributions for a self-associating solute.

4.1.1 Analysis of weight-average molecular weights

The concentration-dependence of weight-average molecular weight deduced from separate sedimentation equilibrium distributions for lysozyme (43) are presented in *Figure 10a*. From this information we now need to manipulate the results to obtain f_1, the weight-fraction of solute in monomeric state. By combining the expression (57, 58):

$$\bar{M}_A/M_1 = d(\ln \bar{c}_A)/d(\ln C_1) \qquad [20]$$

with that for the weight-fraction of monomer:

$$f_1 = c_1/\bar{c}_A = M_1 C_1/\bar{c}_A \qquad [21]$$

where C_1 is the molar concentration of monomer with molecular weight M_1, it follows that:

$$\ln f_1 = \int_0^{\bar{c}_A} \{[(M_1/\bar{M}_A) - 1]/\bar{c}_A\} d\bar{c}_A \qquad [22]$$

A plot of the dependence of $\{[(M_1/\bar{M}_A) - 1]/\bar{c}_A$ upon $\bar{c}_A$ (*Figure 10b*) has the potential to provide their inter-relationship as the precursor of numerical integration to obtain the weight-fraction of monomer from *Equation 22*. Finally, curve-fitting of the dependence of total concentration, c_A, upon monomer concentration, $c_1 = f_1 \bar{c}_A$, to the expression:

$$\bar{c}_A = c_1 + X_2 c_1^2 + X_3 c_1^3 + \ldots \qquad [23]$$

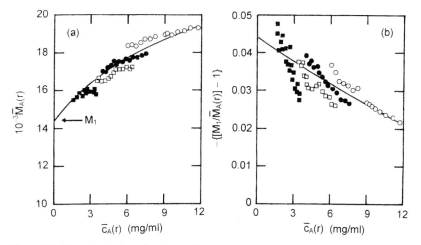

Figure 10 Characterization of solute self-association by analysis of sedimentation equilibrium distributions in terms of weight-average molecular weight. (a) Concentration dependence of $\bar{M}_A(r)$ obtained from four sedimentation equilibrium experiments on lysozyme (pH 6.7, $I = 0.17$, 15 °C), the data being taken from ref. 43. (b) Replot of the results for evaluation of the weight-fraction of monomer via *Equation 22*.

is used to obtain the various equilibrium constants X_i (litre^{i-1} g^{1-i}) describing the formation of *i*-mer from monomer.

A shortcoming of this approach is the differentiation of the sedimentation equilibrium distributions, $\bar{c}_A(r)$ versus r, to obtain $\bar{M}_A(r)$ as $d[\ln \bar{c}_A(r)]/dr^2$ at radial distance r. This procedure magnifies the uncertainty inherent in the experimental distribution, and thereby renders very difficult the delineation of the relationship to be integrated in *Equation 22* (*Figure 10*). Furthermore, the approach is circular in the sense that the molecular weight data obtained by differentiation are then re-integrated to obtain f_1, the weight-fraction of monomer. For over 20 years such use of molecular weight measurements has been rendered redundant by the realization that direct analysis of sedimentation equilibrium distributions affords a simpler and more accurate means of characterizing solute self-association (11, 13, 39, 59).

4.1.2 Direct analysis of sedimentation equilibrium distributions

By analogy with *Equations 13* and *14*, a more direct approach to the analysis of sedimentation equilibrium distributions reflecting solute self-association is to incorporate the psi function (*Equation 14c*) for monomer into *Equation 23*, which then becomes:

$$\bar{c}_A(r) = c_1(r_F)\psi_1(r) + X_2[c_1(r_F)\psi_1(r)]^2 + X_3[c_1(r_F)\psi_1(r)]^3 + \ldots \quad [24]$$

Non-linear least-squares analysis of the dependence of $\bar{c}_A(r)$ upon $\psi_1(r)$ for a given sedimentation equilibrium distribution in terms of this expression with $\psi_1(r)$ as the independent variable thus yields the association constants (X_2, X_3, ...) and $c_1(r_F)$ as the evaluated curve-fitting parameters. In order to accommodate data

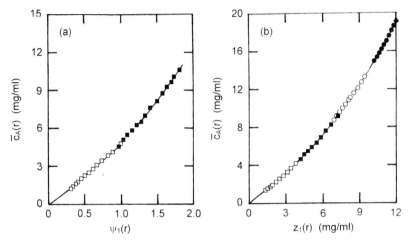

Figure 11 Characterization of lysozyme self-association (pH 8.0, I 0.15, 15 °C) by direct analysis of sedimentation equilibrium distributions. (a) Dependence of total solute concentration upon $\psi_1(r)$ evaluated from two sedimentation equilibrium distributions on lysozyme by choosing r_F in each experiment to achieve a common value of $c_A(r_F)$: also shown is the best-fit description obtained by non-linear regression analysis in terms of *Equation 24*. (b) Dependence of total lysozyme concentration upon the thermodynamic activity of monomer obtained by global analysis of four separate sedimentation equilibrium distributions, including those analysed in (a), on the basis of a common reference radial position (r_F) of 7.05 cm. Data in (a) and (b) are taken from refs 13 and 39 respectively.

from several sedimentation equilibrium distributions into a global analysis, the initial approach (13) was to tie the $\psi_1(r)$ scales for separate data sets by selecting individual reference radial positions (r_F) corresponding to the same total solute concentration $\bar{c}_A(r_F)$—the approach illustrated in *Figure 11a*. An alternative procedure (39) simply entails concomitant analysis of solute distributions from all experiments to obtain a global best-fit value of each equilibrium constant as well as the corresponding $c_1(r_F)$ estimate for each run. Despite differences in the values of $c_1(r_F)$ from the individual experiments, the results are amenable to collective display as a dependence of $\bar{c}_A(r)$ upon the magnitude of $c_1(r) = \psi_1(r)c_1(r_F)$. This feature of the global psi analysis is illustrated in *Figure 11b*, where expression of the abscissa in terms of the thermodynamic activity of monomer (rather than its concentration) emphasizes the ability of the direct approach to make rigorous allowance for the effects of thermodynamic non-ideality on the statistical-mechanical basis of excluded volume (13, 39).

The software packages that are provided with Beckman ultracentrifuges for the analysis of solute self-association are based on the Yphantis method of direct analysis (59), which is equivalent to the above procedure for ideal systems. However, the allowance for thermodynamic non-ideality that is incorporated therein is based on the Adams and Fujita assumption (40) that thermodynamic activities are given by the expression $z_i(r) = iBM_1\bar{c}_A(r)$, where B is an empirical curve-fitting parameter. This assumption implies that thermodynamic non-ideality does not influence the extent of solute self-association because self-cancellation of its effects has been designed in the ratio of activity coefficients that relate the apparent and true equilibrium constants (60, 61). It is therefore hoped that the programs will soon be upgraded to accommodate more realistic allowance for the effects of thermodynamic non-ideality.

4.1.3 Earlier procedures for characterizing two-state self-association

For an ideal monomer–dimer system the application of *Equation 24* in the form:

$$\bar{c}_A(r)/\psi_1(r) = c_1(r_F) + X_2[c_1(r_F)]^2\psi_1(r) \qquad [25]$$

allows $c_1(r_F)$ to be obtained as the ordinate intercept of the dependence of $\bar{c}_A(r)/\psi_1(r)$—a feature illustrated in *Figure 12a* for the dimerization ($2\alpha\beta \leftrightarrow \alpha_2\beta_2$) of aquomethaemoglobin (pH 6.0, $I = 0.10$). Furthermore, combination of the consequent value of $c_1(r_F)$ with the magnitude of the slope, $X_2[c_1(r_F)]^2$, yields a dimerization constant of (2.0 ± 0.2) litre/g under these conditions (56).

Prior to the development of direct analytical procedures for the determination of X_2, weight-average molecular weights were used to evaluate the concentration of monomeric acceptor associated with the total acceptor concentration to which $\bar{M}_A$ referred. From the definition of the weight-average molecular weight for a monomer–dimer system, namely:

$$\bar{M}_A = [c_1 M_1 + (\bar{c}_A - c_1)M_2]/\bar{c}_A \qquad [26a]$$

the concentration of monomeric acceptor may be obtained as:

$$c_1 = \bar{c}_A(M_2 - \bar{M}_A)/(M_2 - M_1) \qquad [26b]$$

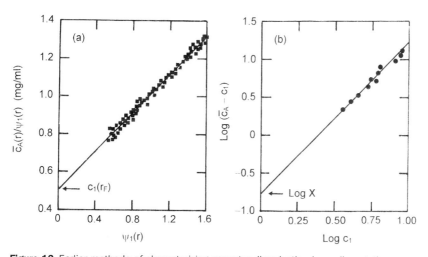

Figure 12 Earlier methods of characterizing acceptor dimerization by sedimentation equilibrium. (a) Direct analysis of sedimentation equilibrium distributions for methaemoglobin (pH 6.0, $I = 0.10$) in terms of *Equation 25*. (b) Use of *Equation 27* to characterize the dimerization of α-chymotrypsin (pH 7.8, $I = 0.28$) from $[\bar{c}_A, c_1]$ data obtained by molecular weight analysis (*Equation 26*) of sedimentation equilibrium distributions. Data in (a) and (b) are taken from refs 56 and 50 respectively.

Evaluation of X_2 from a series of $[\bar{c}_A, c_1]$ data thus generated may then be accomplished by plotting the results according to the expression:

$$\log (\bar{c}_A - c_1) = \log X_2 + 2 \log c_1 \qquad [27]$$

which is the logarithmic form of the law of mass action for a monomer–dimer equilibrium. The application of this procedure to obtain $\log X_2$ from the ordinate intercept of the dependence of $\log (\bar{c}_A - c_1)$ upon $\log c_1$ is illustrated in *Figure 12b*, which refers to the dimerization of α-chymotrypsin in phosphate buffer, pH 7.8, I 0.28 (50).

4.2 Displacement of an acceptor self-association equilibrium by ligand binding

For purposes of illustration we consider a monomer–dimer acceptor system in which ligand (B) binds to p equivalent and independent sites on monomer with intrinsic binding constant k_{1B}, and to q such sites on dimeric acceptor with intrinsic constant k_{2B}. From considerations of mass conservation it follows that the total weight-concentration of acceptor, $\bar{c}_A$, is given by:

$$\bar{c}_A = \bar{c}_1 + \bar{c}_2 = c_1(1 + k_{1B}C_B)^p + c_2(1 + k_{2B}C_B)^q \qquad [28]$$

where $\bar{c}_1$ and $\bar{c}_2$ are the respective constituent concentrations of monomeric and dimeric forms of acceptor. Provided that B is sufficiently small to justify the approximations that $M_{1B_i} \approx M_1$ and $M_{2B_i} \approx M_2$ for all values of i ($0 \le i \le p$ and $0 \le i \le q$ for monomeric and dimeric states respectively), the effect of ligand binding on the monomer–dimer equilibrium position can be monitored by measuring the constitutive dimerization constant, $\bar{X}_2$, defined (51) as:

$$\bar{X}_2 = \bar{c}_2/\bar{c}_1^2 = X_2(1 + k_{2B}C_B)^q/(1 + k_{1B}C_B)^{2p} \qquad [29]$$

where X_2 is the dimerization constant measured in the absence of B. Except for the situation in which ligand binding occurs independently of acceptor self-association ($k_{1B} = k_{2B}$, $q = 2p$), the binding of a small ligand is manifested as a dependence of $\bar{X}_2$ upon free ligand concentration C_B. For data obtained by sedimentation equilibrium of a dimerizing acceptor solution in dialysis equilibrium with a free concentration C_B of ligand, the parameters to emerge from application of *Equation 25* are $\bar{c}_1(r_F)$ and $\bar{X}_2$. Alternatively, analysis of the distributions by means of *Equations 26* and *27* lead to values of $\bar{c}_1$ and $\log \bar{X}_2$.

Preferential binding of ligand to one oligomeric state of acceptor leads to several forms of the dependence of $\bar{X}_2$ upon C_B, which therefore provides a powerful means of probing the relative affinities of monomeric and dimeric acceptor states for ligand. This aspect of sedimentation equilibrium studies is illustrated by considering the effects of ligand binding on several self-associating acceptor systems.

4.3 Illustrative studies of preferential ligand binding by sedimentation equilibrium

The binding of N-acetylglucosamine to lysozyme affords an example of preferential binding despite the fact that monomeric and polymeric states of the

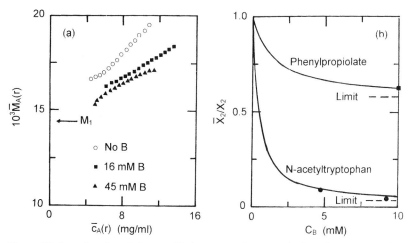

Figure 13 Use of sedimentation equilibrium to monitor perturbation of acceptor self-association as the result of preferential binding of a small ligand to one acceptor state. (a) Effect of N-acetylglucosamine on the weight-average molecular weight of lysozyme, pH 8.0, $I = 0.15$ (52). (b) Dependence of the constitutive dimerization constant for α-chymotrypsin upon the concentration of two competitive inhibitors. Lines denote the theoretical dependencies predicted by *Equation 29* for phenylpropiolate ($k_{2B} = 0.76k_{1B}$) and N-acetyltryptophan ($k_{2B} = 0.19k_{1B}$). Experimental data are taken from refs 50 and 55.

enzyme exhibit equal affinities for ligand ($k_{1B} = k_{2B}$). Because the active site is involved in the head-to-tail association of lysozyme, there is only one binding site for N-acetyltryptophan on each acceptor state (52, 53). In keeping with the qualitative predictions of *Equation 29* for a system with $k_{1B} = k_{2B}$ and $q < 2p$, the extent of self-association decreases with increasing ligand concentration—an effect evident from the dependence of weight-average molecular weight of lysozyme upon N-acetylglucosamine concentration (*Figure 13a*). However, departure of the system from a two-state self-association (*Figure 11*) precludes its use to illustrate quantitatively the prediction (*Equation 29*) that $\bar{X}_2 \to 0$ as $C_B \to \infty$.

Although sites are conserved ($q = 2$, $p = 1$) in the dimerization of α-chymotrypsin, competitive inhibitors such as phenylpropiolate (50) and N-acetyltryptophan (55) bind preferentially to monomeric enzyme because $k_{1B} > k_{2B}$. In keeping with the predictions of *Equation 29*, $\bar{X}_2$ again decreases with increasing ligand concentration (*Figure 13b*). For these systems, however, the predicted limiting magnitude of $\bar{X}_2$ is given by:

$$\lim_{C_B \to \infty} \bar{X}_2 = X_2(k_{2B}/k_{1B})^q \qquad [30]$$

whereupon a limiting value of zero for $\bar{X}_2$ would implicate exclusive binding of ligand to monomer ($k_{2B} = 0$). For the interactions of phenylpropiolate and N-acetyltryptophan with α-chymotrypsin these limits are finite because $k_{2B} = 0.76k_{1B}$ and $0.19k_{1B}$ for the respective ligands (50, 55).

The same limiting expression applies to the preferential interaction of ligands with dimeric acceptor under conditions of site conservation ($q = 2p$, $k_{2B} > k_{1B}$),

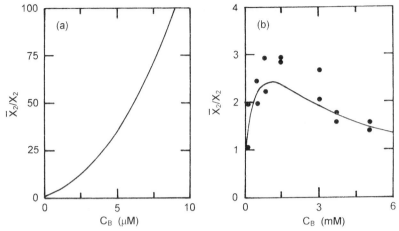

Figure 14 Further examples of the perturbation of acceptor dimerization as the result of preferential ligand binding. (a) Predicted effect of Zn^{2+} concentration on the constitutive binding constant for bacterial α-amylase—a system involving exclusive binding of the metal ion to a single site on dimer (54). (b) Dependence of the constitutive dimerization constant for methaemoglobin upon NADH concentration (56)—a system with $k_{2B} > k_{1B}$ but $q < 2p$ ($p = q = 1$).

but $\bar{X}_2$ now exhibits a positive dependence upon C_B. A ligand-mediated increase in $\bar{X}_2$ without limit is predicted for the binding of Zn^{2+} to bacterial α-amylase (*Figure 14a*), a system for which binding is restricted to a single site on dimer; i.e. a system for which $q = 1$, p and/or $k_{1B} = 0$ (54).

Finally, the interaction of methaemoglobin with NADH (56) and other organophosphate analogues of 2,3-bisphosphoglycerate (51) illustrates the possible existence of a critical point in the dependence of $\bar{X}_2$ upon C_B (*Figure 14b*). Preferential binding is an ambiguous term for this system inasmuch as dimer (the $\alpha_2\beta_2$ species) is the preferred acceptor on the basis of affinity ($k_{2B} > k_{1B}$), whereas monomer ($\alpha\beta$) is the preferred form from the viewpoint of site numbers—one site per base-mole of acceptor compared with one site per two base-moles for dimeric acceptor ($p = 1$, $q = 1$). At low ligand concentrations the magnitude of $\bar{X}_2$ increases with C_B because the binding of a single molecule of organophosphate to the β-cleft of dimer is the dominant phenomenon; but at higher concentrations the denominator of *Equation 29* dominates the magnitude of $\bar{X}_2$, which then decreases with increasing C_B. Non-linear regression analysis of the results in *Figure 14b* according to *Equation 29* with $p = 1$ and $q = 1$ signifies values of (700 ± 100) and (6000 ± 1000) M^{-1} for k_{1B} and k_{2B} respectively (56). A point of interest is that this interplay of equilibria is more amenable to quantitative characterization by sedimentation equilibrium than by classical binding studies.

4 Concluding remarks

Sedimentation equilibrium has much to offer for the quantitative characterization of acceptor–ligand interactions for systems in which both reactants are macromolecular. Its use for the study of the interaction between a small ligand and a macromolecular acceptor is limited unless the ligand has a unique spectral characteristic that allows analysis of the equilibrium distribution for that constituent. On the other hand, sedimentation equilibrium has played a vital role in characterizing the interplay of equilibria responsible for ligand perturbation of acceptor self-association as the result of preferential binding of a small ligand to one oligomeric state. In this review attention has been confined to the study of macromolecular interactions on the basis of thermodynamic ideality on the grounds that this assumption is a reasonable approximation for the relatively dilute solutions that are use in most *in vitro* studies. However, rigorous allowance for the effects of thermodynamic non-ideality on the statistical-mechanical basis of excluded volume has been incorporated into the analysis of sedimentation equilibrium distributions reflecting either solute self-association (13, 29, 39) or interaction between dissimilar macromolecular reactants (29, 39).

References

1. Casassa, E. F. and Eisenberg, H. (1964). *Adv. Protein Chem.*, **19**, 287.
2. Nichol, L. W., Owen, E. A., and Winzor, D. J. (1982). *J. Phys. Chem.*, **86**, 5015.
3. Van Holde, K. E. and Baldwin, R. L. (1958). *J. Phys. Chem.*, **62**, 734.
4. Correia, J. J. and Yphantis, D. A. (1992). In *Analytical ultracentrifugation in biochemistry and polymer science* (ed. S. E. Harding, A. J. Rowe, and J. C. Horton), p. 231. Royal Society of Chemistry, Cambridge.
5. Hexner, P. E., Radford, L. E., and Beams, J. W. (1961). *Proc. Natl. Acad. Sci. USA*, **74**, 2515.
6. Howlett, G. J. and Nichol, L. W. (1972). *J. Phys. Chem.*, **76**, 2740.
7. Svedberg, T. and Pedersen, K. O. (1940). *The ultracentrifuge*. Clarendon Press, Oxford.
8. Goldberg, R. J. (1953). *J. Phys. Chem.*, **57**, 194.
9. Williams, J. W., Van Holde, K. E., Baldwin, R. L., and Fujita, H. (1958). *Chem. Rev.*, **58**, 715.
10. Haschemeyer, R. H. and Bowers, W. F. (1970). *Biochemistry*, **9**, 435.
11. Milthorpe, B. K., Jeffrey, P. D., and Nichol, L. W. (1975). *Biophys. Chem.*, **3**, 169.
12. Wills, P. R. and Winzor, D. J. (1992). In *Analytical ultracentrifugation in biochemistry and polymer science* (ed. S. E. Harding, A. J. Rowe, and J. C. Horton), p. 311. Royal Society of Chemistry, Cambridge.
13. Wills, P. R., Jacobsen, M. P., and Winzor, D. J. (1996). *Biopolymers*, **38**, 119.
14. Jacobsen, M. P., Wills, P. R., and Winzor, D. J. (1996). *Biochemistry*, **35**, 13173.
15. Wills, P. R., Jacobsen, M. P., and Winzor, D. J. (1997). *Prog. Colloid Polym. Sci.*, **107**, 1.
16. Voelker, P. (1995). *Prog. Colloid Polym. Sci.*, **99**, 162.
17. Yphantis, D. A. (1964). *Biochemistry*, **3**, 297.
18. Creeth, J. M. and Harding, S. E. (1982). *J. Biochem. Biophys. Methods*, **7**, 25.
19. Cölfen, H. and Harding, S. E. (1997). *Eur. J. Biophys.*, **25**, 333: see also Harding, S. E., Horton, J. C., and Morgan, P. J. (1992). In *Analytical ultracentrifugation in biochemistry and polymer science* (ed. S. E. Harding, A. J. Rowe, and J. C. Horton), p. 275. Royal Society of Chemistry, Cambridge.
20. Hall, D. R., Harding, S. E., and Winzor, D. J. (1999). *Prog. Coll. Polym. Sci.*, **113**, 62.
21. Richards, E. G., Teller, D. C., and Schachman, H. K. (1968). *Biochemistry*, **7**, 1054.
22. Winzor, D. J., Jacobsen, M. P., and Winzor, D. J. (1998). *Biochemistry*, **37**, 2226.

23. Nichol, L. W. and Ogston, A. G. (1965). *J. Phys. Chem.*, **69**, 4365.
24. Adams, E. T., Jr. (1969). *Ann. N. Y. Acad. Sci.*, **164**, 226.
25. Hancock, D. K. and Williams, J. W. (1969). *Biochemistry*, **8**, 2598.
26. Teller, D. C., Horbett, T. A., Richards, E. G., and Schachman, H. K. (1969). *Ann. N. Y. Acad. Sci.*, **164**, 66.
27. Roark, D. E. and Yphantis, D. A. (1969). *Ann. N. Y. Acad. Sci.*, **164**, 245.
28. Van Holde, K. E., Rossetti, G. P., and Dyson, R. D. (1969). *Ann. N. Y. Acad. Sci.*, **164**, 279.
29. Wills, P. R., Jacobsen, M. P., and Winzor, D. J. (1998). *Biochem. Soc. Trans.*, **26**, 741.
30. Laue, T. M., Senear, D. F., Eaton, S., and Ross, A. J. B. (1993). *Biochemistry*, **32**, 2469.
31. Kim, T., Tsukiyama, T., Lewis, M. S., and Wu, C. (1994). *Protein Sci.*, **3**, 1040.
32. Lewis, M. S., Shrager, R. I., and Kim, S. J. (1994). In *Modern analytical ultracentrifugation: acquisition and interpretation of data for biological and synthetic polymer systems* (ed. T. M. Schuster and T. M. Laue), p. 94. Birkhäuser, Boston, MA.
33. Bailey, M. F., Davidson, B. E., Minton, A. P., Sawyer, W. H., and Howlett, G. J. (1996). *J. Mol. Biol.*, **263**, 671.
34. Wilson, E. K., Scrutton, N. S., Cölfen, H., Harding, S. E., Jacobsen, M. P., and Winzor, D. J. (1997). *Eur. J. Biochem.*, **243**, 393.
35. Nichol, L. W., Jeffrey, P. D., and Milthorpe, B. K. (1976). *Biophys. Chem.*, **4**, 259.
36. Jeffrey, P. D., Nichol, L. W., and Teasdale, R. D. (1979). *Biophys. Chem.*, **10**, 379.
37. Klotz, I. M. (1946). *Arch. Biochem.*, **9**, 109.
38. Behlke, J., Ristau, O., and Schonfeld, H. J. (1997). *Biochemistry*, **36**, 5149.
39. Wills, P. R., Jacobsen, M. P., and Winzor, D. J. (1999). *Progr. Colloid Polym. Sci.*, **113**, 69.
40. Adams, E. T., Jr. and Fujita, H. (1963). In *Analytical ultracentrifugation in theory and experiment* (ed. J. W. Williams), p. 119. Academic Press, New York.
41. Sophianopoulos, A. J. and Van Holde, K. E. (1964). *J. Biol. Chem.*, **239**, 2516.
42. Sarfare, P. S., Kegeles, G., and Kwon-Rhee, S. J. (1966). *Biochemistry*, **5**, 1389.
43. Adams, E. T., Jr. and Filmer, D. L. (1966). *Biochemistry*, **5**, 2971.
44. Adams, E. T., Jr. and Lewis, M. S. (1968). *Biochemistry*, **7**, 1044.
45. Hoagland, V. D. and Teller, D. C. (1969). *Biochemistry*, **8**, 594.
46. Aune, K. C. and Timasheff, S. N. (1971). *Biochemistry*, **10**, 1609.
47. Howlett, G. J., Jeffrey, P. D., and Nichol, L. W. (1972). *J. Phys. Chem.*, **76**, 777.
48. Morris, M. and Ralston, G. B. (1985). *Biophys. Chem.*, **23**, 49.
49. Nichol, L. W., Jackson, W. J. H., and Winzor, D. J. (1967). *Biochemistry*, **6**, 2449.
50. Nichol, L. W., Jackson, W. J. H., and Winzor, D. J. (1972). *Biochemistry*, **11**, 585.
51. Baghurst, P. A. and Nichol, L. W. (1975). *Biochim. Biophys. Acta*, **412**, 168.
52. Sophianopoulos, A. J. (1969). *J. Biol. Chem.*, **244**, 3188.
53. Howlett, G. J. and Nichol, L. W. (1972). *J. Biol. Chem.*, **247**, 5681.
54. Tellam, R., Winzor, D. J., and Nichol, L. W. (1978). *Biochem. J.*, **173**, 185.
55. Tellam, R., de Jersey, J., and Winzor, D. J. (1979). *Biochemistry*, **18**, 5316.
56. Jacobsen, M. P. and Winzor, D. J. (1995). *Biochim. Biophys. Acta*, **1246**, 17.
57. Steiner, R. F. (1952). *Arch. Biochem. Biophys.*, **39**, 333.
58. Adams, E. T., Jr. and Williams, J. W. (1964). *J. Am. Chem. Soc.*, **86**, 3454.
59. Johnson, M. L., Correia, J. J., Yphantis, D. A., and Halvorson, H. R. (1981). *Biophys. J.*, **36**, 575.
60. Ogston, A. G. and Winzor, D. J. (1975). *J. Phys. Chem.*, **79**, 2496.
61. Jacobsen, M. P. and Winzor, D. J. (1992). *Biophys. Chem.*, **45**, 119.
62. Creeth, J. M. and Pain, R. H. (1967). *Prog. Biophys. Mol. Biol.*, **17**, 217.
63. Teller, D. C. (1973). In *Methods in Enzymology*, Vol. 27, p. 346

64. McRorie, D. K. and Voelker, P. J. (1993). *Self-associating systems in the analytical ultracentrifuge*. Beckman Instruments, Fullerton, California.
65. ftp://alpha.bbri.org/rasmb/spin/mac/nonlin-uconn_uaf/
66. Colfen, H. and Winzor, D. J. (1997). *Prog. Colloid Polym. Sci.*, **107**, 36.
67. Harding, S. E., Horton, J. C., Jones, S., Thornton, J. M., and Winzor, D. J. (1999). *Biophys. J.*, **76**, 2432.

Chapter 6
Surface plasmon resonance

P. Anton Van Der Merwe

University of Oxford, Sir William Dunn School of Pathology, South Parks Road, Oxford OX1 3RE, UK.

1 Introduction

Since the development almost a decade ago (1, 2) of the first biosensor based on surface plasmon resonance (SPR), the use of this technique has increased steadily. Although there are several SPR-based systems (3–5), by far the most widely used one is the BIAcore (1, 2), produced by BIAcore AB, which has developed into a range of instruments (*Table 1*). By December 1998 over 1200 publications had reported results obtained using the BIAcore. It is likely that it would be even more widely used were it not for its high cost and the pitfalls associated with obtaining accurate quantitative data (5–8). The latter has discouraged many investigators and led to the perception that the technique may be flawed. This is unjustified because the pitfalls are common to many binding techniques and, once understood, they are easily avoided (4, 8, 9). Furthermore, the BIAcore offers particular advantages for analysing weak macromolecular interactions, allowing measurements that are not possible using any other technique (4, 10). This chapter aims to provide guidance to users of SPR, with an emphasis on avoiding pitfalls. No attempt is made to describe the routine operation and maintenance of the BIAcore, as this is comprehensively described in the BIAcore instrument manual (see Section 9.3). Although written for BIAcore users, the general principles will be applicable to experiments on any SPR instrument.

2 Principles and applications of surface plasmon resonance

2.1 Principles

The underlying physical principles of SPR are complex (see Section 9.1). Fortunately, an adequate working knowledge of the technique does not require a detailed theoretical understanding. It suffices to know that SPR-based instruments use an optical method to measure the refractive index near (within ~ 300 nm) a sensor surface. In the BIAcore this surface forms the floor of a small flow cell,

Table 1 BIAcore instruments currently available (January 1999)

	BIAcore X	BIAcore2000	BIAcore3000
Automated	No	Yes	Yes
Temperature control (°C)	10 below ambient to 40	4 to 40	4 to 40
Flow cell number	2	4	4
Flow cell volume (nl)	60	60	20
Time resolution (Hz)			
single flow cell	10	10	10
two flow cells	5	2.5	5
Refractive index range	1.33–1.40	1.33–1.36	1.33–1.40
Sample recovery	Manual	Automatic	Collects surface bound material only, but in very small volume (3–7 µl)
Fraction collection	Manual	Automatic	
Baseline noise (RU, root mean square)	0.3	0.3	0.1
Online subtraction of background response	Yes	Yes (flow cells 2-1, 3-1, 4-1)	Yes (flow cells 2–1, 3–1, 4-1, 3–4)

20–60 nl in volume (*Table 1*), through which an aqueous solution (henceforth called the *running buffer*) passes under continuous flow (1–100 µl.min^{-1}). In order to detect an interaction one molecule (the *ligand*) is immobilized onto the sensor surface. Its binding partner (the *analyte*) is injected in aqueous solution (*sample buffer*) through the flow cell, also under continuous flow. As the analyte binds to the ligand the accumulation of protein on the surface results in an increase in the refractive index. This change in refractive index is measured in real time, and the result plotted as response or resonance units (RUs) versus time (a *sensorgram*). Importantly, a response (*background response*) will also be generated if there is a difference in the refractive indices of the running and sample buffers. This background response must be subtracted from the sensorgram to obtain the actual binding response. The background response is recorded by injecting the analyte through a control or reference flow cell, which has no ligand or an irrelevant ligand immobilized to the sensor surface.

One RU represents the binding of approximately 1 pg protein/mm^2. In practise > 50 pg/mm^2 of analyte binding is needed. Because is it very difficult to immobilize a sufficiently high density of ligand onto a surface to achieve this level of analyte binding, BIAcore have developed sensor surfaces with a 100–200 nm thick carboxymethylated dextran matrix attached. By effectively adding a third dimension to the surface, much higher levels of ligand immobilization are possible. However, having very high levels of ligand has two important drawbacks. First, with such a high ligand density the rate at which the surface binds the analyte may exceed the rate at which the analyte can be delivered to the surface (the latter is referred to as mass transport). In this situation, mass transport becomes the rate-limiting step. Consequently, the measured association rate constant (k_{on}) is slower than the true k_{on} (see Section 7.2). A second, related problem is that, following dissociation of the analyte, it can re-bind to the unoccupied ligand before diffusing out of the matrix and being washed from the flow cell. Consequently, the measured dissociation rate constant (apparent k_{off}) is slower than the true k_{off} (see Section 7.2). Although the dextran matrix may exaggerate these kinetic artefacts (mass transport limitations and re-binding) they can affect all surface-binding techniques.

2.2 Applications

This section outlines the applications for which SPR is particularly well suited. Also described are some applications for which it is probably not the technique of choice. Of course, future technical improvements are likely to extend the range of applications for which the SPR is useful.

2.2.1 What SPR is good for

i. Evaluation of macromolecules

Most laboratories studying biological problems at the molecular or cellular level need to produce recombinant proteins. It is important to be able to show that the recombinant protein has the same structure as its native counterpart. With the possible exception of enzymes, this is most easily done by confirming that the protein binds its natural ligands. Because such interactions involve multiple residues, which are usually far apart in the primary amino acid sequence, they require a correctly folded protein. In the absence of natural ligands monoclonal antibodies (mAbs) that are known to bind to the native protein are an excellent means of assessing the structural integrity of the recombinant protein. The BIAcore is particularly well suited to evaluating the binding of recombinant proteins to natural ligands and mAbs. Setting up an assay for any particular protein is very fast, and the data provided are highly informative.

ii. Equilibrium measurements (affinity and enthalpy)

Equilibrium analysis requires multiple *sequential* injections of analyte at different concentrations (and at different temperatures). Because this is very time-consuming it is only practical to perform equilibrium analysis on interactions that attain equilibrium within about 30 min. The time it takes to reach equilibrium is determined primarily by the dissociation rate constant or k_{off}; a useful rule of thumb is that an interaction should reach 99% of the equilibrium level within $4.6/k_{off}$ sec. High affinity interactions ($K_D < 10$ nM) usually have very slow k_{off} values and are therefore unsuitable for equilibrium analysis. Conversely, very weak interactions ($K_D > 100$ μM) are easily studied. The small sample volumes required for BIAcore injections (< 20 μl) make it feasible to inject the very high concentrations (> 500 μM) of protein required to saturate low affinity interactions (11).

Equilibrium affinity measurements on the BIAcore are highly reproducible. This feature and the very precise temperature control makes it possible to estimate binding enthalpy by van't Hoff analysis (12). This involves measuring the (often small) change in affinity with temperature (Section 7.4). Although not as rigorous as calorimetry, much less protein is required.

iii. Kinetic measurements

The fact that the BIAcore generates real-time binding data makes it well suited to the analysis of binding kinetics. There are, however, important limitations to kinetic analysis. Largely because of mass transport limitations it is difficult to measure accurately k_{on} values faster than about 10^6 $M^{-1}s^{-1}$. This upper limit is dependent on the size of the analyte. Faster k_{on} values can be measured with analytes with a greater molecular mass. This is because the larger signal produced by a large analyte allows the experiment to be performed at lower ligand densities, and lower ligand densities require lower rates of mass transport. For different reasons measuring k_{off} values slower than 10^{-5} s^{-1} or faster than ~ 1 s^{-1} is difficult. Because the BIAcore is easy to use and the analysis software is user-friendly, it is deceptively easy to generate kinetic data. However *obtaining accurate kinetic data is a very demanding and time-consuming task, and requires a thorough understanding of binding kinetics and the potential sources of artefact* (Section 7.2).

iv. Analysis of mutant proteins

It is possible using BIAcore to visualize the capture of proteins from crude mixtures onto the sensor surface. This is very convenient for analysing mutants generated by site-directed mutagenesis (13, 14). Mutants can be expressed as tagged proteins by transient transfection and then captured from crude tissue culture supernatant using an antibody to the tag, thus effectively purifying the mutant protein on the sensor surface. It is then simple to evaluate the effect of the mutation on the binding properties (affinity, kinetics, and even thermodynamics) of the immobilized protein. This provides the only practical way of quantifying the effect of mutations on the thermodynamics and kinetics of weak protein/ligand interactions (15).

2.2.2 What SPR is not good for

i. High throughput assays

The fact that with BIAcore only one sample can be analysed at a time, with each analysis taking 5–15 min, means that it is neither practical nor efficient for high throughput assays. Automation does not solve this problem because the sensor surface deteriorates over time and with reuse. Blockages or air bubbles in the micro-fluidic system are also common in long experiments, especially when many samples are injected.

ii. Concentration assays

The BIAcore is also unsuitable for concentration measurements, because these require the analysis of many samples in parallel, including the standard curve. A second problem is that, for optimal sensitivity, concentration assays require long equilibration periods.

iii. Studying small analytes

Because the SPR measures the mass of material binding to the sensor surface, very small analytes ($M_r < 1000$) give very small responses. The recent improvements in signal to noise ratio have made it possible to measure binding of such small analytes. However a very high surface concentration of active immobilized ligand ($\sim$ 1 mM) is needed, and this is difficult to achieve. Furthermore, at such high ligand densities accurate kinetic analysis is not possible because of mass-transport limitations and re-binding (Section 7.2). Thus only equilibrium analysis is possible with very small analytes, and then only under optimal conditions. This assessment may need to be revised as and when future improvements are made in the signal to noise ratio.

3 General principles of BIAcore experiments

3.1 A typical experiment

A typical SPR experiment involves several discrete tasks.

(a) Prepare ligand and analyte.

(b) Select and insert a suitable sensor chip.

(c) Immobilize the ligand and a control ligand to sensor surfaces.

(d) Inject analyte and a control analyte over sensor surfaces and record response.

(e) Regenerate surfaces if necessary.

(f) Analyse data.

While the ligand and analyte could be almost any type of molecule, they are usually both proteins. This chapter focuses on the analysis of protein–protein interactions.

3.2 Preparation of materials and buffers

BIAcore experiments are frequently disrupted by small air bubbles or other particles passing through the flow system. Usually these can be flushed out and the experiment repeated, wasting only time and reagents. Occasionally, however, the damage is irreversible necessitating the replacement of an expensive sensor chip or integrated fluidic cartridge. This can be minimized by following a few simple rules (Section 9.2.1).

3.3 Monitoring the dips

The output from the photo-detector array that is used to determine the surface plasmon resonance angle (θ_{spr}, Section 9.1) can be viewed directly as 'dips'. The current BIAcore documentation does not describe how to view and interpret dips, and so this information is supplied here (*Protocol 1*).

Protocol 1
Normal and abnormal 'dips'

A Viewing the dips

1. Enter the service mode on the BIAcore control software by simultaneously pressing the *control*, *alt*, and *s* keys.
2. When the dialog box appears requesting a password ignore this and press the *enter* or *return* key. An additional *Service* menu will appear on the menu bar.
3. Select *View dips* from this menu.
4. A graph appears similar to the one in *Figure 1* showing the amplitude of light reflected off the sensor surface (reflectance) measured over a small range of angles. The angle of the minimum reflectance (θ_{spr}) is calculated by fitting a curve to all this data, thereby enabling θ_{spr} to be measured at a far greater resolution than the resolution at which the data are actually collected.

B Normal dip

The important feature of a normal dip is its depth (*Figure 1*, Dip 1). Generally it bottoms out at a reflectance of ~ 10000. When the refractive index above the sensor surface increases (e.g. because of protein binding), the dip shifts to the right while maintaining its shape and depth and (Dip 2).

C Abnormal dip

There are two main types of abnormal dip.

1. The shallow dip (Dip 3). The θ_{spr} is measured along a section of the sensor surface rather than at a single point. Refractive index heterogeneity on the surface gives heterogeneous θ_{spr} values. When averaged the result is a shallow dip that does not reach below 10000. Heterogeneity can be the result of differences in the amount of material immobilized along the surface, in which case the dip is usually slightly shallow, or the result of small air bubbles or particles in the flow cell, in which case the dip is very shallow.
2. No dip (Dip 4). A large change in the refractive index beyond the instrument dynamic range (*Table 1*) will shift the θ_{spr} so much that no dip is evident. Usually this is the result of air in the flow cell.

While slight shallowness of the dip is acceptable, and is common after coupling large amounts of protein, more severe abnormalities should not be ignored. Attempts should be made to return dips to normal by flushing the flow cells with buffer and/or regenerating the sensor surfaces. If this is ineffective a new sensor surface should be used.

SURFACE PLASMON RESONANCE

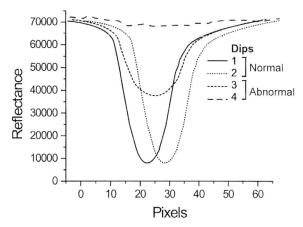

Figure 1 Analysis of dips. Several examples are shown of normal (1 and 2) and abnormal (3 and 4) dips. These are discussed in *Protocol 1*.

4 Ligand

4.1 Direct versus indirect immobilization

The most challenging step when setting up SPR experiments is immobilizing of one of the proteins (the *ligand*) to the sensor surface without disrupting its activity. Immobilization can either be direct, by covalent coupling, or indirect, through capture by a covalently coupled molecule.

The major advantage of *direct covalent immobilization* is that it can be used for any protein provided that it is reasonably pure ($> 50\%$) and has a pI > 3.5. However, it has three important drawbacks. First, because they usually have multiple copies of the functional group that mediates immobilization, proteins are coupled heterogeneously and sometimes at multiple sites. Secondly, direct coupling often decreases or completely abrogates binding to analyte. And thirdly, directly-coupled proteins are difficult to regenerate (see below).

Indirect immobilization has the disadvantage that it can only be used for proteins that have a suitable binding site or tag for the covalently coupled molecule. However, it has four important advantages, which make it the method of choice in most cases. First, proteins are seldom inactivated by indirect coupling. Secondly, the protein need not be pure. It can be captured from a 'crude' sample. Thirdly, all the molecules are immobilized in a known and consistent orientation on the surface. Finally, using appropriate buffers it is often possible to dissociate selectively the non-covalent ligand/analyte bond, thereby enabling the ligand surface to be reused—a process termed 'regeneration'.

4.2 Covalent immobilization

4.2.1 A general approach

Instead of replicating the detailed protocols in the BIAcore literature (see Section 9.3), I will suggest a general approach (*Protocol 2*) to covalent coupling, discussing several aspects in detail.

Protocol 2
An approach to covalent coupling of a protein

1. Select the coupling chemistry (Section 4.2.2).
2. Prepare the protein (Section 4.2.3).
3. Optimize the pre-concentration step (Section 4.2.4).
4. Couple the protein (Section 4.2.5).
5. Evaluate the activity of the immobilized protein (Section 4.4).
6. Establish conditions for regeneration (Section 4.2.6).
7. Adjust the immobilization conditions (Section 4.2.7).

4.2.2 Choice of chemistry

There are three main types of coupling chemistry, which utilize, respectively, amine (e.g. lysine), thiol (cysteine), or aldehyde (carbohydrate) functional groups on glycoproteins (*Table 3*). All covalent coupling methods utilize free carboxymethyl groups on the sensor chip surface. They can therefore be used for any of the sensor chips that have such carboxymethyl groups (see *Table 2*). If the protein to be immobilized has a surface-exposed disulfide or a free cysteine, ligand–thiol coupling is probably the method of choice. Failing this, amine coupling should be tried in the first instance. If amine coupling inactivates the protein (as assessed by ligand and/or mAb binding), aldehyde coupling can be attempted, provided that the protein is glycosylated. Detailed protocols are available from BIAcore for all coupling techniques (see Section 9.3). Only amine coupling is described here in some detail.

4.2.3 Prepare the protein

Only a modest amount (5–10 μg) of protein is needed. The major requirement is that the protein is pure and has a high level of activity. Because direct coupling is relatively indiscriminate, all protein in the preparation will be coupled. Thus if the preparation is contaminated by other proteins or is partially active, the level of binding observed will be proportionally decreased. Because the protein needs to be diluted into the pre-concentration buffer (Section 4.2.4) to a final concentration of 20–50 μg/ml, the stock should be fairly concentrated (> 0.5 mg/ml). If the protein is in solutions that are strongly buffered or contain high salt concentrations, primary amine groups or sodium azide, then these must either be dialysed out or much larger dilutions (1:100) made. In the latter case the protein will need to be at higher concentrations (> 2 mg/ml).

4.2.4 Pre-concentration

The purpose of pre-concentration is to concentrate the protein to very high levels (> 100 mg/ml) within the dextran matrix, thereby driving the coupling

Table 2 Sensor chips available for use on the BIAcore

	Surface chemistry	Free carboxy-methyl groups	Comments
Sensor chip CM5 (research grade)	Carboxymethylated dextran matrix	Yes	Most widely used sensor chip. Suitable for most applications.
Sensor chip CM5 (certified grade)	As above	Yes	Less chip-to-chip variation than research grade chip but much more expensive.
Sensor chip SA	As above with streptavidin pre-coupled	No	Quite expensive, especially since it cannot be reused. It is cheaper to couple streptavidin to CM5 research grade chip (11).
Sensor chip NTA	As above with NTA pre-coupled	No	Comes with nitrilotriacetic acid (NTA) coupled. For capturing proteins with oligo-histidine tags. Expensive but can be reused many times.
Sensor chip HPA	Flat hydrophobic surface	No	For immobilizing lipid monolayers. Problem with non-specific binding.
Pioneer chip C1	No dextran matrix. Carboxymethylated	Yes	For binding particles (e.g. cells) too large to enter dextran matrix. Also helps eliminate potential matrix-related artefacts.
Pioneer chip B1	Low level of carboxymethylation	Yes	Its lower charge density makes it useful reducing non-specific interactions with charged analytes.
Pioneer chip F1	Thin dextran matrix	Yes	For binding to bulky particles. Reduces matrix-related artefacts.
Pioneer chip J1	Unmodified gold surface	No	Can be used to design novel coupling chemistry.
Pioneer chip L1	Dextran derivatized with lipophilic compounds	No	For immobilizing liposomes via interactions with lipophilic compounds. Apparently lower levels of non-specific binding than HPA chip.

reaction. Without pre-concentration far higher concentrations of protein would need to be injected to get equivalent levels of coupling. Pre-concentration is driven by an electrostatic interaction between the negatively charged carboxylated dextran matrix and positively charged protein. To this end the protein is diluted into a buffer with a low ionic strength (to minimize charge screening) with a pH below its isoelectric point or pI (to give the protein net positive charge). However amine coupling is most efficient at a high pH because activated carboxyl groups react better with uncharged amine groups. Thus the highest pH compatible with pre-concentration is determined empirically (*Protocol 3*). Electrostatically-bound protein should dissociate rapidly and completely when injection of running buffer resumes, both because the proteins net positive charge will decrease and because electrostatic interactions will be screened by

Table 3 Covalent coupling chemistry

	Surface activation[a]	Group required on protein	Protein preparation	Comments
Amine	EDC/NHS[b]	Amine groups (lysine and unblocked N-termini).	No preparation needed.	Can be used with most proteins. Risk of multiple coupling.
Surface-thiol	EDC/NHS followed by cystamine/DTT to introduce a thiol group	Amine or carboxyl groups.	Need to introduce reactive disulfide using heterobifunctional reagent such as PDEA (carboxyl) or SPDP (amine).	Not widely used
Ligand-thiol	EDC/NHS followed by PDEA to introduce a reactive disulfide group	Surface exposed free cysteine or disulfide.	Reduce disulfides under non-denaturing conditions to generate free cysteine.	Useful for proteins with exposed disulfides or free cysteines. Multiple coupling unlikely.
Aldehyde coupling	EDC/NHS followed by hydrazine	Aldehyde groups.	Create aldehydes by oxidizing cis-diols (in sugars) with periodate. Works especially well with sialic acid.	Useful for polysaccharides and glycoproteins.

[a] All coupling reactions utilize carboxymethyl groups on the sensor surface and can be used with any sensor chip which has free carboxymethyl groups (Table 2). The first step is to activate these groups with N-hydroxysuccinimide, thus creating a highly reactive succinimide ester which reacts with amine groups on protein, or can be further modified.

[b] Abbreviations: EDC, N-ethyl-N'-(dimethylaminopropyl)carbodiimide; NHS, N-hydroxysuccinimide; DTT, dithiothreitol; PDEA, 2-(2-pyridinyldithio)ethane-amine.

the high ionic strength of the running buffer. Incomplete dissociation suggests that the interaction was not purely electrostatic, perhaps because of the binding, at low pH, of a denatured form of the protein.

Protocol 3
Determining the optimum pre-concentration conditions for a protein

Equipment and reagents
- BIAcore
- Protein solution
- Pre-concentration buffers

Method

1. Dilute the protein to a final concentration of 20–50 µg/ml (final volume 100 µl) into pH 6.0, 5.5, 5.0, 4.5, and 4.0 pre-concentration buffers (see Section 9.2.2).
2. Start a manual BIAcore run using a single flow cell (flow rate 10 µl/min).
3. Inject 30 µl of each sample, beginning at pH 6.0 and working down.
4. If no electrostatic interaction is observed continue with protein samples diluted in pH 3.5 and 3.0 pre-concentration buffers.
5. Use the highest pH at which > 10 000 RU of protein binds electrostatically during the injections.
6. Check that all the bound protein dissociates after the injection. If not it suggests that not all the protein was bound electrostatically and that the protein denatures irreversibly at that pH.

4.2.5 Amine coupling

There is a standard protocol for amine coupling (*Protocol 4*). The first step is to activate the carboxymethyl groups with N-hydroxysuccinimide (NHS), thus creating a highly reactive succinimide ester which reacts with amine and other nucleophilic groups on proteins. The second (coupling) step is to inject the protein in pre-concentration buffer, thereby achieved high protein concentrations and driving the coupling reaction. The third (blocking) step, blocks the remaining activated carboxymethyl groups by injecting very high concentrations of ethanolamine. The high concentration of ethanolamine also helps to elute any non-covalent bound material. The final regeneration step is optional. When a protein is coupled for the first time it is advisable not to include any regeneration step. If it is included and the protein has poor activity it will not be clear whether covalent coupling or regeneration was responsible for disrupting the protein. The structural integrity of the protein should be evaluated (Section 4.2.6) before regeneration is attempted. Once regeneration conditions have been established these can be added on to the coupling protocol.

Protocol 4

Amine coupling

Reagents

- 120 μl of protein at 20–50 μg/ml in suitable pre-concentration buffer (*Protocol 3*)
- 120 μl of EDC (0.4 M) mixed 1:1 with NHS (0.1 M); make up just before coupling
- 120 μl of 1 M ethanolamine–HCl pH 8.0
- 120 μl of regeneration solution (if regeneration conditions known)

Method

1. Establish pre-concentration conditions (*Protocol 3*).
2. Set flow rate to 10 μl/min.
3. Activation: inject 70 μl of EDC/NHS.
4. Coupling: inject 70 μl of protein.
5. Blocking: inject 70 μl of ethanolamine.
6. Regeneration: inject 30 μl of regeneration solution (if regeneration conditions known, see *Protocol 5*).
7. View the 'dips' (*Protocol 1*) to confirm that the immobilization is homogeneous.

4.2.6 Regeneration

Once a covalently immobilized protein has been shown to be active with respect to binding its natural ligand or a monoclonal antibody (Section 4.4), regeneration can be attempted. A general approach to establishing regeneration conditions can be used (*Protocol 5*). The goal here is to elute any non-covalently bound analyte without disrupting the activity of the ligand. Regeneration allows surfaces to be reused many times, saving both time and money. However establishing ideal regeneration conditions can be a very time-consuming, and in many cases impossible, task. Thus it may be more cost-effective to opt for imperfect or no regeneration, using new sensor surfaces instead.

Protocol 5

Establishing regeneration conditions

Regeneration solutions

These fall into four main classes: divalent cation chelator, high ionic strength, low pH, high pH (Section 9.2.2).

- If the interaction is likely to be dependent on divalent cations try buffers with EDTA.
- If the monomeric interaction is weak, try high ionic strength buffers.

Protocol 5 continued

- Otherwise start with a low pH buffer.
- If this is without effect, try high pH buffer.

Method

1. Covalently couple ligand to the surface.
2. Make up enough analyte for several injections.
3. Inject analyte over the immobilized ligand and measure the amount of binding.
4. Inject selected regeneration buffer for 3 min, and measure the amount of analyte that remains bound after this.
5. If the decrease is < 30% switch to a different class of regeneration buffer and return to step 4.
6. If the decrease is > 30% but < 90%, try repeated injections. If this fails to elute > 90% select a stronger regeneration buffer of the same type and return to step 4.
7. If this fails, try a different class of buffer and return to step 4.
8. When > 90% of bound analyte is eluted, return to step 3, using identical analyte and injection conditions. If binding post-regeneration remains > 90% of binding before regeneration, the regeneration conditions may be adequate.
9. If residual activity is < 90%, select a different class of regeneration buffer. If residual binding is very low it may be necessary to return to step 1, starting with a new surface.
10. If two different types of buffer give partial elution, try using both sequentially.

4.2.7 Adjusting the immobilization conditions

Since the level of ligand coupling achieved is unpredictable it is usually necessary to modify the initial protocol in order to achieve the desired immobilization level. It is also often necessary to create several surfaces on the same chip with different levels of coupling. The best way to achieve different levels of coupling is to change the duration of the activation step, by varying the volume of NHS/EDC injected. The level of coupled ligand varies in proportion with the duration of the activation step. Thus, a twofold reduction in activation period will usually lead to a twofold reduction in coupling. It is possible using an option in the inject command to couple ligand simultaneously in multiple flow cells, varying only the length of activation step.

4.3 Non-covalent immobilization (ligand capture)

There are two requirements for non-covalent or indirect immobilization of a ligand (henceforth called 'ligand capture'). First, it must be possible to obtain or create a sensor surface that can capture a ligand. Ideally, it should be possible to regenerate this surface so that repeated capture is possible. Second, the ligand

Table 4 Techniques for ligand capture

	Covalently coupled molecule	Captured ligand	Valency of ligand	Removal of tag	Comments
Antibodies	Rabbit-anti mouse Fc polyclonal (26)	Any mouse IgG monoclonal antibody	Divalent	No	Antibody available from BIAcore. Because it is polyclonal very high levels of binding to a particular isotype are difficult to achieve.
	Mouse anti-human Fc, monoclonal (R10Z8E9) (13, 27)	Any protein fused to the Fc portion of human IgG$_1$. This includes most Fc chimeras.	Divalent	Not routinely.	Antibody available from Recognition Systems, University of Birmingham Science Park, Birmingham B15 2SQ, UK.
	Mouse anti rat CD4, monoclonal (OX68) (17)	Any protein fused to rat CD4 domains 3 and 4.	Monovalent	No	Antibody available from Serotec.
	Anti-GST[b]	Any protein expressed as a GST chimera.	Divalent	Yes	Antibody available from BIAcore. Because GST dimerizes, the fusion proteins are dimers.
Other molecules	Streptavidin	Any biotinylated molecule. Biotinylation can be indiscriminate or (preferably) targeted[a].	Monovalent, can be made tetravalent with streptavidin	No	Streptavidin sensor chips are available from BIAcore. However streptavidin can be coupled to standard CM5 chips at lower cost (11). Note that streptavidin cannot be regenerated.
	Ni-NTA	Any protein with oligo-histidine tag.	Monovalent	Yes	Sensor chips can be purchased from BIAcore with Ni-NTA already coupled.

[a] Targeted biotinylation involves introducing a single biotin group on the molecule in a position in which it will not interfere with binding to analyte. This can be achieved either by chemical synthesis (e.g. oligonucleotides), or by enzymatic biotinylation. An established method uses the enzyme BirA to biotinylated a specific peptide which has been added on to the N or C terminus of a recombinant protein (11).

[b] Abbreviations GST, glutathione-S-transferase; Ni-NTA, nickel-nitrilotriacetic acid.

needs to have a suitable binding site or modification that allows it to be captured.

4.3.1 Using an existing strategy

A number of established techniques are available for ligand capture (*Table 4*). When expressing a recombinant protein to be used in SPR studies it is advisable to consider adding a suitable domain or peptide motif so that one of these techniques can be used. The precise choice of tag will depend on whether the tag is needed for purification and what other uses are envisaged for the ligand.

(a) If it is envisaged that the ligand is to be used as an analyte in SPR studies it should be monovalent or, if multivalent (Fc- and GST-chimeras), the tag should be readily removable.

(b) If the ligand is to be used for structural studies, the tag should also be removable.

(c) If the ligand is to be used to probe for binding partners on cells or tissues it should be multivalent, or it should be possible to make it multivalent.

(d) If the ligand is to be captured from crude mixtures (i.e. after expression without purification) the capturing agent needs to have a high affinity and to be highly specific. For example, CD4 (14) and anti-human IgG$_1$ (13) mAbs have been shown to be suitable for this purpose.

4.3.2 Developing a new strategy

The widespread availability of purified monoclonal or polyclonal mAbs, and their ability to tolerate harsh regeneration conditions, make them suitable reagents for non-covalent immobilization. Typically there is a need to immobilize a number of related ligands, with an invariant and a variant portion. If several antibodies against the invariant portion are available, it is likely that at least one of these will be suitable for indirect coupling. A basic approach to developing such a method is described in *Protocol 6*.

Protocol 6

Developing a new antibody-mediated indirect coupling method

Reagents

- Panel of antibodies
- See *Protocols* 4 and 5

Method

1. Obtain a panel of antibodies that can bind a suitable ligand and are available in pure form.
2. Analyse the binding properties and select a high affinity antibody with a slow k_{off}.

> **Protocol 6 continued**
>
> 3 Covalently couple the antibody by amine coupling (*Protocol 4*) and check for activity. If inactive, try a different antibody.
> 4 Establish regeneration conditions (*Protocol 5*).
> 5 If suitable regeneration conditions cannot be found, try a different antibody.

4.4 Activity of immobilized ligand

It is important to evaluate the functional integrity of the immobilized protein. This is best achieved by using a protein which binds to correctly-folded ligand. An ability to bind its natural ligand is reassuring evidence that an immobilized ligand is functionally intact. Since monoclonal antibodies (mAbs) usually bind to 'discontinuous epitopes' on a protein, they are excellent probes of protein structure. It is preferable to check the binding of several mAbs, including ones that bind to the same binding sites or, failing this, the same domain of the natural ligand. mAbs are particularly useful if the natural binding partner has not been identified and/or a candidate binding partner being assessed for an interaction with the ligand. In is important to know not just whether immobilized ligand is 'active' but also what proportion is active (*Protocol 7*).

Protocol 7
Quantitating binding levels

The ligand activity (*ActL*), stoichiometry (*S*), molecular mass of ligand (M_L) and analyte (M_A), and the analyte binding level at saturation (*A*) are related as follows:

$$ActL * S = \left(\frac{M_L}{M_A}\right) * \frac{A}{L}$$

where L is the level of ligand immobilized,
   S is the molar ratio of analyte to ligand in analyte/ligand complex.
The product ActL*S is readily calculated once A and L have been measured. Either ActL or S need to be independently determined in order to calculate the other. A convenient way to measure ActL by SPR is to use Fab fragments of mAbs specific for the ligand. Intact mAbs are less useful because of the uncertainty as to their binding stoichiometry.

4.5 Control surfaces

A control surface should be generated which is as similar as possible to the ligand surface, including similar levels of immobilization. This is to measure non-specific binding and to record the background response. The immobilization levels need to be similar because this affects the background response measured with analytes that have a high refractive index (Section 5.3).

4.6 Reusing sensor chips

Because each sensor chip has several flow cells, it is common to have unused flow cells at the end of an experiment. In addition many covalently coupled ligands are very stable, enabling surfaces to be reused over several days. It is therefore convenient to be able to remove and reinsert sensor chips in the BIAcore (*Protocol 8*).

Protocol 8
Reusing sensor chips

1. Undock the sensor chip, choosing the empty flow cell option in the dialogue box.
2. Store the sensor chip in its cassette at 4 °C.
3. When it is to be reused, re-insert and dock the sensor chip.
4. After priming the system check the 'dips' (Section 3.3).

5 Analyte

The extent to which the analyte needs to be characterized depends on the nature of the experiment. Quantitative measurements (Section 7) require that the analyte is very well characterized and of the highest quality. In contrast, this is less important for qualitative measurements (Section 6).

5.1 Purity, activity, and concentration

In order to determine affinity and association rate constants it is essential that the *concentration* of the injected material is known with great precision. The only sufficiently accurate means of measuring concentration is to use a spectrophotometer to measure the optical density of a solution of *pure* protein, usually at 280 nm (OD_{280}). In order to calculate the concentration from the absorption at OD_{280}, two additional measurements are required. First, it is necessary to determine the extinction coefficient. Although this can be calculated from the primary sequence it is best determined directly by amino acid analysis of a sample of the protein with a known OD_{280}. Secondly, it is important to assess what proportion of the purified protein is 'active', i.e. able to bind to the ligand. This can be done by depletion experiments in which the ligand-coated Sepharose beads are used to deplete the analyte from solution (16). mAb-coated Sepharose beads can be used instead if ligand-coated beads are impractical. If all the analyte can be depleted in this way it is 100% active.

5.2 Valency

In general, affinity and kinetic measurement require that each analyte molecule has a single binding site, i.e. is monovalent. If the protein has a single binding site it is only necessary to show that it exists as a monomer in solution. This is most readily achieved by size exclusion chromatography (Chapter 3) or analytical ultracentrifugation (Chapters 4 and 5). It is important to emphasize that these analytical techniques will not detect the presence of very low concentrations of multivalent aggregated material (17, 18). In order to ensure that this material does not contribute to the binding it is ESSENTIAL to purify the monomeric peak by size exclusion chromatography immediately before BIAcore experiments and to analyse it before concentrating or freezing it. Only when it has been shown that concentration, storage, or freezing do not affect the measured affinity and kinetic constants is it wise to deviate from this strict principle. Fortunately the presence of multivalent aggregates is readily excluded by analysis of the binding kinetics (Section 7.2). If the dissociation of bound analyte is monophasic (mono-exponential) multivalent binding can be ruled out. If dissociation is bi-exponential with > tenfold difference in the two k_{off} values, multivalent binding is likely. Bi-exponential dissociation with smaller differences in the two k_{off} values could have several explanations (see Section 7.2).

5.3 Refractive index effect and control analytes

When an analyte is injected over a surface it is important, for two reasons, to perform a second injection with a control analyte. First, this helps rule out non-specific binding. Secondly, it controls for any refractive index artefacts. These occur when the background signal measured during the injection of an analyte sample differs between flow cells. Clearly such an artefact will create problems for affinity measurements. It can occur whenever there is a substantial difference between surfaces. For example, if very different levels of material are immobilized on each surface. In this case, because the immobilized material displaces volume, the volume accessible to the injected analyte sample will differ between flow cells. If the analyte sample has a higher refractive index than the running buffer, a larger background signal will be seen from the surface with less immobilized material. This artefact is greater when:

(a) There are big differences in the levels of immobilized material (e.g. > 2000 RUs).

(b) The background signal is very large (e.g. > 1000 RUs).

(c) The binding response is much smaller (< 10%) than the background response.

These conditions are common when measuring very weak interactions, because high concentrations of analyte are injected, and with low molecular weight analytes, which give a very small response. A refractive index artefact can be detected by injecting a control solution with a similar refractive index to the analyte sample. If the control solution gives the same response in both flow cells a refractive index artefact can be excluded. Refractive index artefacts are most easily avoided by taking care to immobilize the same amount of total material on both the control and ligand sensor surfaces.

5.4 Low molecular weight analytes

Recent improvements in the signal to noise ratios of BIAcore instruments have enabled binding to be detected of analytes with M_r as low as 180 (19). There are two major problems associated with such studies. First, very high densities of ligand must be immobilized in order to detect binding. The levels can be calculated using equation in *Protocol 7*. For an analyte of $M_r \sim 200$ that binds a ligand with an M_r of $\sim 40\,000$, approximately 10 000 RU of active ligand needs to be immobilized to see 50 RU of analyte binding. Achieving this level of immobilization is very difficult. A second problem is that with such small binding responses refractive index effects become significant. The latter can be avoided by dissolving and/or diluting the analyte in the running buffer and using a control flow cell with very similar levels of immobilization.

6 Qualitative analysis; do they interact?

6.1 Positive and negative controls

The main purpose of a qualitative analysis is to establish whether or not there is an interaction between a given analyte and ligand. If binding is detected it is necessary to test negative controls to exclude a false positive. These include negative ligand controls and negative analyte controls. Blocking experiments, using molecules known to block the interaction, are also useful negative controls. In contrast, if no binding is detected it becomes necessary to run positive controls to establish whether this reflects the absence of an interaction, a very low affinity, or an artefact resulting from defective ligand or analyte. The ligand is readily assessed by showing that it can bind to one or more mAbs or additional analytes should they exist. The analyte can be tested by injecting it over a surface with either a known interacting ligand or mAb immobilized to the surface.

6.2 Qualitative comparisons using a multivalent analyte

The binding of a multivalent analyte can be heavily influenced by the level of immobilized ligand, since the latter will influence the valency of binding. Thus when comparing the binding of a multivalent analyte to different ligands it is important that these ligands are immobilized at comparable surface densities.

7 Quantitative measurements

Quantitative measurements are far more demanding than qualitative measurements because of the quality and amount of materials required and the difficulties associated with designing the experiments and analysing the data. When undertaking these measurements it is particularly important to understand the various pitfalls and how these can be avoided (4, 5, 8). *Any quantitative analysis on the BIAcore requires that the analyte binds in a monovalent manner.* Since many binding parameters are temperature-dependent *it is important to perform key measurements* at *physiological temperatures (i.e. 37°C in mammals)*. Considering how easy it is to regulate the temperature on the BIAcore it is surprising how seldom this is done.

7.1 Affinity

7.1.1 Concepts

As has been described earlier in this volume (see, e.g. Chapters 1, 3-5), there are a number of ways to represent the affinity of an interaction.

(a) The 'association constant' (K_A) or affinity constant is simply the ratio at equilibrium of the 'product' and 'reactant' concentrations. Thus for the interaction $A + B \longleftrightarrow AB$:

$$K_A = \frac{C_{AB}}{C_A * C_B} \quad [1]$$

Note that K_A has units M^{-1} (i.e. $L.mol^{-1}$).

(b) Many prefer to express affinity as the 'dissociation constant' or K_D, which is simply the inverse of the K_A, and therefore has the units M.

(c) Affinity can also be expressed as the binding energy or, more correctly, the standard state molar free energy ($\Delta G°$). This can be calculated from the dissociation constant as follows:

$$\Delta G° = RT \ln \frac{K_D}{C°} \quad [2]$$

where T is the absolute temperature in Kelvin (298.15 K = 25°C), R is the universal gas constant (1.987 $cal.K^{-1}.mol^{-1}$), $C°$ is the standard state concentration (i.e. 1 M).

7.1.2 Experimental design

In principle the affinity constant can be measured directly by equilibrium binding analysis, or calculated from the k_{on} and k_{off}. However, because of the difficulties associated with obtaining definitive kinetic data on the BIAcore, equilibrium binding analysis is more reliable. It involves injecting a series of analyte concentrations and measuring the level of binding at equilibrium. The relationship between the binding level and analyte concentration enables the affinity constant to be calculated (7). A basic approach to such measurements is outlined on *Protocol 9*.

i. Preliminary steps

Because equilibrium measurements are unaffected by mass-transport or rebinding artefacts, high levels of immobilization can be used to increase the binding response. This is particularly useful when the background signal is high or the analyte very small.

Ideally, equilibrium affinity measurements require that the level of active ligand on the surface is the same for each concentration of analyte injected. This is usually straightforward when the ligand is covalently coupled (and so does not dissociate) and the analyte dissociates spontaneously within a few minutes (so that regeneration is not required). In cases where regeneration *is* required it must be shown that ligand activity is unaffected by repeated regeneration. Where captured ligands dissociate spontaneously or require regeneration, it may be difficult to maintain the level of the ligand constant. It is possible to correct for this if the level of active ligand can be accurately monitored (20).

It is important to ensure that the analyte injections reach equilibrium. While the approximate time it takes to reach equilibrium can be calculated (see footnote to *Protocol 9*), it is advisable to measure this directly in preliminary experiments under the same conditions (flow rate, analyte concentration, ligand density) as those to be used for the affinity measurements. Enough time must be allowed following the injection for the bound analyte to dissociate completely from the sensor surface (see footnote to *Protocol 9*). If dissociation is incomplete, or takes too long, it may be necessary to enhance dissociation by injecting regeneration solution.

ii. The experiment

Ideally the analyte concentration should be varied over four orders of magnitude, from $0.01 * K_D$ to $100 * K_D$. However it is often only practical to vary the concentration over 2–3 orders of magnitude. This can be achieved with 10 twofold dilutions starting at between $10 * K_D$ and $100 * K_D$.

An assumption in these affinity measurements is that the level of active immobilized ligand remains constant. This should be checked by showing that a reference analyte binds to the same level at the beginning and end of the experiment. A second internal control is to reverse the order of injections. An efficient way of doing this is to work up from the lowest concentration of analyte, give one injection at the highest concentration, and then work back down to the lowest concentration. The same affinity should be obtained irrespective of the order of injections.

7.1.3 Data analysis

In order to derive an affinity constant from the data a particular binding model must be used. The simplest (Langmuir) model ($A + L \leftrightarrow AL$) is applicable in the vast majority of cases. It assumes that the analyte (A) is both monovalent and homogeneous, that the ligand (L) is homogeneous, and that all binding events are independent. Under these conditions data should conform to the Langmuir binding isotherm:

$$Bound = \frac{C^A * Max}{C^A + K_D} \quad [3]$$

where 'Bound' is measured in RUs and 'Max' is the maximum response (RUs). C^A is the concentration of injected analyte, and K_D is in the same units as C^A (normally mol L^{-1}, i.e. M).

The K_D and *Max* values are best obtained by non-linear curve fitting of the equation to the data using suitable computer software such as *ORIGIN* (MicroCal) or *SIGMAPLOT*.

A Scatchard plot of the same data (see Chapter 3), obtained by plotting Bound/C^A against Bound, is useful for visualizing the extent to which the data conform to the Langmuir model. A linear Scatchard plot is consistent with the model. *Scatchard plots alone should not be used to estimate affinity constants* since they place inappropriate weighting on the data obtained with the lowest concentrations of analyte, which are generally the least reliable.

Non-linear Scatchard plots indicate that the data do not fit the Langmuir model. Before considering models that are more complex it is important to exclude trivial explanations (Section 7.1.5).

7.1.4 Controls

Several artefacts can result in erroneous affinity constants. These include an effect of ligand immobilization on the binding, an error estimating the active concentration of analyte, and an incorrect assumption that the analyte is monovalent. The most rigorous control is to confirm the affinity constant in the reverse or 'upside down' orientation since this excludes all three artefacts. If this is not possible the experiment should be repeated with the ligand immobilized in a different way, which addresses the possible effects of immobilization on binding. The affinity should also be confirmed with two independently-produced batches of protein and with different recombinant forms of the same proteins.

7.1.5 Non-linear Scatchard plots

A non-linear Scatchard plot indicates that binding does not conform to the Langmuir model. Many binding models can be invoked to explain non-linear Scatchard plots. Distinguishing between these models can be very difficult and is beyond the scope of this chapter. The priority should be to exclude trivial explanations for non-linear Scatchard plots. A 'concave up' Scatchard plot is the most common deviation from linearity. It may be a consequence of heterogeneous ligand, multivalent analyte, or (rarely) negative cooperativity between binding sites. A trivial cause of analyte heterogeneity is the presence of multivalent analyte. Ligand heterogeneity may be a consequence of immobilization. A 'concave down' Scatchard plot is unusual and indicates either positive cooperativity between binding sites or self-association of the analyte, either in solution or on the sensor surface.

Protocol 9
Affinity measurements

A Preliminary steps

1. Immobilize the ligand and a control. High levels of immobilization are acceptable.
2. Ensure that the analyte is monomeric and binds monovalently. Determine accurately the concentration of the analyte and the proportion that is active.

Protocol 9 continued

3. Determine the time it takes to reach equilibrium and the time it takes for the bound analyte to dissociate completely from the sensor surface. While this should be done empirically, under the conditions to be used for the equilibrium measurements, approximate times can be calculated from the k_{off}.[a]
4. If necessary, establish regeneration conditions (*Protocol 5*).
5. Obtain a rough estimate of the K_D by injecting a series of fivefold dilutions.

B Measurements

1. Prepare a dilution series of analyte starting at 10–100 times the K_D with at least 9 twofold dilutions thereof. There should be enough for two injections at each concentration except the highest concentration, where only enough for one injection is required. A minimum of 17 µl of sample is required per injection.[b]
2. Make up separate control sample of analyte (at concentration ~ K_D) with enough for two injections.
3. Set the flow rate. To conserve sample this can be as slow as 1 µl.min^{-1}.
4. Inject the control sample.
5. Inject the dilution series starting from low and moving up to the highest concentration (low-to-high) and then moving back down to the low concentrations (high-to-low). Inject the highest concentration only once. It is important to inject for a period sufficiently long to reach equilibrium. Either enough time must be allowed for spontaneous dissociation or the analyte must be eluted with regeneration buffer.
6. Repeat the injection of the control sample.
7. For all injections measure the equilibrium response levels in the ligand and control flow cells. The difference between these two is the amount of binding at each concentration.

C Data analysis

1. Plot the binding versus the concentration for both the low-to-high and the high-to-low series (20).
2. Fit the Langmuir (1:1) binding isotherm to the data by non-linear curve fitting. Use this to determine the K_D and maximal level of binding.
3. Do a Scatchard plot and check if it is linear. The points on the plot where binding is less than 5% of maximum are highly inaccurate and should be ignored.
4. If the Scatchard plot is not linear do further experiments to establish cause (Section 7.1.5).

D Controls

1. The affinity constant should be confirmed in the reverse orientation.
2. Should this be impossible (e.g. because the ligand is multivalent) the affinity con-

> **Protocol 9 continued**
>
> stant should be confirmed with the ligand immobilized by a different mechanism, preferably by ligand capture.
>
> 3 Use at least two independently produced batches of protein.
>
> 4 Use different recombinant forms of the same proteins.
>
> [a] Both the time taken to reach equilibrium and the time it takes for the bound analyte to dissociate are governed primarily by the k_{off}. For the simple 1:1 model binding will reach 99% of the equilibrium level within $4.6/k_{off}$ seconds. Similarly, it will take $4.6/k_{off}$ seconds for 99% of the analyte to dissociate. Thus for $k_{off} \sim 0.02\ s^{-1}$, equilibrium will be reached within ~ 230 sec, and the bound analyte will take ~ 230 sec to dissociate.
>
> [b] If the Quickinject command is used as little as 15 μl of analyte sample is used.

The control experiments outlined in 7.1.4 will help to eliminate some trivial explanations. For example, if the analyte has a multivalent component the non-linear Scatchard plot will not be evident in the reverse orientation. The shape of the Scatchard plot will also depend on the surface density of ligand. If the ligand immobilization is responsible for heterogeneity, this should be eliminated in the reverse orientation and if the ligand is immobilized indirectly.

7.2 Kinetics

7.2.1 Concepts

The period during which analyte is being injected is termed the 'association phase' whereas the period following the end of the injection is termed the 'dissociation phase'. During the association phase there is simultaneous association and dissociation. Equilibrium is reached when the association rate equals the dissociation rate. Under ideal experimental conditions only dissociation should take place during the dissociation phase. In reality some re-binding (see below) often occurs. The main factors affecting the association rate are the concentration of analyte near the ligand (C^A), the concentration of ligand (C^L), and the association rate constant (k_{on}). Because of the high surface density of ligand on the sensor surface, the rate at which analyte binds ligand can exceed the rate at which it is delivered to the surface (referred to as mass transport). In this situation binding is said to be mass transport limited. Analysis of association rate under mass transport limited conditions will yield an *apparent* k_{on} that is slower than the true k_{on}. It is difficult to determine the k_{on} under these circumstances, and so experimental conditions must be sought in which mass transport is not limiting. Analyte is transported to the surface by both convection and diffusion. Convection transport can be increased simply by increasing the flow rate. However even at the maximal flow rates permissible mass transport can still be limiting (10, 16) because there is an unstirred 'diffusion' layer near the sensor surface through which transport is solely by diffusion (5). In this case mass transport limits can only be avoided by decreasing the surface density of immobilized ligand.

SURFACE PLASMON RESONANCE

The main factors affecting the analyte dissociation rate are the surface density of bound analyte, the dissociation rate constant (k_{off}), and the extent to which dissociated analyte re-binds to ligand before leaving the sensor surface (termed 're-binding'). The latter is also a consequence of mass transport deficiency but here it is transport away from the surface that is limiting. Convection transport can be increased by increasing the flow rate. However re-binding will still occur because diffusion out of the unstirred layer is little affected by convection transport. Re-binding is most easily avoided by decreasing the level of ligand immobilized on the surface. An alternative method is to inject during the dissociation phase a competing molecule that can rapidly bind to free analyte or ligand and block re-binding (21). If it binds ligand the competing molecule needs to be small so that it does not influence the SPR signal. Finally, when the ligand is saturated the initial part of the dissociation phase will not be affected by re-binding, since no free ligand is available for re-binding. However such selective analysis of a part of the dissociation phases should be avoided; it provides no indication as to whether the data conforms to any particular binding model and can give highly misleading results.

In summary, mass transport limitations, which lead to an underestimation of the intrinsic kinetics, are aggravated by low flow rates, high levels of immobilized ligand, and high intrinsic association rate constants. They can be reduced by increasing the flow rate and, most importantly, lowering the level of immobilized ligand.

7.2.2 Experimental design

Because mass transport may limit binding it is essential to use the lowest density of ligand that gives an adequate level of analyte binding. Depending on the background response 100 RU of binding should be adequate. In order to determine whether binding is limited by mass transport the kinetics should be measured in several flow cells with different levels of immobilized ligand. The immobilization level should vary at least twofold.

Protocol 10
Kinetic measurements

A Preliminary steps

1. Immobilize the ligand in three flow cells at different levels (e.g. 500, 1000, and 2000 RU).
2. Immobilize a control ligand in the remaining flow cell at a level midway between the range of immobilization levels in the other three flow cells.
3. Ensure that the analyte is monomeric and binds monovalently. Determine accurately the concentration of the analyte.
4. Determine the time it takes reach equilibrium and for the bound analyte to dissociate completely from the sensor surface (see footnote in *Protocol 9*). If less than

Protocol 10 continued

4–5 sec (fast) it will be necessary to collect at the maximal rate possible (10 Hz). This is only possible if data is collected from one flow cell at a time. Because the sample needs to be injected once for each flow cell studied, more sample is needed.

5 Establish regeneration conditions if necessary. With kinetic determinations it is not as important to maintain the same level of active ligand for each injection.

B Measurements

1 Prepare a twofold dilution series of the analyte ranging from concentrations of $8*K_D$ to approximately $0.25*K_D$. Take care to prepare enough sample for the special kinetic injection command (KINJECT), which utilizes more material. If kinetics are fast and a high data collection rate is needed, enough analyte needs to be prepared for separate injections in each flow cell.

2 Set the flow rate to 40–100 μl/min in order to maximize analyte mass transport. The duration of the injection is not critical since binding does not need to reach equilibrium. However, equilibrium should be approached at the higher analyte concentrations.

3 Inject the dilution series in any particular order. It is usual to start from lower concentrations.

C Data analysis

1 Use the BIAevaluation software supplied by BIAcore.

2 Subtract the response in the control flow cell from the responses in each of the ligand flow cells.

3 Group and analyse together the binding curves obtained with each dilution series, one flow cell at a time (with control responses subtracted).

4 If equilibrium is reached within 1 sec the association phase will not produce useful data. In this case only the dissociation phase should be analysed.

5 Attempt a global fit of the simple 1:1 binding model to the entire series of curves. Include in the fit as much of the association and dissociation phase as possible.

6 Repeat the analysis with data obtained at the other levels of ligand immobilization. In order to prove that binding is not limited by mass transport it is necessary to show that the same rate constants are obtained at two different ligand immobilization levels. *If this is not possible the measured rate constants should be considered to be lower limits of the true rate constants (16, 18).*

7 If poor fits are obtained using the simple 1:1 binding model, the binding is considered complex. After excluding trivial explanations (*Table 5*) an attempt should be made to establish the cause (*Table 6*).

D Controls

1 The kinetics constants should always be confirmed in the reverse orientation and/or with the ligand immobilized in a completely different manner.

Protocol 10 continued

2 Always confirm the results using separate batches of recombinant protein.

3 If possible, confirm the results using different recombinant forms of the same protein.

A second important point is that more of the analyte is needed for kinetic analysis. This is because the experiments are performed at a high flow rate, the KINJECT command wastes more material, and separate injections may be required for each flow cell.

Table 5 Trivial causes of complex binding kinetics

Cause	How to detect and/or eliminate
Mass transport limitations	Increase the flow rate and lower the density of immobilized ligand. Alternatively, include mass transport term in fitting equation.
Drifting baseline	Should be evident in the control sensorgrams. Try to eliminate this by subtracting the control sensorgram. Alternatively, the simple 1:1 binding model can be modified to include a drifting baseline.
Bulk refractive index artefacts	Should be evident in the control sensorgrams. Try to eliminate this by subtracting the control sensorgram. Alternatively, the simple 1:1 binding model can be modified to include bulk refractive index artefacts.
Re-binding	The fit will be worse at higher levels of immobilization. Increase the flow rate and lower the density of immobilized ligand. Inject a competing small analyte during the dissociation phase (21).
Heterogeneous immobilization	Immobilize ligand in a different way (preferably indirectly, by ligand capture).
Analyte is multimeric	Perform size exclusion chromatography and/or analytical ultracentrifugation to ensure that analyte is monomeric. If the ligand is monomeric, perform kinetic analysis in reverse orientation.
Analyte is monomeric but contaminated by multivalent aggregates	Repeat the experiment using the monomeric fraction immediately after size exclusion chromatography, avoiding concentrating or freezing the sample.

7.2.3 Data analysis

Analysis of kinetic data is best performed using the *BIAevaluation* software supplied with the instruments as this has been designed especially for the purpose (9). Another programme *CLAMP* (available at http://www.hci.utah.edu/groups/interaction/) has also been designed specifically for analysis of kinetic data generated on the BIAcore (22). While a complete discussion of kinetic theory is beyond the scope of this review a basic approach to kinetic analysis is provided instead. After subtracting the background responses (obtained in the control flow cells) an attempt should be made to fit the simple 1:1 Langmuir binding model to the data. For any particular sensorgram as much of the data as possible

Table 6 Distinguishing some non-trivial causes of complex kinetics

Binding mechanism	Description	Distinguishing features
Heterogeneous analyte	More than one form of the analyte binds to a single ligand.	Dissociation will slow as the duration of the injection is increased. This effect will not be seen in the reverse orientation.
Heterogeneous ligand	One analyte binds to more than one ligand.	Dissociation will not be affected by the duration of the injection. However, in the reverse orientation dissociation will slow as the duration of the injection is increased.
Two-state binding (conformational change)	After forming, the ligand/analyte complex interconverts between two or more forms with different kinetic properties.	The dissociation phase will slow as the duration of the injection is increased. This affect will also be seen in the reverse orientation.

should be included in the fit. This normally includes the entire association and dissociation phases, omitting only the 'noisy' few seconds at the beginning and end of the analyte injection. Noise in the dissociation phase is reduced by using the KINJECT command. It is good practise to fit both the association and dissociation phases simultaneously rather than separately. However the association phase cannot be analysed if equilibrium is attained within 2–4 sec, which is usually the case if the k_{off} is > 1 s^{-1}. In contrast, the dissociation phase can analysed even if the k_{off} is >1 s^{-1} (11, 20). A rigorous test of the binding model is to fit it simultaneously to multiple binding curves obtained with different analyte concentrations. This *global fitting* (8, 9) establishes whether a single 'global' k_{on} and k_{off} provide a good fit to all the data. An important internal test of the validity of the kinetic constants is to determine whether the calculated K_D ($K_{Dcalc} = k_{off}/k_{on}$) is equal to the K_D determined by equilibrium analysis.

When a poor fit is obtained to the data using the simple 1:1 binding model the binding kinetics are considered *complex*. Since the most likely explanation for this is experimental artefact, initial efforts should be directed at excluding trivial causes (*Table 5*). Only when trivial explanations have been excluded should any effort be expended on trying to establish what complex binding model explains the kinetics. This is a difficult and often impossible task (8, 9). The simple 1:1 binding model predicts that both the association and dissociation phases are mono-exponential, i.e. described by an equation with a single exponential term. When a poor fit is obtained excellent fits can usually be obtained using equations with two exponential terms (a bi-exponential fit). Because all complex binding models generate equations with two or more exponential terms, it is usually impossible to distinguish between different models by curve fitting alone. Instead further experiments need to be performed (*Table 6*).

7.2.4 Controls

The same controls should be performed as for the affinity measurements (see Section 7.1.4 and *Protocol 10*).

7.3 Stoichiometry

The binding stoichiometry can be determined if the molecular mass of ligand and analyte are known and the activity of the ligand is known. The basic approach is to immobilize a defined amount of ligand and then saturate this with analyte. The stoichiometry can then be calculated according to *Protocol 7*. Because it is very difficult to saturate with analyte the maximum level of analyte binding is best obtained by doing a standard equilibrium affinity determination. A fit of the simple 1:1 binding model to this data yields the maximum level of analyte binding as well as a K_D. The key problem is establishing the activity of the immobilized ligand. If the ligand has 100% activity in solution and is immobilized by ligand capture, it is reasonable to assume that it is all active. Activity levels can also be determined using a Fab fragment of a mAb specific for the ligand. Finally, the stoichiometry should be identical when measured in the reverse orientation.

7.4 Thermodynamics

The binding energy or affinity includes contributions from changes in enthalpy (heat absorbed or ΔH) and entropy (increased disorder or ΔS).

$$\Delta G = \Delta H - T^*\Delta S \quad [4]$$

While ΔS cannot be measured, ΔH (or the heat absorbed upon binding) can be measured directly, by microcalorimetry (ΔH_{cal}), or indirectly, by van't Hoff analysis (ΔH_{vH}). If it is assumed that ΔH and $\Delta S°$ are temperature-independent, the linear form of the van't Hoff equation can be used.

$$\ln \frac{K_D}{C°} = \frac{\Delta H_{vH}}{R*T} - \frac{\Delta S°}{R} \quad [5]$$

where $C°$ is the standard state concentration (1 M) and $\Delta S°$ is the change in entropy in the standard state. K_D is measured over a range of temperatures and $\ln(K_D/C°)$ plotted against $1/T$. If linear, the slope of this plot equals $\Delta H_{vH}/R$. A drawback of this approach is that ΔH varies with temperature for protein–ligand interactions and so the plot is not linear. Consequently, K_D needs to be measured over a small range around the temperature of interest, and the slope determined within this range. This is technically difficult and likely to be inaccurate. A more rigorous approach is to measure the affinity ($\Delta G°$, see Section 7.1.1) over a wider range of temperature and then fit an integrated (non-linear) form of the van't Hoff equation to the data (23).

$$\Delta G° = \Delta H_{T_0} - T*\Delta S°_{T_0} + \Delta C_p(T - T_0) - T*\Delta C_p*\ln\left(\frac{T}{T_0}\right) \quad [6]$$

where T is the temperature in Kelvin (K), T_0 is an arbitrary reference temperature (e.g. 298.15 K), ΔH_{T_0} is the enthalpy change upon binding at T_0 (kcal.mol^{-1}), $\Delta S°_{T_0}$ is the standard state entropy change upon binding at T_0 (kcal.mol^{-1}), and ΔC_p is the change in heat capacity (kcal.mol^{-1}.K^{-1}), and is assumed to be temperature-independent.

The ΔC_p is a measure of the dependence of ΔH (and ΔS) on temperature. It is almost invariably negative for protein–protein interactions (24), indicating that enthalpic effects become more favourable and entropic effects less favourable as temperature increases. This negative heat capacity is believed to be the result of the disruption at high temperatures of the ordered 'shell' of water that forms over the non-polar surfaces of a macromolecule. Consequently, the favourable entropic effect of displacing the shell upon binding is reduced. And because fewer solvent bonds are disrupted at the higher temperature the net enthalpy change becomes more favourable. ΔC_p is a useful measure of the extent of non-polar surface that is buried upon binding (25). However determining ΔC_p by van't Hoff analysis is likely to be inaccurate. A second drawback of van't Hoff analysis is that changes in temperature may also affect the interactions between the proteins and the solution components, including water (26). If these equilibria are coupled to the protein–protein interaction they will contribute to the ΔH_{vH}, which will therefore differ from the ΔH determined by calorimetry. Because of these drawbacks, it is advisable to confirm ΔH and ΔC_p determinations by calorimetry. Unfortunately even recently developed microcalorimeters require about 100-fold more protein than the BIAcore. Thus the BIAcore may be the only means of obtaining enthalpy and heat capacity data when limited amounts of material are available.

7.5 Activation energy

The k_{on} and k_{off} will generally increase with temperature. The extent of this increase is a measure of the amount of thermal energy required for binding or dissociation, and is referred to as the activation energy of association (E_a^{ass}) or dissociation (E_a^{diss}). E_a can be determined using the Arrhenius equation. Assuming that E_a is constant over the temperature range examined, then:

$$\ln k = \ln A - \frac{E_a}{R*T} \qquad [7]$$

where k is the relevant rate constant (e.g. k_{on} and k_{off}), R is the gas constant, and A is a constant known as the pre-exponential factor. E_a is determined from the slope of a plot of $\ln k$ versus $1/T$. Importantly, because E_a^{ass} and E_a^{diss} can be considered activation enthalpies, the reaction enthalpy can be calculated from the relationship:

$$\Delta H = E_a^{ass} - E_a^{diss} \qquad [8]$$

An unusually high E_a value indicates that binding and/or dissociation require the surmounting of high potential energy barriers, suggesting that conformational rearrangements are required.

8 Conclusion

SPR provides a powerful tool for the analysis of protein–protein interactions. This is particularly true for low affinity interactions, which are difficult to study using any other technique. One of the most useful features of SPR is that it provides, through binding analysis, a quick way of checking the structural integrity of recombinant molecules. SPR is also useful for measuring the affinity, enthalpy, stoichiometry, kinetics, and activation energy of an interaction. A major advantage of SPR over other techniques such as calorimetry is that much smaller amounts of protein are required. The pitfalls associated with SPR are easily avoided once they are understood. SPR is not well suited to high throughput assays, or the analysis of small molecules ($M_r < 1000$).

9 Appendix

9.1 Physical basis of SPR

When a beam of light passes from material with a high refractive index (e.g. glass) into material with a low refractive index (e.g. water) some light is reflected from the interface. When the angle at which the light strikes the interface (the angle of incidence or θ) is greater than the critical angle (θ_c), the light is completely reflected (total internal reflection). If the surface of the glass is coated with a thin film of a noble metal (e.g. gold), this reflection is not total; some of the light is 'lost' into the metallic film. There then exists a second angle greater than the critical angle at which this loss is greatest and at which the intensity of reflected light reaches a minimum or 'dip'. This angle is called the surface plasmon resonance angle (θ_{spr}). It is a consequence of the oscillation of mobile electrons (or 'plasma') at the surface of the metal film. These oscillating plasma waves are called surface plasmons. When the wave vector of the incident light matches the wavelength of the surface plasmons, the electrons 'resonate', hence the term surface plasmon resonance. The 'coupling' of the incident light to the surface plasmons results in a loss of energy and therefore a reduction in the intensity of the reflected light. It is because the amplitude of the wave vector in the plane of the metallic film depends on the angle at which it strikes the interface that an θ_{spr} is observed. An evanescent (decaying) electrical field associated with the plasma wave travels for a short distance ($\sim$ 300 nm) into the medium from the metallic film. Because of this, the resonant frequency of the surface plasma wave (and thus θ_{spr}) depends on the refractive index of this medium. If the surface is immersed in an aqueous buffer (refractive index or μ $\sim$ 1.0) and protein ($\mu \sim$ 1.33) binds to the surface, this results in an increase in refractive index which is detected by a shift in the θ_{spr}. The instrument uses a photo-detector array to measure very small changes in θ_{spr}. The readout from this array can be viewed on the BIAcore as 'dips' (Section 3.3). The change is quantified in resonance units or response units (RUs) with 1 RU equivalent to a shift of 10^{-4} degrees. Empirical measurements have shown that the binding of 1 ng/mm^2 of protein to the sensor surface leads to a response of $\sim$ 1000 RU. Since the matrix is $\sim$ 100 nm thick, this represents a protein concentration within the matrix of 10 mg/ml. Apart from the refractive index, the other physical parameter which affects θ_{spr} is temperature. Thus a crucial feature of any SPR instrument is precise temperature control.

9.2 Samples and buffers

9.2.1 Guidelines for preparing samples and buffers for use on the BIAcore

All buffers should be filtered through 0.2 μm filters and degassed at room temperature. The latter can be achieved by filtering under vacuum or by using a vacuum chamber.

(a) All samples > 3 ml should be filtered and degassed in the same way.

(b) Samples < 3 ml should be spun at high speed in a microcentrifuge for 10–20 min at 4 °C and then degassed in a vacuum chamber.

(c) Before samples are placed in the sample rack they should be pulsed briefly in a microcentrifuge. This dislodges air bubbles from the bottom of the container, and ensures that the meniscus is horizontal.

(d) Vials should be capped to prevent sample evaporation.

(e) When running long experiments consider cooling the sample rack base using a thermostatic re-circulator.

9.2.2 Standard buffers

i. Running buffers

(a) HBS or HBS-EP: 10 mM Hepes pH 7.4, 150 mM NaCl, 3 mM EDTA, 0.005% P20.

(b) HBS-P: 10 mM Hepes pH 7.4, 150 mM NaCl, 0.005% P20.

(c) HBS-N: 10 mM Hepes pH 7.4, 150 mM NaCl.

ii. Pre-concentration buffers

(a) pH 3.0–4.5: 10 mM formate.

(b) pH 4.0–5.5: 10 mM acetate.

(c) pH 5.5–6.0: 5 mM maleate.

iii. Regeneration buffers

Grouped according to chemical properties. They increase in strength from left to right.

(a) Cation chelator: HBS with 20 mM EDTA pH 7.5.

(b) High ionic strength: 1 M NaCl; 4 M KCl; 2 M $MgCl_2$.

(c) Low pH: 100 mM glycine–HCl pH 2.5; 10 mM HCl; 100 mM HCl; 100 mM H_3PO_4.

(d) High pH: 5 mM NaOH; 50 mM NaOH.

9.3 Additional information

BIAcore maintains a useful **web site** [http://www.biacore.com/]. This announces new product developments and includes a continuously updated list of SPR publications and an electronic version of the BIAjournal.

The **handbooks** that come with the instrument are an essential resource. These can also be purchased from BIAcore along with the following books, which describe the technology and its applications in more detail.
BIAtechnology Handbook
BIAapplication Handbook
BIAevaluation software Handbook

Acknowledgements

I would like to thank Liz Davies, Neil Barclay, and Don Mason for critical comments on the manuscript. This work was supported by the Medical Research Council.

References

1. Jönsson, U., Fägerstam, L., Ivarsson, B., Johnsson, B., Karlsson, R., Lundh, K., et al. (1991). *BioTechniques*, **11(5)**, 620.
2. Jönsson, U. and Malmqvist, M. (1992). *Adv. Biosensors*, **2**, 291.
3. Hodgson, J. (1994). *Bio/Technology*, **12**, 31.
4. van der Merwe, P. A. and Barclay, A. N. (1996). *Curr. Opin. Immunol.*, **8**, 257.
5. Schuck, P. (1997). *Annu. Rev. Biophys. Biomol. Struct.*, **26**, 541.
6. O'Shannessy, D. J. (1994). *Curr. Opin. Biotechnol.*, **5**, 65.
7. van der Merwe, P. A., Brown, M. H., Davis, S. J., and Barclay, A. N. (1993). *Biochem. Soc. Trans.*, **21**, 340S.
8. Myszka, D. G. (1997). *Curr. Opin. Biotechnol.*, **8(1)**, 50.
9. Karlsson, R. and Falt, A. (1997). *J. Immunol. Methods*, **200(1-2)**, 121.
10. van der Merwe, P. A. and Barclay, A. N. (1994). *Trends Biochem. Sci.*, **19**, 354.
11. Wyer, J. R., Willcox, B. E., Gao, G. F., Gerth, U. C., Davis, S. J., Bell, J. I., et al. (1999). *Immunity*, **10**, 219.
12. Willcox, B. E., Gao, G. F., Wyer, J. R., Ladbury, J. E., Bell, J. I., Jakobsen, B. K., et al. (1999). *Immunity*, **10**, 357.
13. van der Merwe, P. A., Crocker, P., Vincent, M., Barclay, A. N., and Kelm, S. (1996). *J. Biol. Chem.*, **271**, 9273.
14. van der Merwe, P. A., McNamee, P. N., Davies, E. A., Barclay, A. N., and Davis, S. J. (1995). *Curr. Biol.*, **5**, 74.
15. Davis, S. J., Davies, E. A., Tucknott, M. G., Jones, E. Y., and van der Merwe, P. A. (1998). *Proc. Natl. Acad. Sci. USA*, **95**, 5490.
16. van der Merwe, P. A., Bodian, D. L., Daenke, S., Linsley, P., and Davis, S. J. (1997). *J. Exp. Med.*, **185**, 393.
17. van der Merwe, P. A., Brown, M. H., Davis, S. J., and Barclay, A. N. (1993). *EMBO J.*, **12**, 4945.
18. van der Merwe, P. A., Barclay, A. N., Mason, D. W., Davies, E. A., Morgan, B. P., Tone, M., et al. (1994). *Biochemistry*, **33**, 10149.
19. Karlsson, R. and Ståhlberg, R. (1995). *Anal. Biochem.*, **228**, 274.
20. Nicholson, M. W., Barclay, A. N., Singer, M. S., Rosen, S. D., and van der Merwe, P. A. (1998). *J. Biol. Chem.*, **273**, 763.
21. Felder, S., Zhou, M., Hu, P., Ureña, J., Ullrich, A., Chaudhuri, M., et al. (1993). *Mol. Cell. Biol.*, **13(3)**, 1449.
22. Myszka, D. G. and Morton, T. A. (1998). *Trends Biochem. Sci.*, **23**, 149.

23. Yoo, S. H. and Lewis, M. S. (1995). *Biochemistry*, **34**, 632.
24. Stites, W. E. (1997). *Chem. Rev.*, **97**, 1233.
25. Ladbury, J. E. and Chowdhry, B. Z. (1996). *Chem. Biol.*, **3(10)**, 791.
26. Naghibi, H., Tamura, A., and Sturtevant, J. M. (1995). *Proc. Natl. Acad. Sci. USA*, **92(12)**, 5597.

Chapter 7
Capillary electrophoresis

Niels H. H. Heegaard

Department of Autoimmunology, Statens Serum Institut, 5 Artillerivej, 2300 Copenhagen S, Denmark.

1 Introduction

Electrophoresis is the transport of charged molecules in solution by the action of an electrical field (1). Electrophoretic techniques are widely used to separate and characterize biomolecular mixtures and very high resolution may be achieved by the use of denaturing one- or two-dimensional gel methods. The separation efficiency in these techniques is normally achieved at the expense of analyte functions such as enzyme and ligand binding activities that are adversely affected by the denaturing conditions required. With the advent of capillary electrophoresis (CE) (2), however, it has become possible to preserve molecular function whilst still maintaining separation efficiency for the reason that extremely high resolution can be obtained in buffers that do not interfere with the native structure of the analyte. This has expanded the use of affinity electrophoresis methods considerably.

The basic aspects of high resolution electrophoresis in capillaries were worked out in the 1930s by Tiselius and co-workers (3, 4). Later important contributions to the field were made by Hjertén and others (2, 5, 6). The use of CE in binding studies were subsequently initiated by reports in the 1990s (7-16). Several recent reviews cover theory and applications in affinity CE (17-25).

Particularly valuable methods for the study of affinity interactions by CE use immobilized ligands in a manner much similar to affinity chromatographic methods (21, 26). However, the present chapter focuses on techniques taking place in solution and describes the methods and their limitations for quantitative and qualitative biomolecular binding studies in both low and high affinity systems.

2 The capillary electrophoresis technique

The basic components of electrophoretic systems using capillaries for the separation path are shown in *Figure 1*. Apart from the capillary itself the components include a sample introduction system based on hydrostatic pressure or electrical current, a high voltage power supply (usually capable of delivering up to 30 kV),

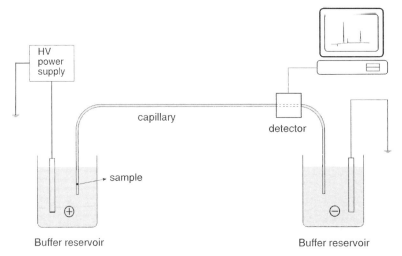

Figure 1 Schematic drawing of a capillary electrophoresis system. The capillary typically holds 1–2 μl buffer and the sample mixture is usually introduced in a nl volume as a small zone at the anode end of the capillary. Subsequently, the separation is toward the cathode after application of voltage.

reservoirs for buffers and cleaning solutions, a detector (usually a UV absorbance detector but important alternatives exist, cf. Section 4.1.2), and a system for data acquisition, analysis, and reporting. The following options may also be possible: sample cooling, switching between different detectors, capabilities for automatic buffer replenishment, fraction collection, and coupling to mass spectrometers. Software is usually instrument specific but alternative software may be preferred because it is more useful, e.g. by offering automated area corrections. A number of important features of CE are compared with the characteristics of conventional slab gel electrophoretic techniques in *Table 1*. For more theory and treatment of technical details of CE separations the reader is referred to several recent publications (27–31). However, in the context of binding studies by electrophoresis two specific points merit particular attention here:

(a) In contrast to the situation in conventional gel electrophoresis where the electro-osmotic flow is usually of no significance for molecular movement the force that moves an analyte molecule in CE is the sum of the electrophoretic mobility (μ) and the electro-osmotic flow. The latter is of considerable magnitude in CE under neutral conditions where it provides a higher velocity than the electrophoretic mobility of the analytes (32). This means that all analytes are transported against the cathode irrespective of charge. The resulting mobility of an analyte is mainly dependent on its charge-to-mass ratio and is highly dependent on the pH and ionic strength conditions under which it is electrophoresed because these parameters influence analyte charge as well as the electro-osmotic flow that, e.g. decreases with lowered pH and increased ionic strength.

Table 1 Comparison of conventional gel electrophoresis and capillary electrophoresis

Feature	Gel electrophoresis	Capillary electrophoresis
Analyte detection	After separation	During separation
Analyte quantitation	Indirect	Direct
Max. resolution (theoretical plates)	10^3–10^4	10^5–10^6
Detection limit (proteins)	nM	µM[a]
Typical sample volume	µl	nl
Parallel sample analysis	Yes	No,[b] sequential separation of individual samples
Speed	Hours	Minutes
Migration data	Distance (fixed separation time)	Time (fixed path length)
Separation field strength (V/cm)	2–20	200–500
Preparative capability	Possible	Poor
Analyte versatility	Organic macromolecules	Ions, small molecules, and macromolecules
Physiological buffer conditions	Usually not compatible	Yes

[a] Using UV-based detection systems.
[b] Capillary arrays have been devised for DNA sequencing (63) but are not an option in standard instruments.

(b) The data output in CE is in the form of peak appearance times (t), i.e. the time it takes for an analyte to travel the fixed distance from the injection end of the capillary to the detector window. This contrasts to conventional gel electrophoresis where migration data output is in the form of the distance (d) travelled at a fixed time. Since the electrophoretic mobility µ (see *Table 2*) is directly proportional to d and inversely proportional to t (corrected for the contribution from the electro-osmotic flow to t) this has consequences for transforming the equations used in conventional gel affinity electrophoresis (33) to the situation in CE (17, 34).

3 Why use electrophoresis for the study of reversibly interacting molecules?

The existence of numerous approaches for binding studies (cf. e.g. the present volume, refs 35, 36, and *Table 3*) reflects the fact that reversible binding interactions of widely different nature are of interest in biology and that no universally applicable method exists. Electrophoresis and CE methods are usually not first choices for binding assays due to the existence of more traditional and well-proven methods that often work very well. Also, there are limitations in the applicability of CE to binding studies and occasional difficulties occur in the interpretation of electrophoresis data especially with regard to migration shift experiments where the system response to binding (migration changes) is quite different from other analytical methods. Finally, capillary electrophoresis requires relatively specialized and expensive equipment.

Table 2 Equations used in electrophoretic binding studies[a]

Migration shift experiments:
Assumption: [L] >> [R].
One-to-one binding (see also *Table 6*).
Mobility change: $\Delta\mu = \Delta\mu_{max} - K_d \times \Delta\mu/[L]$.
Mobility change and corrected peak appearance time: $\Delta\mu = L/E \times [(1/t - 1/t_r) - (1/t_0 - 1/t_{r0})] = L/E \times \Delta(1/t)$.
Mobility change expressed using corrected peak appearance times: $\Delta(1/t) = \Delta(1/t)_{max} - K_d \times \Delta(1/t)/[L]$.
Direct plot $\Delta\mu$ as a function of [L] or $\Delta(1/t)$ as a function of [L] should show a definite curvature (17).
Non-linear curve fitting to the direct plot using a one site hyperbola function yields the K_d if binding behaves according to a 1:1 molecular association binding isotherm that has the equation: $\Delta(1/t) = (\Delta(1/t)_{max} \times [L])/(K_d + [L])$.
Linearizing plots: $\Delta\mu$ vs $\Delta\mu/[L]$ or $\Delta(1/t)$ vs $\Delta(1/t)/[L]$. Linear regression by least squares dubious (35). Double reciprocal plots are more biased towards the data points obtained at low ligand concentration (17, 64).
Linear plot slope: $-K_d$ (see *Figure 3A*).

Pre-incubation experiments:
Samples pre-equilibrated: $[R] + [L] \Leftrightarrow [RL]$ (see also *Table 7*).
[R] determined experimentally by CE at different R:L ratios in pre-incubation mixtures.
Direct binding curve: [RL] as a function of [L] or log[L] (61) to ensure saturability.
A 1:1 molecular association binding isotherm is a rectangular hyperbola that can be fitted to the experimental data in the direct binding curve using non-linear regression. From this curve the K_d is obtained since it is described by:
$[RL] = ([R]_{tot} \times [L])/(K_d + [L])$.
Linearizing plot (Scatchard plot): [RL]/[L] vs [RL], linear regression by least squares dubious (35).
Linear plot slope: $-1/K_d$ (see *Figure 3B*).

[a] E, field strength; L, total capillary length; l, length to detector window; [L], equilibrium concentration of free ligand. Migration shift experiments only valid when [L] >> [R] and/or when binding is weak. K_d, apparent dissociation constant; [R], equilibrium concentration of free receptor (analyte) molecule; $[R]_{tot}$, total receptor (analyte) concentration; [RL], equilibrium concentration of RL complex; t, peak appearance time; t_0, peak appearance time in the absence of ligand; t_r, reference (marker) peak appearance time; $\Delta(1/t)$, difference in corrected inverse peak appearance time in experiment with and without added ligand; μ, electrophoretic mobility. $\mu = (L \times l)/(V \times t)$. $\Delta\mu$, difference in μ in experiments with and without added ligand; V, applied voltage.

Table 3 Comparison of some commonly used methods for quantitative binding studies

Technique	Time	Measurement	Purified	Label	Material consump.	Low affinity
Equilibrium dialysis	Days	Size	Yes	(Yes)	Moderate	Yes
Calorimetry	Min	Transition temp.	No	No	High	No
Spectroscopy	Min	Quench/enhancem.	Yes	No	Moderate	No
Solid phase	Hours	Retention or mass[a]	Yes	(Yes)	Moderate	No
Chromatography	Hours	Size[b]	Yes	(No)	High	Yes
Capillary	Min–h	Charge/mass electrophoresis	No	No	Low[c]	Yes

[a] Surface plasmon resonance techniques.
[b] Gel filtration experiments.
[c] Low affinity binding shifts may require quite high amounts of purified ligand.

CE-based methods for binding studies (affinity CE) should be considered when:

(a) Purified analyte is not available and/or an unknown fraction of the analyte is devoid of binding activity.

(b) Complex formation only leads to small changes in the system response that is relevant for a given binding assay.

(c) Analyte or ligand cannot be analysed by gel electrophoresis or affinity gel electrophoresis offers too low a resolution.

(d) Material is available in scarce amounts.

(e) Weak binding is expected.

(f) The formation of complexes of different stoichiometries are expected.

(g) It is critical not to disrupt molecular structure.

(h) Many putative ligands must be screened for binding activity.

Thus, when analyte material is scarce, not necessarily pure but in solution and identifiable in the separation pattern, and when complexation leads to a distinct change in electrophoretic mobility, the electrophoretic techniques may be very well suited to measure strong as well as weak binding while some intermediate cases may not be amenable to quantitative analysis.

It was realized before CE became more widely accessible that electrophoresis carried out under non-denaturing conditions had some potential as an alternative to some of the other methods for binding studies (Table 3). The fundamental basis of using electrophoresis to examine binding between soluble molecules is that complexed molecules are distinguished by changed positions in the separation profile (20, 22, 37) (with the important exception of competition experiments where more than one ligand is used) (34).

The quite numerous approaches that have been developed for classical gel affinity electrophoretic methods (see e.g. ref. 33) are generally also applicable in CE and often offer considerable advantages partly because of the versatility of the CE technique. This versatility derives from CE facilitating the analysis of a very wide range of analytes using a wide range of buffer conditions and partly because of its high resolving power, speed, minimal sample consumption and waste generation, and finally because of the automation inherent in this technique.

4 Instrumentation and experimental variables

4.1 Apparatus

A number of commercial instruments exist. For binding studies 'home-made' equipment may certainly suffice but the automated sample and data processing in commercial instruments are a definite advantage. The most critical parts of the instrumentation are the separation capillaries that are used in them, the detector, and the efficiency of the capillary temperature control.

4.1.1 Capillary columns

Fused silica capillaries coated with polyimide on the outside for flexibility but unmodified at the inner surface are presently the standard choice for the separation path in CE and are available in pre-assembled instrument-specific cassettes from the instrument companies or in spools with for example 10 m capillaries from several manufacturers (e.g. PolyMicro Technologies, Supelco, and Composite Metal Service). Typical internal diameters (i.d.) are 20–100 μm with lengths of between 10–100 cm. A first choice for binding experiments would be a 50 μm i.d. capillary with as short a length as possible. Temperature effects are less in small diameter capillaries but so is detection sensitivity. The position of the detector window relative to the capillary ends is instrument-specific. The simplest way to make a detector window is to burn the polyimide coating away on a 0.3–0.5 cm length of the capillary and remove debris with ethanol-wetted tissue.

Binding studies do not confer special requirements *per se* on the choice of columns but it is important to be aware that uncoated fused silica offers a vast negatively charged surface at neutral pH that will attract and bind positively charged molecules or positively charged areas in macromolecules. Therefore, an important first step in binding studies (cf. *Protocol 1*) is to establish proper buffer conditions for reproducible analyte and ligand recovery (see Section 4.2.1). It is recommended to start with the simplest solution, i.e. to use uncoated fused silica and some of the electrophoresis buffers suggested in *Table 4*. Also, new capillaries should be conditioned before use. We routinely use alternating 5 min rinses of water and 1 M NaOH for 30 min followed by a rinse with electrophoresis buffer. The use of standard analyte mixtures can ensure satisfactory capillary performance and may be used at regular intervals to check the quality of the capillary. Peak tailing, irregular current, unstable and noisy baseline, and/or drifting peak appearance times even when using freshly prepared buffers indicate that capillary replacement is due.

4.1.2 Detectors

The most widely used type is single wavelength UV detection. Diode-array detection is also standard on some instruments but may not be advantageous when sensitivity is an issue. Due to the short path length across a capillary the concentration detection limit of UV detection over a capillary is not much below μM for peptides detected at 200–210 nm (8, 38). The direct consequence of this for binding studies is that strong binding characterized by, e.g. nM dissociation constants is usually only amenable to study by CE if more sensitive systems than UV detectors are used. Capillaries that are enlarged at the detection window site are fractionally more sensitive than conventional capillaries with a uniform diameter. However, laser-induced fluorescence detection which is the most widely used alternative can offer substantially more sensitive detection and an easier monitoring of the fate of the specific labelled molecules. Unfortunately, the labelling procedure (e.g. fluoresceinylation) may change the molecular structure of the analyte enough to affect its binding interactions.

4.2 Experimental variables

The choice of capillaries has been treated above. Given suitable binding kinetics, the most important variables for the success of electrophoretically based binding studies are the electrophoresis buffer selection and the sample preparation.

4.2.1 Electrophoresis buffers

Usually buffers of physiological pH and ionic strength are used for binding studies, in some cases with the additions of Ca^{2+}, Mg^{2+}, or other ions if required for binding or mild non-ionic detergents such as β-octyl glucoside or Tween 20 at 0.1–1% (v/v) if required for keeping analytes soluble (39). The possible detrimental effects of any buffer addition on the binding interaction must always be considered. Buffer components may show unwanted interactions with the sample as well as with the capillary wall. Ca^{2+} may for example bind to the negatively charged capillary wall and thereby drastically change the electro-osmotic flow. The same anomalous binding may be the case for a putative ligand: anomalous

Table 4 Examples of electrophoresis buffers used to establish optimal separation conditions in CE-based binding studies

All buffers are prepared with MilliQ-quality water and filtered through 0.22 μm pore size filters before use. Degassing is optional. May be kept at 4 °C for months.

0.1 M phosphate pH 7.4
40.5 ml of 0.2 M Na_2HPO_4 (35.61 g/l of $Na_2HPO_4.2H_2O$)
9.5 ml of 0.2 M NaH_2PO_4 (27.6 g/l of $NaH_2PO_4.H_2O$)
50 ml H_2O

Isotonic borate pH 7.4
10 ml of 0.05 M $Na_2B_4O_7$ (19.11 g/l of $Na_2B_4O_7.10H_2O$) (A)
90 ml of 0.2 M HBO_4 (12.40 g/l) (B)
270 mg NaCl (C)
To scan for analyte recovery at a range of pH values (see e.g. *Figure 2*) borate buffers of the following compositions are helpful (65):
pH 6.8: 3 ml (A), 97 ml (B), 270 mg (C)
pH 7.8: 20 ml (A), 80 ml (B), 260 mg (C)
pH 8.1: 30 ml (A), 70 ml (B), 240 mg (C)
pH 8.4: 45 ml (A), 55 ml (B), 210 mg (C)
pH 8.6: 55 ml (A), 45 ml (B), 190 mg (C)
pH 8.8: 70 ml (A), 30 ml (B), 140 mg (C)
pH 9.1: 90 ml (A), 10 ml (B), 70 mg (C)

Hepes pH 7.4
10 mM *N*-2-hydroxyethylpiperazine-*N'*-ethanesulfonic acid (Hepes) (2.38 g/l)
adjusted with NaOH to pH 7.4
150 mM NaCl (8.77 g/l)

Tricine pH 8.15
20 mM *N*-Tris(hydroxymethyl)methylglycine (Tricine) (3.58 g/l)
adjusted with NaOH to pH 8.15
150 mM NaCl (8.77 g/l)

Tris-buffered saline pH 7.4
5 mM Tris(hydroxymethyl)aminomethane (Tris) (0.61 g/l)
adjusted with HCl to pH 7.4
150 mM NaCl (8.77 g/l)

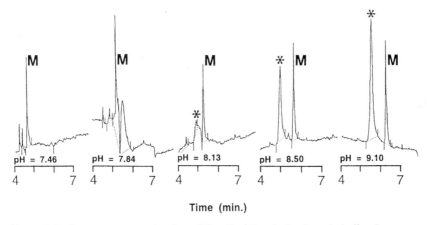

Figure 2 Analyte recovery as a function of the pH of the electrophoresis buffer. A monoclonal antibody (anti-phosphotyrosine IgG, marked with an *) at 0.05 mg/ml in isotonic borate pH 8.9 together with 0.1 mg/ml of a synthetic peptide (M) (H-Asp–Ala–Glu–Phe–Arg-OH) was injected for 1 sec into a 57 cm uncoated fused silica capillary (50 μm diameter) with the detector window at 50 cm. Electrophoresis at 25 kV with detection at 200 nm has taken place in different borate buffers (cf. *Table 4*) at the pH values indicated.

binding activity can be checked by adding it to the electrophoresis buffer. Anomalous binding may induce peak appearance time changes that are entirely due to the secondary and non-specific effects on the electro-osmotic flow if the ligand coats the inner capillary walls or it may change the buffer viscosity or conductivity. Analyte–wall interactions will typically result in impaired analyte recovery and broad, asymmetric, or missing peaks. High salt concentrations, change of pH, and the use of buffer additives such as zwitterionic detergents may counteract detrimental analyte–wall interactions but may also interfere with the binding interactions under study. As an example the experiment shown in *Figure 2* shows the complete lack of recovery of a monoclonal antibody analysed at neutral pH. The antibody becomes fully recoverable when raising the pH of the electrophoresis buffer to above 8.2.

Alternative measures to inhibit solute–wall interactions are the use of covalently bonded columns that can be produced in the laboratory or are available commercially with neutral or positively charged surfaces. Stably bonded capillaries may be the only solution when working, e.g. with very basic proteins (40) (*Table 5*). When modified capillaries and/or buffer additives are used the consequences for the electro-osmotic flow rate must be judged in each individual case.

Examples of starting buffers that are used to establish optimal conditions for analyte separation and recovery are listed in *Table 4*. A suitable buffer would have a pK value within less than 1 unit away from the desired pH. Most running buffers can be reused several times but this depends on the running voltage and type of buffer including its buffer capacity. The buffer ion depletion that occur during a run will eventually result in drifting migration times in repeated analyses. When this happens simply change to a fresh buffer. The potential

Table 5 Examples of methods used to counteract solute–capillary wall interaction

pH decrease or increase (31, 41).
Buffer ionic strength increase (66).
Buffer additives (37, 67).
Dynamic capillary coatings (68, 69).
Stably bonded capillaries (70–74).

problem of Joule heating is very dependent on the ionic composition of the buffer and therefore minimized with the use of low conductivity buffers where zwitterions control pH, e.g. Hepes and Tricine buffers (41). We now consider its control in some detail.

4.2.2 Temperature and voltage selection

The dissipation of the Joule heat from the electrophoresis buffer that takes place when a current passes through it is the main upper limiting factor for choice of field strength in CE (2). Similar problems affect the electro-optic methods considered in the following chapter. The highest possible field strength gives the highest separation efficiency and therefore low conductivity buffers in as narrow capillaries as possible are desirable to limit the induced current and consequently the temperature gradients that induce detrimental viscosity changes and convection currents (28, 41). However, the suppression of capillary wall–analyte interactions and the creation of near physiological conditions relevant for binding studies favour the use of higher ionic strength buffers. Also, it is not necessarily advantageous in all types of binding experiments to use the highest possible voltage (cf. Section 5.3.1).

Capillary cooling is achieved by either forced air flow or by a cooling liquid. The latter is probably the most effective way but both methods cool the capillary from the outside and a temperature gradient across the capillary cannot be avoided by any of these methods. Rather, the main point of the temperature control is to ensure reproducible temperature profiles from run to run. This is important for binding studies using migration shift experiments since temperature affects the buffer viscosity to a degree that influences analyte migration considerably. Reproducible cooling capabilities and the use of reasonable field strengths are therefore important. A rule-of-thumb for current limits is that the value of [current (μA) × voltage (V)]/[total length of capillary (cm) × 1000] must be below 0.05. A more precise value for a given buffer may be obtained by constructing an Ohm's Law plot (current reading as a function of voltage). The breakpoint of the curve indicates the limiting voltage above which Joule heating begins to exceed the heat dissipating ability of the system (42).

4.2.3 Sample preparation

The sample buffer should be compatible with the running buffer and normally of the lowest concentration possible compared to the running buffer concentration

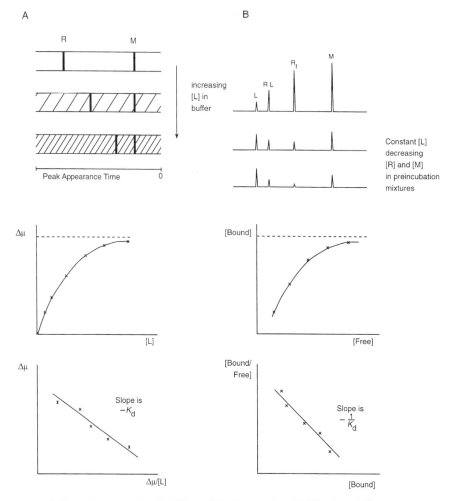

Figure 3 The main approaches in CE-based binding studies. (A) Migration shift electrophoresis. Analyte zone is R, marker zone is M. Changes in analyte electrophoretic mobility ($\Delta\mu$) caused by addition of increasing concentrations of ligand [L] to the electrophoresis buffer are studied. Dashed line indicates the level of maximum mobility shift which is the difference between the mobility of the molecular complex itself and the free analyte mobility. See also *Table 2*. (B) Pre-incubation approach. R_f are free receptor (analyte) molecules. Dashed line indicates the maximum binding (total concentration of receptor binding sites). Linearizations in (A) and (B) according to equations given in *Table 2*, assumptions and prerequisites as set forth in *Tables 6* and *7*.

so as to sharpen the analyte band. Highly concentrated samples and large sample volumes may cause a local electrical field distortion because the sample itself acts as a conducting electrolyte and this may disturb peak shape (43). Furthermore, adsorbing sample constituents that are not washed away during the washing steps in between runs will result in gradually changing migration times due to the resulting changes in the electro-osmotic flow. Since these apparent migration changes are not due to specific interactions the inclusion of an internal marker in the sample is a very important way of detecting such phenomena.

5 Capillary electrophoretic binding experiments

5.1 Experimental approaches for electrophoretic binding studies

In order to show binding in any electrophoretic system (except for some competitive assays where more than one ligand is used, cf. Section 5.3.1) it is a requirement that the electrophoretic mobility must be different for the complexed and free analyte.

The two main approaches to determining the affinity binding constant using CE are schematically illustrated in *Figure 3* and the important requirements for their use are briefly outlined in *Tables* 6 and 7 (see also Sections 5.3.1 and 5.3.2). The requirements characteristically and uniquely do not include differences in size between analyte and ligand molecule, do not depend on the availability of secondary reagents such as antibodies or precipitating chemicals, and do not normally include labelling or immobilization of any of the molecules involved in the interactions. Other CE-based approaches for binding studies that use

Table 6 Migration shift affinity electrophoresis—requirements and assumptions for binding constant determination

Distinct mobility of free and complexed molecules.
Migration data adjusted by use of an internal non-interacting marker molecule.
Binding is not influenced by electrical field.
Molecular concentration of ligand is 10–500 times higher than analyte concentration.
Ligand concentration known precisely.
No stacking effects.
On-and-off kinetics fast as compared with time for electrophoresis, typically $k_{off} > 0.1$ s^{-1} (75), in practice this is seen as peak migration shifts that do not affect peak shape and/or size (*Figure 5A*).
1:1 binding stoichiometry and homogeneous and totally accessible distribution of independent binding sites.
No interactions of analyte with other buffer components or with capillary wall contribute to separation pattern changes.

Table 7 Pre-equilibration approach—requirements for electrophoretic binding constant determination

Distinct mobility of complexed molecules (change in peak appearance time greater than peak width).
Stable complexes (complex dissociation half-time > peak appearance time (t), or: $k_{off} <$ 0.105/t).
Equilibrium reached.
Quantitation of free molecules is possible.

frontal analysis (15, 44) or Hummel and Dreyer-like methods (see Chapter 3) as the vacancy peak method (45) may be useful in special cases but generally consume quite large amounts of material and will not be considered in this chapter.

5.2 Demonstration of ligand binding

The simplest and most efficient way of using CE to screen for the binding of ligand to a specific analyte or a mixture of analytes is to separate the sample with and without the ligand added to the buffer as detailed in *Protocol 1*. With a

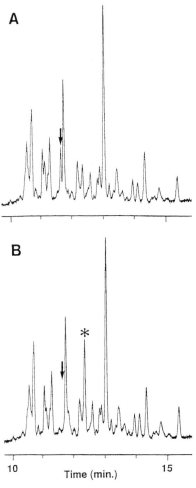

Figure 4 Demonstration of heparin binding in a peptide mixture (76). Part of a CE separation of a mixture of tryptic peptides from the protein serum amyloid P component in the absence (A) or the presence (B) of 1 mg/ml of heparin added to the electrophoresis buffer (isotonic borate pH 8.4, cf. *Table 4*). Conditions otherwise as for *Figure 2* except that sample was injected for 5 sec and electrophoresis took place at 12.5 kV. Peptide marked with an arrow is selectively affected in the experiment with heparin present in the electrophoresis buffer where it appears as a later peak (marked with an *).

charged ligand this will lead to specific changes in the separation pattern both if binding is weak and if it is strong as illustrated in *Figure 4* where a heparin binding peptide is identified in a mix of peptides. By binding to the highly anionic heparin molecules during electrophoresis the peak appearance time for the peptide is considerably and specifically prolonged.

The nature of the observed changes gives a direct clue to the approach most suitable for subsequent electrophoretically based determination of apparent binding constants (cf. Sections 5.3.1 and 5.3.2). If a neutral ligand is examined for binding and the system cannot be inverted (i.e. so that the ligand is used as the analyte, as used for example in SPR—see Chapter 6) a competitive binding approach can be used by including another charged ligand with which the neutral ligand can compete (34).

As CE methods and instrumentation get scaled down and separations become more rapid (separations lasting less than 30 sec on chips) (46, 47) it is worth noting that some classes of interactions may have k_{on} (association rate constants) values that are too slow to effect discernible binding during a short electrophoresis run. Longer electrophoresis times will be more advantageous in these cases. Very short analysis times are advantageous for measurement of multiple samples with strong binding interactions such as, for example, is the case when CE is used to measure specific analytes in biological fluids with high affinity antibodies (46, 48).

Table 8 Reasons for apparent non-reactivity in capillary electrophoresis-based binding studies

No binding at chosen conditions.
Complexed molecules have identical mobility to free ones.
Binding of ligand blocked by binding to other components in system including capillary wall, self-aggregation, and interaction with buffer constituents.
Binding blocked by other component(s) that bind to analyte but do not change mobility.
Weak binding (fast complex dissociation) (in pre-equilibration experiments).
Very slow association (in migration shift experiments).

Protocol 1
Capillary electrophoretic procedure to demonstrate binding

Equipment and reagents
- Electrophoresis equipment (see Section 4)
- Electrophoresis buffers (see *Table 4*)
- Ligand molecule
- Marker molecule, e.g. synthetic peptide

Method
1 Establish electrophoretic conditions for reproducible analysis of the analyte molecule. Capillary wall interactions are usually seen as analyte peak broadening, tailing,

Protocol 1 continued

or even total disappearance. High ionic strength and/or pH extremes suppress electrostatic interactions between wall and analytes. Other measures are mentioned in *Table 5*. Electrophoresis buffers that are physiologically relevant with respect to pH and ionic strength should normally be used. Start e.g. with 0.1 M phosphate pH 7.4. Other possibilities are listed in *Table 4*. In the choice of electrophoresis parameters also consider the electrical current limitations identified in Section 4.2.2.

2 Establish if the ligand molecule is soluble and recoverable under the same conditions and in a similar or higher concentration than the approximate analyte concentration.

3 Include a suitable marker molecule, e.g. a synthetic peptide that is very soluble negatively charged and can be added to the sample if the analyte is not part of a mixture.

4 Perform electrophoresis with two or three different concentrations of the charged putative ligand (in molar excess of the approximate concentration of analyte) added to the electrophoresis buffer. Look for changes in the ensuing separation patterns (cf. *Figure 5*). NB. If the putative ligand is more negatively charged than the analyte molecule it will be possible to save on the amount of ligand required by filling the volume of the capillary with ligand-containing buffer and then performing the electrophoresis using running buffer without added ligand. Also note that if carry-over of ligand into the sample vial may be a problem then a plug of running buffer injected from a large volume prior to the sample injection or reversed direction filling of the capillary may be helpful.

5 If changes in the electrophoresis patterns appear as specific mobility shifts proportional with the ligand concentration in the buffer and peaks are otherwise unchanged with respect to area and shape (*Figure 5A*) the interaction is amenable to migration shift analyses as described in Section 5.3.1. If peak is tailing, split, diminished, or even disappearing (*Figure 5B-5E*) a pre-equilibration approach may be used to estimate binding constants, see *Protocol 3*.

6 If no changes in the separation patterns can be observed the possible causes of non-reactivity listed in *Table 8* should be considered. If it is suspected that complex formation does occur but does not lead to a mobility change such as when a large analyte binds a small uncharged ligand the system may be reversed so that the ligand is used as the analyte and binding is checked by adding the larger molecule to the electrophoresis buffer once it is remembered that larger molecules are more prone to capillary wall interactions.

5.3 Determination of binding constants

5.3.1 Migration shift affinity capillary electrophoresis (fast on-and-off rates)

In the case of weak interactions where electrophoresis of pre-equilibrated samples fail to reveal binding because the complexes dissociate too quickly, the

CAPILLARY ELECTROPHORESIS

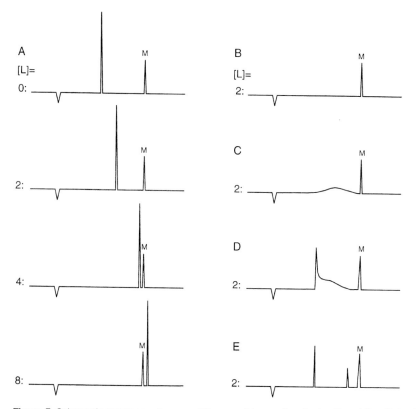

Figure 5 Schematic summary of some of the possible results of migration shift affinity CE experiments with additions of ligand to the electrophoresis buffer. [L] is the concentration of ligand added to the electrophoresis buffer. M is a non-interacting internal marker added to the sample.

electrophoretic approach is analogous to methods used in affinity chromatography (Chapter 3) except that the ligand does not have to be physically immobilized. Binding-induced shifts in analyte appearance times are linked to ligand affinity constants in experiments using different concentrations of ligand as an additive to the electrophoresis buffer. The link between the experimentally observed changes in peak appearance times and the binding constant is expressed in relations (49, 50) (see *Table 2*) that originate from theory describing binding parameters involved in reversible binding examined by column chromatography (51–53) which has been used in conventional affinity gel electrophoresis for many years (33, 54). The general assumptions concerning complex mobility, analyte solubility, molar excess of ligand, and the importance of a suitable marker molecule are listed in *Table 6*. The assumptions also include that the interactions take place between monovalent reactants. When a 1:1 complex formation cannot be assumed such as when there is more than one binding site present on the analyte the binding constants that are extracted from migration shift assays may be regarded as intrinsic constants (55). For antibody-monovalent antigen interactions relations have been worked out for evaluating the two dissociation constants of the antibody–antigen complex (37).

The weak-to-low affinity interactions suited for analysis by migration shift affinity CE can be operationally defined as interactions with binding kinetics causing clean migration shifts when ligand is added to the buffer (*Figures 5A* and *6*). This situation exists when establishment of the molecular equilibrium between free and complexed states is fast. Then the experimentally determined electrophoretic analyte mobility in a solution containing the ligand will be the weighted average of the mobilities of the analyte in the free and complexed states (17). The quantitative use of migration shift CE is normally restricted to interactions with off-rates that give equilibrium conditions during electrophoresis. If the interaction rates are slower as would be typical for a stronger interaction (k_{off} (the dissociation rate constant) is smaller) the migration shift experiment would result in analyte peak disappearance, broadening, tailing, or splitting (*Figure 5*). As a rule, values of $1/k_{off}$ that approach the electrophoresis time for the analyte peak will result in non-equilibrium conditions and consequently induce peak broadening (56, 57). Average weighted peaks can only be expected if their dissociation half-time, $\ln 2/k_{off}$ is equal to or less than 1% of the peak appearance time (56). Very long runs would in theory permit the analysis of more slowly dissociating interactions by migration shift electrophoresis (56) but in practice a CE separation rarely exceeds 30 minutes. In some cases the peak shape changes caused by intermediate binding can be used to estimate the binding rate constants (k_{on} and k_{off}) (58) but the approach is based on an iterative process that requires a prior knowledge of the binding constant (59). However, given suitable kinetics (see other important requirements in *Table 6*) migration shift affinity CE can be used to calculate apparent dissociation constants as detailed in *Protocol 2* and in *Table 2* and illustrated by *Figure 6B*. In that figure a monoclonal antibody against DNA is analysed by performing the separations in electrophoresis buffers containing various concentrations of a double-stranded synthetic DNA molecule. The migration shifts to later peak appearance times are caused by the complexation of the analyte with this negatively charged ligand and are observed easily (*Figure 6A*). The data make it possible to calculate a molar dissociation constant, K_d value of 0.1 μM as shown in *Figure 6B*. Additionally, the analysis shows that part of the monoclonal antibody preparation that is otherwise pure by usual criteria is non-binding, i.e. probably irreversibly denatured. This illustrates the power of combining separation and binding analysis in a single operation.

An additional important advantage of the approach is that a knowledge of the exact concentration of analyte is not necessary for the calculations (as long as the ligand molecules are added in a molar surplus, cf. *Table 6*). The exact concentration of specific ligand must, however, be known. Uniquely, information is extracted from the ligand-induced changes in the peak appearance pattern and is not based on determination of component concentrations after the establishment of equilibrium. The upper limit of K_d that can be determined by migration shift electrophoresis is in practice limited only by ligand availability, solubility,

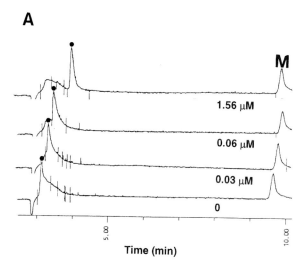

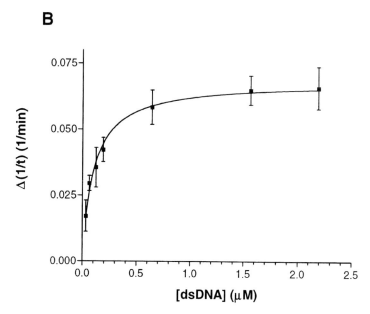

Figure 6 Migration shift affinity CE for quantitative assessment of a binding interaction (77). (A) Monoclonal anti-DNA antibody (0.7 mg/ml in 0.01 M phosphate pH 8.13 with 0.03 mg/ml tyrosine phosphate added as an internal marker (M)) was injected for 2 sec into a 27 cm, 50 μm internal diameter, untreated fused silica capillary with 20 cm to the detector. Electrophoresis subsequently took place at 8.5 kV in 0.1 M phosphate pH 8.13 with additions of double-stranded 32mer biotin–DNA at the concentrations given in the figure. Detection at 200 nm. • marks the position of the monoclonal antibody peak. (B) Data from binding experiments such as those presented in (A) expressed as outlined in *Figure 3A* and *Table 2*. Data points represent the mean and the standard deviation of migration shift (corrected inverse peak appearance times) differences from triplicate experiments. The curve represent a non-linear curve fit using a one site binding hyperbola (GraphPadPrism). R^2 = 0.99 and the equation for the curve yields a K_d of 0.10 μM.

and by viscosity increase, conductivity change, detector signal saturation, and other buffer effects at very high concentrations. The lower limit of dissociation constants quantifiable by affinity CE is actually higher than that by traditional affinity gel electrophoresis due to the higher speed of the CE analyses.

When ligand is added directly to the running buffer the filling (pre-rinse) of the capillary which will typically hold 1–2 µl can be done from minute (10–50 µl) volumes. The capillary can then be switched back to empty buffers for the run in the cases where a negatively charged ligand is used. This approach avoids the relatively quick depletion of ligand from solutions where anionic material will be deposited on the anode when the electric field is applied.

Alternative CE-based procedures for quantitative binding studies of weak interactions employ mobility ratios (60) but have the prerequisite that the complex electrophoretic mobility is known from independent experiments and this parameter can only be accurately measured in special cases (17).

Protocol 2
Capillary electrophoresis migration shift binding assay

Equipment and reagents
- See *Protocol 1*

Method
1 Establish reproducible and suitable (e.g. physiological) analysis conditions for analyte, marker molecule, and ligand separately and ensure that they migrate differently.

2 Perform electrophoresis in the presence of various known concentrations of ligand added to the electrophoresis buffer. Depending on the availability it will be advantageous to use the most charged molecule as the ligand (the buffer additive). Mix analyte in a suitable proportion with the marker molecule and perform the CE analysis. Look for signs of migration shift not affecting peak shape and size.

3 Perform affinity electrophoresis in the presence of ligand molar concentrations from 10–500 times the expected dissociation constant value while keeping the approximate analyte concentration at least 10 times lower than the lowest ligand concentration.

4 Process peak appearance shift data according to the relations given in *Table 6*. A direct binding curve of $\Delta(1/t)$ as a function of [L] is plotted to estimate the saturability of the system and to fit the binding isotherm to the experimental data using non-linear curve fitting methods (*Figure 6B*). This yields the K_d from the formula for a one site binding hyperbola. Data linearization also gives a measure of the K_d from the slope of the line (*Figure 3A*).

5.3.2 Capillary electrophoresis of pre-incubated samples (slow off-rates)

In the case of strongly interacting molecules (i.e. slowly dissociating complexes) there is no difference between choosing electrophoresis for studying binding or using the more classical methods that are mentioned, e.g. in the present book or in other reviews (35, 36) where the bound or free ligand concentrations at different receptor:ligand ratios are directly measured (*Table 3*) (35, 61).

Slower interactions result in stable complexes that will lead to an uneven distribution of bound and free states among analyte molecules during a run in the presence of a given concentration of ligand and therefore result in peak shape disturbances (*Figure 5B*). The consequence of this is that a migration shift approach can only rarely be used in practice when analysing strong binding for quantitative purposes. On the other hand, the stability of the complexes often means that samples can be analysed after pre-equilibration without complication through interference from dissociating molecules during sample introduction and electrophoresis into the empty electrophoresis buffer. It follows that in this situation it is desirable to maximize the speed and efficiency of the separation as well as the difference in mobility between complexed and free molecules. The goal is that the area measurements of peaks corresponding to free molecules from the equilibrated sample will not be disturbed by dissociating material after the peak has been separated from the peak representing the complex. As a rule, the stability of a complex represented by a dissociation rate constant of less than $0.105/t$, where t is the peak appearance time is sufficient if no more than 10% of the specifically bound ligand is to be allowed to dissociate during the separation (62). Since the rate constants will rarely be known, the best way to assess the possible influence of the interaction kinetics is to observe the separation pattern for signs of peak trailing or peak broadening for the peak representing free molecules. In cases with weak interactions there will be no difference in free peak areas because all material in complexes will dissociate quickly upon initiation of electrophoresis. The limits of usefulness of the approach thus are determined by the interplay between k_{off} and the speed of separation (for intermediate affinity binding) and the lower limit of detection (for high affinity binding). The detection limitation arises from the necessity of getting data points from the range where the binding is not saturable. The use of sensitive detection systems such as laser-induced fluorescence detection makes the quantitative study of stronger binding possible and makes it straightforward to carry out inhibition studies using labelled and unlabelled ligands. This is a particularly valuable approach for the study of antigen–antibody interactions (48).

Thus, for different types of interactions it will to a certain extent be possible by adjusting the analysis conditions appropriately to perform binding assays based on pre-equilibrated samples covering a range of binding constants. When the necessary requirements are met (see also *Table 7*) the electrophoresis step is simply a method to separate bound from free molecules and to quantify this behaviour on the basis of peak areas. If different binding stoichiometries exist they may be directly visualized from the CE separation (47). From these data direct or derived binding curves can be constructed (*Figures 3* and *7B*). The steps of the approach are outlined in *Protocol 3* and *Figure 7* illustrate the determination of a K_d value for the interaction of a peptide with heparin. Due to limita-

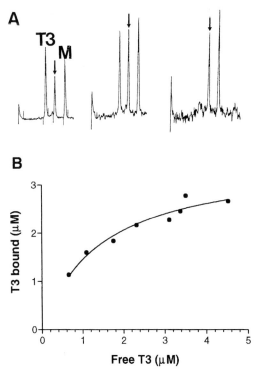

Figure 7 Pre-incubation approach for electrophoretic determination of binding constants (78). (A) A heparin binding peptide (T3) derived from serum amyloid P component was incubated at 11, 7.4, and 3 μM (from left to right) together with a marker peptide (M) with 0.44 mM low molecular weight heparin in 10 mM phosphate pH 7.81. After 2 h at room temperature 8 sec injections of the samples were analysed by CE in 75 mM phosphate pH 7.81 at 10 kV with detection at 200 nm. Capillary length was 37 cm with 30 cm to the detector window. Inner diameter was 50 μm. Arrows mark the increasing relative fraction of T3 that is bound. Absorbance readings at 200 nm (different scales are used in the three experiments). (B) Direct binding curve of experimental data (peak area measurements converted to concentrations) obtained for the T3–heparin interaction by analysis of pre-incubated samples at different analyte (T3):ligand (heparin) ratios as shown in (A). A best fit ($R^2 = 0.95$) one site binding hyperbola yields a K_d of 1.5 μM.

tions in the availability of the peptide the binding curve in *Figure 7B* only partly covers the optimal range of the binding curve; however, a one site binding hyperbola fits well to the experimental data and yields a K_d of 1.5 μM. Even though this does not reflect a very strong binding the complex formation was stable enough for pre-incubation experiments while migration shift experiments in this case did not give gradual migration shifts but tailing (not shown). This is indicative of an interaction characterized by both slow association and dissociation rates.

Intermediate binding typified by cases where the half-lives of analyte–ligand complexes are in the same range as the electrophoresis time (typical k_{off} between 0.001–0.1 s^{-1} (34) corresponding to 0.1–11 min half-lives) may not be amenable to any of the electrophoretic approaches described in the present chapter unless the electrophoresis parameters can be changed enough to afford sufficiently longer or shorter electrophoresis times.

Protocol 3
Measurement of strong binding by capillary electrophoresis

Equipment and reagents
- See *Protocol 1*

Method

1   Analyte, ligand, and marker recovery should be ensured as given in *Protocol 1*.

2   A quick screening of the nature of the binding can be performed by doing a migration shift affinity electrophoresis experiment (*Protocol 1*). If migration shifts of unchanged peaks do not occur such as e.g. when peak tailing, splitting, or broadening are observed the pre-equilibration approach as outlined here must be used instead of migration shift experiments.

3   Establish standard curves of the analyte peak area as a function of analyte concentration at the buffer and electrophoresis conditions that are going to be used in the binding assays.

4   The necessary time for establishing equilibrium is conveniently monitored by CE by performing repeated injections of sample at defined time intervals from an analyte–ligand mixture. When the peak corresponding to complexed and/or free ligand is invariant with time equilibrium has been reached. It is only necessary that one of the components, preferably the peak representing the free molecules, is measured.

5   Set up the binding experiment by incubating a number of mixtures of analyte and ligand molecules at fixed volumes and fixed ligand concentration with analyte concentrations spanning from 0.1–100 times the expected K_d value under the conditions that are relevant for the binding interaction under study. It is not a requirement that these conditions correspond to the electrophoresis buffer (*Figure 7A*).

6   Perform electrophoretic separation of bound and free analyte and measure peak areas with necessary corrections for small injection volume and migration time changes by using the internal marker.

7   Construct a direct binding curve based on the equations given (*Table 2*, *Figure 7B*). A non-linear curve fitting of the binding isotherm to the experimental data based on the assumption of one site binding gives K_d from the equation for the curve. A linearization of data may reveal deviations from simple binding as curvature.

5.3.3 Pitfalls

A major issue in binding experiments is the specificity of the response that is observed in the experiments. In affinity CE it must be demonstrated that the observed mobility changes are due exclusively to the interaction of analyte and ligand. Any of the above approaches for binding studies by CE must support specificity at least by considering the following points:

(a) **Saturability**: Migration changes (i.e. changes in corrected inverse peak appearance times) or the deduced amount of bound ligand, respectively, must be shown to be saturable at higher levels of ligand concentrations and linear transformations of binding data should not be performed if this cannot be demonstrated.

(b) **Analyte selectivity**: Marker molecules that do not participate in binding but mark non-specific differences from run to run in electrophoretic mobility (caused by electrolyte depletion, temperature-induced pH shifts, viscosity and conductivity changes) and in electro-osmosis flow velocity are essential to adjust the contribution from these factors to the observed changes in migration shift electrophoresis. In pre-incubation experiments the internal marker also supports selectivity and correct for small injection volume and sample dilution errors.

(c) **Ligand selectivity**: If possible, the specificity of the interaction is highly supported by demonstrating that closely related, similarly charged and sized molecules used as a putative ligand are without effects on analyte migration or analyte peak area.

If a binding behaviour in CE satisfies the above controls it is most likely real. The opposite situation where no evidence of binding is found even though it is expected may arise because of some of the reasons given in *Table 8*.

6 Conclusions

CE is useful for binding assays when low sample consumption, the use of unmodified molecules, speed, high resolution, and versatility with respect to usable analytes and analysis conditions are important features. The system output (migration changes) is exceedingly sensitive to complex formation, is very reproducible, and occurs simultaneously with the separation process so that for example the non-binding fraction of an analyte can be visualized directly. The low sample volumes that are required means that the technique is also very well suited for the monitoring of binding in pre-incubated samples over time. A 10–20 µl sample volume is normally sufficient for hundreds of injections. Equilibration time and optimal binding conditions may thus be examined with the requirement of less material than most other binding assays. Also, the independence of size changes as a measure of complex formation and the high resolution inherent in CE makes it possible to measure binding interactions between unlabelled, small molecules in solution, often a problem with other methods. Thus the method is by no means restricted to protein–ligand interactions.

A number of requirements concerning detectability, solubility, and absence of irrelevant interactions, however, must be met and generally complexed molecules should have an electrophoretic mobility that is different from free molecules, that is to say at least one of the interacting molecules must be charged. Even when these requirements are met there are intermediate affinity ranges that are not suited for quantitative analysis by either form of affinity CE described in this chapter; in these cases, other methods described in these volumes would be more appropriate.

Acknowledgements

I thank Ms. Liselotte Stummann for the drawings and Dr Casper Paludan for helpful criticism.

References

1. Cantor, C. R. and Schimmel, P. R. (1980). *Biophysical chemistry. Part II: Techniques for the study of biological structure and function* (1st edn). W. H. Freeman and Company, New York.
2. Jorgenson, J. W. and Lukacs, K. D. (1981). *Anal Chem.*, **53**, 1298.
3. Tiselius, A. (1930). *Nova acta regiae societatis scientiarum upsaliensis.*, **Ser. IV**, **7**, **No. 4**, 1.
4. Tiselius, A. (1937). *Trans. Faraday Soc.*, **33**, 524.
5. Hjertén, S. (1959). *Protides Biol. Fluids Proc. Colloq.*, **7**, 28.
6. Mikkers, F. E. P., Everaerts, F. M., and Verheggen, T. P. E. M. (1979). *J. Chromatogr.*, **169**, 11.
7. Nielsen, R. G., Rickard, E. C., Santa, P. F., Sharknas, D. A., and Sittaampalam, G. S. (1991). *J. Chromatogr.*, **539**, 177.
8. Heegaard, N. H. H. and Robey, F. A. (1992). *Anal. Chem.*, **64**, 2479.
9. Carpenter, J. L., Camilleri, P., Dhanak, D., and Goodall, D. (1992). *J. Chem. Soc. Chem. Commun.*, 804.
10. Chu, Y.-H. and Whitesides, G. M. (1992). *J. Org. Chem.*, **57**, 3524.
11. Chu, Y.-H., Avila, L. Z., Biebuyck, H. A., and Whitesides, G. M. (1992). *J. Med. Chem.*, **35**, 2915.
12. Kajiwara, H. (1991). *J. Chromatogr.*, **559**, 345.
13. Kajiwara, H., Hirano, H., and Oono, K. (1991). *J. Biochem. Biophys. Methods*, **22**, 263.
14. Honda, S., Taga, A., Suzuki, K., Suzuki, S., and Kakehi, K. (1992). *J. Chromatogr.*, **597**, 377.
15. Kraak, J. C., Busch, S., and Poppe, H. (1992). *J. Chromatogr.*, **608**, 257.
16. Guttman, A. and Cooke, N. (1991). *Anal. Chem.*, **63**, 2038.
17. Rundlett, K. L. and Armstrong, D. W. (1997). *Electrophoresis*, **18**, 2194.
18. Chu, Y.-H. and Cheng, C. C. (1998). *Cell Mol. Life Sci.*, **54**, 663.
19. Heegaard, N. H. H. (1998). In *New methods for the study of biomolecular complexes* (ed. W. Ens and K. Standing), p. 305. Kluwer Academic Publishers, Dordrecht, The Netherlands.
20. Heegaard, N. H. H. and Shimura, K. (1998). In *Quantitative analysis of biospecific interactions* (ed. P. Lundahl, E. Greijer, and A. Lundqvist), p. 15. Harwood Academic Publishers gmbh, Amsterdam, The Netherlands.
21. Heegaard, N. H. H., Nilsson, S., and Guzman, N. A. (1998). *J. Chromatogr. B Biomed. Sci. Appl.*, **715**, 29.

22. Colton, I. J., Carbeck, J. D., Rao, J., and Whitesides, G. M. (1998) *Electrophoresis*, **19**, 367.
23. Heegaard, N. H. H. (1998). *J. Mol. Recogn.*, **11**, 1.
24. Heegaard, N. H. H. (ed.). (1998). *Electrophoresis*, Paper Symposium on Affinity Capillary Electrophoresis, **19**, 367.
25. Chu, Y.-H., Avila, L. Z., Gao, J., and Whitesides, G. M. (1995). *Acc. Chem. Res.*, **28**, 461.
26. Schweitz, L., Andersson, L. I., and Nilsson, S. (1997). *Anal. Chem.*, **69**, 1179.
27. Guzman, N. A. (1993). *Capillary electrophoresis technology*. Marcel Dekker, Inc., New York, NY, USA.
28. Grossman, P. D. (1992). In *Capillary electrophoresis* (ed. P. D. Grossman and J. C. Colburn), p. 3. Academic Press, Inc., San Diego, CA, USA.
29. Landers, J. P. (1997). *Handbook of capillary electrophoresis*, 2nd edn. CRC Press.Boca Raton, FL, USA.
30. Li, S. F. Y. (1992). *Capillary electrophoresis*. Elsevier Science Publishers, Amsterdam, The Netherlands.
31. Grossman, P. D. and Colburn, J. C. (ed.) (1992). *Capillary electrophoresis*. Academic Press, Inc., San Diego, USA.
32. Hjertén, S. (1990). *Electrophoresis*, **11**, 665.
33. Takeo, K. (1987). In *Advances in electrophoresis*, Vol. I (ed. A. Chrambach, M. J. Dunn, and B. J. Radola), p. 229. VCH Verlag, Weinheim, Germany.
34. Shimura, K. and Kasai, K. (1995). *Anal. Biochem.*, **227**, 186.
35. Hulme.E. C. (ed.) (1992). *Receptor-ligand interactions*. IRL, Oxford University Press. Oxford, England.
36. Phizicky, E. M. and Fields, S. (1995). *Microbiol. Rev.*, **59**, 94.
37. Mammen, M., Gomez, F. A., and Whitesides, G. M. (1995). *Anal. Chem.*, **67**, 3526.
38. Grossman, P. D., Colburn, J. C., Lauer, H. K., Nielsen, R. G., Riggin, R. M., Sittampalam, G. S., *et al.* (1989). *Anal. Chem.*, **61**, 1186.
39. Heegaard, N. H. H., Hansen, B. E., Svejgaard, A., and Fugger, L. H. (1997). *J. Chromatogr.*, **781**, 91.
40. Heegaard, N. H. H. and Brimnes, J. (1996). *Electrophoresis*, **17**, 1916.
41. Lauer, H. H. and McManigill, D. (1986). *Anal. Chem.*, **58**, 166.
42. Wheat, T. E. (1994). In *Methods in molecular biology* (ed. B. M. Dunn and M. W. Pennington), p. 65. Humana Press, Inc., Totowa, NJ, USA.
43. Jorgenson, J. W. and Lukacs, K. D. (1983). *Science*, **222**, 266.
44. Ohara, T., Shibukawa, A., and Nakagawa, T. (1995). *Anal. Chem.*, **67**, 3520.
45. Chu, Y.-H., Lees, W. J., Stassinopoulus, A., and Walsh, C. T. (1994). *Biochemistry*, **33**, 10616.
46. Koutny, L. B., Schmalzing, D., Taylor, T. A., and Fuchs, M. (1996). *Anal. Chem.*, **68**, 263.
47. Chiem, N. H. and Harrison, D. J. (1998). *Electrophoresis*, **19**, 3040.
48. Tao, L. and Kennedy, R. T. (1997). *Electrophoresis*, **18**, 112.
49. Nakamura, S. and Wakeyama, T. (1961). *J. Biochem. Tokyo*, **49**, 733.
50. Takeo, K. and Nakamura, S. (1972). *Arch. Biochem. Biophys.*, **153**, 1.
51. Wilson, J. N. (1940). *J. Am. Chem. Soc.*, **62**, 1583.
52. Weiss, J. (1943). *J. Chem. Soc.*, **81**, 297.
53. DeVault, D. (1943). *J. Am. Chem. Soc.*, **65**, 532.
54. Heegaard, N. H. H. and Bøg-Hansen, T. C. (1990). *Appl. Theor. Electrophoresis*, **1**, 249.
55. Winzor, D. J. (1995). *J. Chromatogr. A*, **696**, 160.
56. Matousek, V. and Horejsí, V. (1982). *J. Chromatogr.*, **245**, 271.
57. Lim, W. A., Sauer, R. T., and Lander, A. D. (1991). In *Methods in enzymology*, Vol. 208, (ed. R. T. Saver) p. 196. Academic Press, Inc., San Diego, CA, USA.

58. Horejsí, V. (1979). *J. Chromatogr.*, **178**, 1.
59. Avila, L. Z., Chu, Y.-H., Blossey, E. C., and Whitesides, G. M. (1993). *J. Med. Chem.*, **36**, 126.
60. Bose, S., Yang, J., and Hage, D. S. (1997). *J. Chromatogr. A*, **697**, 77.
61. Klotz, I. M. (1989). In *Protein function: a practical approach* (ed. T. E. Creighton), p. 25. IRL Press, Oxford, England.
62. Hulme, E. C. (1990). In *Receptor biochemistry: a practical approach* (ed. E. C. Hulme), p. 303. IRL Press, Oxford, England.
63. Huang, X. C., Quesada, M. A., and Mathies, R. A. (1992). *Anal. Chem.*, **64**, 2149.
64. Dowd, J. E. and Riggs, D. S. (1965). *J. Biol. Chem.*, **240**, 863.
65. Perrin, D. D. and Dempsey, B. (1974). *Buffers for pH and metal ion control.* Chapman & Hall, London, England.
66. Chen, F. A., Kelly, L., Palmieri, R., Biehler, R., and Schwartz, H. (1992). *J. Liq. Chromatogr.*, **15**, 1143.
67. Gong, B. Y. and Ho, J. W. (1997). *Electrophoresis*, **18**, 732.
68. Towns, J. K. and Regnier, F. E. (1991). *Anal. Chem.*, **63**, 1126.
69. Gilges, M., Husmann, H., Klemiss, M.-H., Motsch, S. R., and Schomburg, G. (1992). *J. High Resol. Chromatogr.*, **15**, 452.
70. Schmalzing, D., Piggee, C. A., Foret, F., Carrilho, E., and Karger, B. L. (1993). *J. Chromatogr. A*, **652**, 149.
71. Zhao, Z., Malik, A., and Lee, M. L. (1993). *Anal. Chem.*, **65**, 2747.
72. Cobb, K. A., Dolnik, V., and Novotny, M. (1990). *Anal. Chem.*, **62**, 2478.
73. Huang, M., Mitchell, D., and Bigelow, M. (1996). *J. Chromatogr. B*, **677**, 77.
74. Bentrop, D., Kohr, J., and Engelhardt, H. (1991). *Chromatographia*, **32**, 171.
75. Shimura, K. and Kasai, K. (1997). *Anal. Biochem.*, **251**, 1.
76. Heegaard, N. H. H., Heegaard, P. M. H., Roepstorff, P., and Robey, F. A. (1996). *Eur. J. Biochem.*, **239**, 850.
77. Heegaard, N. H. H., Olsen, D. T., and Larsen, K.-L. P. (1996). *J. Chromatogr.*, **744**, 285.
78. Heegaard, N. H. H. (1998). *Electrophoresis*, **19**, 442.

Chapter 8
Molecular electro-optics

Dietmar Porschke
Max Planck Institut für biophysikalische Chemie, D-37077 Göttingen, FR Germany.

1 Introduction

Molecular electro-optical techniques provide a high potential for the acquisition of a unique set of data on macromolecules in solution, including hydrodynamic, spectroscopic, and electrostatic parameters. The information obtained by molecular electro-optics is very useful for assigning macromolecular structures. A major advantage is the particular high sensitivity. Molecular electro-optics has been used for the characterization of many different types of macromolecules, including large numbers of proteins, and is also very useful for the analysis of protein–ligand interactions.

The literature on molecular electro-optical methods and applications is extensive. Among the books (1–4), *Electric dichroism and electric birefringence* by Fredericq and Houssier (1) is still the most useful as an introduction into the theory and the general approach, whereas the description of the instruments and the examples are a little out-of-date.

The principle of molecular electro-optical experiments is quite simple: short pulses of electric fields are applied to solutions of macromolecules; because virtually all macromolecules have some electric dipole, the field pulses induce alignment of the molecules along the direction of the electric field vector (*Figure 1*); this molecular alignment is recorded either by measurements of the

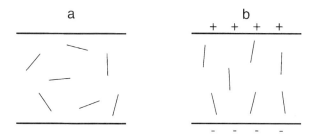

Figure 1 Scheme of the orientation of rod-like molecules by an external electric field.
(a) In the absence of an external field the distribution of the molecules in space is random.
(b) In the presence of an external field the molecules are partially oriented in the direction of the field vector.

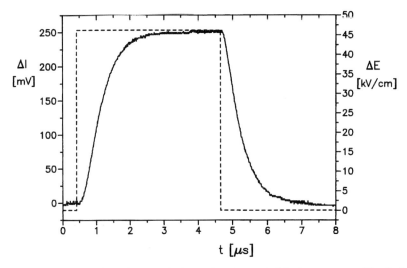

Figure 2 Example of an electro-optical experiment recorded for lac repressor. The dashed line shows the pulse of the electric field E in kV/cm (right ordinate); the continuous line shows the change of the intensity of light polarized parallel to the field vector ΔI in mV (left ordinate). Experimental conditions: 10 °C; λ = 280 nm, absorbance of the sample 0.42; buffer: 10 mM NaCl, 5 mM Tris–HCl pH 8.0, 0.1 mM $MgCl_2$, 0.1 mM DTE. The curve represents a single shot recorded with the detector adjusted to a rise time of 0.45 μs; i.e. the transients are convoluted with the detector rise curve; the rise and decay time constants of the sample are obtained from these data by deconvolution; because numerical deconvolution is reliable (5), the time constants are obtained with high accuracy owing to a high signal to noise ratio.

absorbance of polarized light (electric dichroism) or by measurements of the anisotropy of the refractive index (electric birefringence) or by any other optical technique like fluorescence or light scattering.

A simple form of an electro-optical signal induced by a rectangular pulse is shown in *Figure 2* for a system with a simple, single exponential dichroism decay. Separate relaxation processes are induced first upon pulse application and second upon pulse termination.

(a) Application of the electric field pulse induces partial alignment of the molecules; the relaxation process reflecting this alignment is affected by the electric field.

(b) Upon pulse termination the molecules relax back to their initial random distribution. This relaxation process is observed in the absence of external electric forces and provides the decay time constant τ_d which is a measure of rotational diffusion of the molecules. The time constant of rotational diffusion is very sensitive to molecular dimensions. Changes of the molecular dimensions due to ligand binding, for example, can be assigned by measurements of dichroism decay time constants with a particularly high accuracy. Association between protein molecules may also be detected at an unusually high sensitivity.

During application of an electric field pulse the change of the optical parameter approaches a limiting value, which is characteristic of a stationary state with partially aligned molecules. The change of the optical parameter, reflecting the degree of alignment, increases with increasing electric field strength. Extrapolation of these changes measured at various field strengths to infinite electric field strength using an appropriate 'orientation' function provides the optical anisotropy of the fully aligned state, which can be used to determine the orientation of the molecular subunit responsible for the optical signal.

The dependence of the optical signal on the electric field strength can also be used to evaluate the nature and the magnitude of the dipole. According to standard physics there are two different types of molecular dipole: induced dipoles are formed by 'polarization' under the influence of an external electric field, whereas permanent dipoles have a permanent anisotropy of their charge distribution, which is present without any external electric field. Macromolecules usually bear large numbers of charged residues and in most cases are associated with a mobile ion cloud. Under these conditions the interactions with external electric fields may lead to special effects, which are not consistent with the simple types of dipole models (6).

As already mentioned, the field induced alignment of molecules may be recorded by various spectroscopic techniques. For simplicity, the following description will be given for the case of electric dichroism, because this can be calculated in a relatively simple and straightforward manner from molecular structure. The data obtained by other spectroscopic techniques, like electric birefringence, are closely related (for a comparison of the relative advantages of electric dichroism and of electric birefringence see Section 6.1).

It may be useful here to summarize the individual types of information obtained by electro-optical techniques for the case of electric dichroism:

(a) The dichroism decay time constant(s) reflect rotational diffusion and provide a very sensitive measure of hydrodynamic dimensions.

(b) Extrapolation of the stationary values of the dichroism measured at various electric field strengths to infinite field strength provides the limiting value of the electric dichroism, which can be used to determine the orientation of the particular chromophore used for the measurements with respect to the dipole vector.

(c) The stationary values of the dichroism measured at various field strengths can also be used to assign the nature and the magnitude of electric dipole moments.

Induction of electro-optical effects requires:

(a) The existence of some electric anisotropy, i.e. a permanent or induced dipole.

(b) The existence of some optical anisotropy, i.e. preferential absorbance of polarized light for a given molecular orientation.

Most protein molecules are anisotropic with respect to both electric and optical parameters and, thus, electro-optical effects can be measured and analysed.

2 Equations for representing the field-induced orientation of molecules

2.1 Dichroism amplitude

The orientation of molecules in the presence of an external electric field results from the interaction of their dipole moments with the field. In the case of induced dipoles with a preferential polarizability α along one axis of the molecule, the energy of interaction is given by:

$$U_i = -(1/2)\alpha E^2 \cos^2\theta \qquad [1]$$

where E is the electric field strength and θ is the angle between the induced dipole and the electric field vector. In the case of permanent dipoles with a dipole moment μ_p the energy of interaction is given by:

$$U_p = -\mu_p E \cos\theta \qquad [2]$$

The distribution of molecular orientations is determined by the interaction energy with respect to the thermal energy kT (k = the Boltzmann constant, T = absolute temperature) and is described by the Boltzmann function:

$$f(\theta) = \frac{\exp(-U/kT)}{\int_0^\pi \exp(-U/kT) 2\pi \sin\theta d\theta} \qquad [3]$$

where $U = U_i + U_p$.

When the molecules are aligned in the direction of the electric field, the absorbance of light is changed relative to the natural state, where the molecules are in the usual random spatial distribution. The change of the absorbance of light polarized parallel to the field vector $\Delta A_\parallel$ is a measure of the degree of orientation. The theory predicts that the change of the absorbance of light polarized perpendicular to the field vector $\Delta A_\perp$ measured under the same conditions fulfils the relation:

$$\Delta A_\parallel = -2 \cdot \Delta A_\perp \qquad [4]$$

The relative change of the absorbance defined by:

$$(\Delta A_\parallel - \Delta A_\perp)/\bar{A} = (1.5 \cdot \Delta A_\parallel)/\bar{A} = \xi \qquad [5]$$

is the reduced electric dichroism, where $\bar{A}$ is the isotropic absorbance measured in the absence of an electric field.

The degree of molecular orientation and thus the magnitude of electric dichroism increases with the electric field strength E. Complete orientation in the direction of the electric field may be expected only in the limit of infinitely high E. The distribution of the orientational states is described quantitatively by the Boltzmann function. The dependence of the dichroism on the electric field strength requires integration of the following form of the Boltzmann function, also denoted as the 'orientation function':

$$\phi = \frac{\int_0^\pi [\exp(-U/kT)] \cdot (3\cos^2\theta - 1)(\pi \sin\theta d\theta)}{\int_0^\pi [\exp(-U/kT)] \cdot 2\pi \sin\theta d\theta} \qquad [6]$$

and the dependence of the dichroism on the electric field strength is given by:

$$\xi = \phi \cdot \xi_\infty \qquad [7]$$

where ξ_∞ is the limiting value of the electric dichroism at infinite field strength.

In the case of induced dipoles integration of *Equation 6* provides the orientation function:

$$\phi = \left\{ \frac{3}{4} \left[\left(e^\gamma / \sqrt{\gamma} \int_0^{\sqrt{\gamma}} e^{x^2} dx \right) - 1/\gamma \right] - \frac{1}{2} \right\} \qquad [8]$$

where $\gamma = (\alpha E^2)/(2kT)$.

In the case of permanent dipoles the orientation function is given by:

$$\phi = \left[1 - \frac{3(\coth \beta - 1/\beta)}{\beta} \right] \qquad [9]$$

where $\beta = \mu_p \cdot E/kT$.

The orientation functions may be used to determine the limiting value of the electric dichroism corresponding to complete molecular orientation, by least squares fitting of dichroism values measured at different field strengths. The limiting value of the electric dichroism provides direct information about the orientation of the chromophores with respect to the dipole vector according to the following relation:

$$\xi_\infty = (3/2) \cdot (3\cos^2(\varphi) - 1) \qquad [10]$$

where φ is the angle of the transition dipole moment of the chromophore relative to the dipole vector. When the transition dipole moment of the chromophore is oriented parallel to the dipole vector, corresponding to $\varphi = 0$, the limiting value of the dichroism is +3. In the other limit case, where the transition dipole moment of the chromophore is in a perpendicular direction to the dipole vector $\varphi = 90$), the limiting value of the dichroism is −1.5. Thus, the limiting value of the electric dichroism can be used to calculate the angle φ of the optical transition dipole with respect to the direction of the electric dipole (cf. *Figure 3*).

A quantitative analysis of stationary values of the dichroism measured at different field strengths using appropriate orientation functions may also be used to determine the nature and the magnitude of the dipole moment. The electrostatic parameters provide information on the state of the charged residues of the macromolecule.

2.2 Dichroism decay

When an electric field pulse is switched off, the molecules revert to their random distribution by the process of rotational diffusion. Rotational diffusion is very strongly dependent on the form and dimensions of the molecules.

Analytical solutions for the time constants of rotational diffusion are only

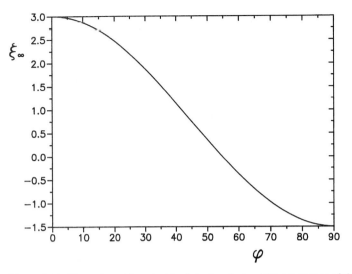

Figure 3 Limiting value of the electric dichroism ξ_∞ for different values of the angle φ between the directions of the transition dipole moment and the electric dipole moment, according to *Equation 10*. The direction of the transition dipole moment determines the plane of polarization, where absorption of light is maximal.

available for simple geometries. In the case of spherical molecules the dichroism decay time constant is given by (7):

$$\tau_d = V_h \eta / kT \quad [11]$$

where V_h is the effective hydrated volume and η the viscosity of the solvent.

Rotational diffusion of rigid, rod-like molecules is mainly determined by the length ℓ of the rod, whereas the radius of the rod r is of marginal influence only, except for very short rods. The dichroism decay time constant τ_d^{rr} for such rods may be described by (8, 9)

$$\tau_d^{rr} = \frac{\pi \eta \ell^3}{18kT \cdot [\ln(q) - 0.662 + 0.917/q - 0.050/q^2]} \quad [12]$$

where $q = \ell/2r$. This equation may be used to calculate the length ℓ of a rigid rod-like molecule from its dichroism decay time constant.

Another special case are thin circular discs, encountered, e.g. in membrane fragments of bacteriorhodopsin (10) or Na^+/K^+-ATPase (11). In this case the dichroism decay time constant is given by (12):

$$\tau_d = (2\eta/9kT)d^3 \quad [13]$$

where d is the diameter of the disc.

Both rigid, rod-like molecules and thin circular discs are limiting cases of ellipsoids. In general the rotational diffusion of ellipsoids is described by two friction coefficients, representing diffusion about the two different semi-axes and leading to two different time constants. In practice it is quite difficult for

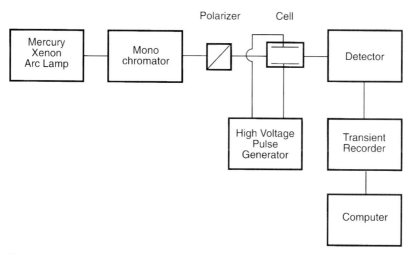

Figure 4 Block diagram with the components of an instrument for measurement of electric dichroism.

various reasons to extract these individual time constants from experimental data (cf. ref. 7). Accurate descriptions of the sum of individual relaxation processes including both time constants and amplitudes are possible by 'quantitative molecular electro-optics' based on molecular models or crystal structures (cf. Section 5.4). For this numerical approach, there are no restrictions with respect to shape.

3 Instruments

3.1 General

The general design of an instrument for electro-optical measurements is shown in *Figure 4*. The main parts required are a pulse generator, a system for spectrophotometric detection, a device for analog-digital conversion, and a computer for data processing. The details of the instrumentation depend very much on the particular systems to be analysed. In general, investigations with high time resolution and high electric field strengths require more sophisticated and also more expensive equipment. Because of the progress in electronics, the fast time range is now much more easily accessible than previously. Nevertheless, construction of instruments for measurements in the nanosecond time range, required for analysis of small proteins, is still a challenge and also requires potent financial resources. Small proteins do not only exhibit short rotational diffusion time constants but also require relatively high electric field pulses to induce sufficient degrees of orientation. α-chymotrypsin, for example, with a molecular weight of ~ 25 kDa and an almost spherical shape, shows a dichroism decay time constant of 31 ns at 2 °C and requires electric field strengths of ~ 70 kV/cm to approach saturation of the dichroism (13, 14). Obviously, the difficulties and the expenses associated with the construction of pulse generators increases with the magnitude of the electric field strength and with decreasing rise/decay times of the pulses.

3.2 Pulse generators

Most of the electro-optical data in the literature have been obtained with commercial pulse generators. The maximum voltages of these generators are in the range of 2.5 kV at rise/decay times of ~ 20 ns. The pulse voltages can be increased by external transformers up to 30 kV at the expense of a reduction of rise/decay times to the range of 10 µs (maximal currents are also reduced). Generators of this type are available from, e.g. Velonex, Santa Clara, CA, USA.

Higher voltage pulses with short rise/decay times have been generated by the cable discharge technique (15–17): coaxial cables are charged up by standard high voltage power supplies up to the range of 100 kV; the cable is then connected to the measuring cell by a spark gap, which is used as a switching device. The essential parts of spark gaps are two electrodes, usually of spherical shape, which are kept at a distance just above that allowing spontaneous discharge; the spark gap is converted into a highly conductive state within a very short time by moving the electrodes to a distance sufficient for spark discharge. The pulse may be terminated by a second spark gap, which is triggered by a helper spark. The schematic illustration of an instrument based on this principle is shown in *Figure 5*. Short rise/decay times require optimal adaptation of the connections to the cable impedance.

The 'cable discharge technique' has also been used for electro-optical measure-

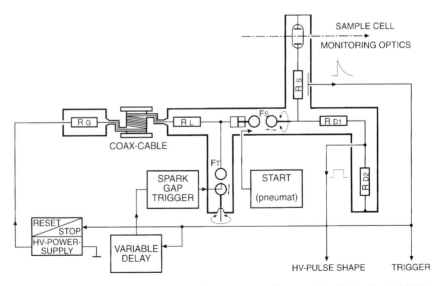

Figure 5 Diagram of a coaxial high field pulse generator. The coaxial cable is charged up by a HV power supply; then the pulse is applied to the sample by pneumatic closing of the spark gap F_s, finally the pulse is terminated by the spark gap F_T which is triggered by a helper spark initiated by a variable delay (modified form of a generator constructed by Grünhagen) (16).

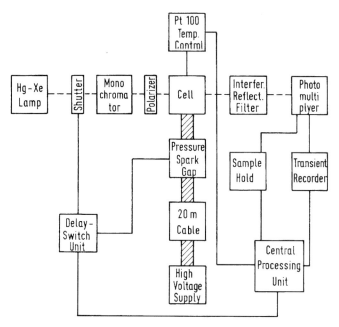

Figure 6 Block diagram with the components of an automatic field jump apparatus for electro-optical measurements at physiological salt concentrations (17).

ments at physiological salt concentrations. A schematic representation of this instrument is shown in *Figure 6*.

3.3 Measuring cells

For application of relatively low field pulses (up to a few kV/cm), simple cell constructions are sufficient. An example with a Teflon insert holding Pt-electrodes at a defined distance within a standard quartz cuvette is shown in *Figure 7a*. A more careful design is required for application of high field pulses. A typical construction used in the laboratory of the author is shown in *Figure 7b*. The electrodes are machined from brass and are covered by a $\sim$ 1 mm layer of Pt. The quartz windows for spectrophotometric detection are carefully inserted into the body machined from synthetic polymer (e.g. polymethylmethacrylate) with a thin layer of silicon grease, such that the conical windows are without strain.

3.4 Spectrophotometric detection

Standard commercial spectrophotometers are not appropriate for molecular electro-optical measurements, because their time resolution is usually limited to the msec/sec time range. However, the components required for fast spectrophotometric detection are available from several sources. The light source should have as high intensity as possible, in order to get a maximal signal to noise ratio. Particularly high light intensities are emitted by high pressure Hg/Xe arc lamps at special Hg emission wavelengths. The advantage resulting from

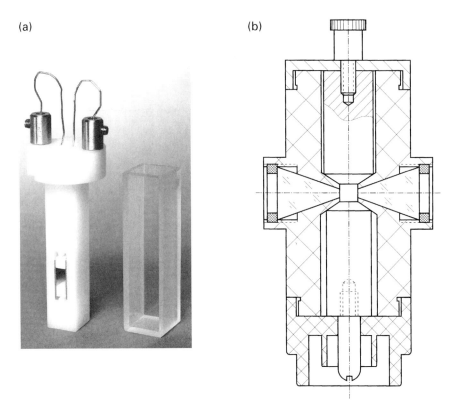

Figure 7 (a) Cell for electro-optical measurements at low electric field strengths: a holder machined from Teflon for the Pt-electrodes and the supply wiring is inserted into a standard cuvette containing the solution. (b) Construction of a cell for measurements at high field strengths: the lower electrode is connected to the cable (cf. text) and the upper one to ground. The quartz windows for spectroscopic detection are truncated cones, which are inserted with a thin layer of silicon grease into the cell body machined from synthetic polymer.

high light intensities in most cases overcompensates any disadvantage that may be associated with restriction to the Hg emission lines. The light from the source is focused to the entrance slit of a monochromator. Efficient grating monochromators are supplied by various companies. The monochromatic light is then polarized, preferentially by some crystal polarizer. Among the various polarizer constructions, 'Glan' polarizing prisms prepared from calcite with prism halves separated by an air gap are recommended; for applications in the far-UV, the calcite has to be selected carefully for optimal transmission. The most effective detectors in the UV spectral range still appear to be photomultipliers with appropriate photocathodes. All these components can be obtained from various commercial sources.

A major problem for molecular electro-optical measurements results from the fact that mV signals have to be recorded simultaneously to the application of kV pulses. Under these conditions induction effects may lead to serious perturbations of measured signals. These problems can be avoided by careful shielding of the detection device. The photomultiplier may be shielded in special covers made from µ-metal, protecting mainly against magnetic perturbation. In addition, the multiplier head including the amplifier should be shielded in a solid metal cover for protection against electric induction effects. Finally, the connection between the multiplier head and the digitizer should be shielded as well. The electric supply lines of all instruments should be checked for 'earth circuits', i.e. circular connections of ground wires, which may lead to large perturbations and, thus, should be eliminated.

3.5 Transient data storage and data processing

Thanks to the progress in electronics, fast analog to digital conversion of data is not a problem anymore, except for the very short time range, where analog–digital converters are still expensive. The digitized data are transmitted to appropriate computers and then are analysed by suitable software. In most cases the required software is prepared individually. However, standard software, e.g. for evaluation of exponentials, may be obtained free from, for example, Stephen Provencher, e-mail address: sp@S-provencher.COM; download from http://S-provencher.COM

3.6 Automatic data collection

In many cases electro-optical signals are large enough, such that single shots are sufficient for a satisfactory signal to noise ratio. However, in other cases it is necessary to increase signal to noise ratios by averaging of transients. Electro-optical instruments can be adapted relatively easily for automatic collection of many transients (17). Standard PC's are sufficient as control units for timing of electric field pulses, activation and reading of transient data storages, and finally averaging of data. Problems resulting from mutual perturbation of different parts of the instrument should be avoided by galvanic separation using optoelectronic coupling as much as possible. Repetition of field pulses of a given polarity leads to motion of molecules with a net charge towards one of the electrodes. To avoid this problem, the polarity of subsequent pulses should be inverted. Some biological macromolecules may be damaged by photoreactions resulting from high light intensities; this problem may be reduced by using photo-shutters for closing the light beam except during data acquisition.

4 Experimental procedures

4.1 Sample preparation

Electro-optical measurements are usually performed at relatively low ionic concentrations in order to reduce Joule heating of the solutions during the pulses as much as possible. Low conductivities of solutions are also important, in order to minimize reduction of the electric field strength during pulses. Furthermore, electro-optical signals are often strongly affected by details of the ion composition of the solution. Thus, it is important to control the ion concentrations as carefully as possible. A convenient and reliable procedure for adjustment of ion concentrations in solutions of macromolecules is dialysis. Membranes for dialysis are available with various pore sizes, such that dialysis is possible even for proteins of relatively low molecular weight.

Bivalent ions may have a particularly strong influence on electro-optical signals and, thus, their presence should be controlled with special care. It is possible to remove these ions by dialysis against EDTA. Furthermore, it is useful either to add EDTA (e.g. 100 μM) or to add some concentration of bivalent ions (e.g. 100 μM Mg^{2+}) to solutions under investigation, in order to ensure well defined conditions.

Protocol 1
Dialysis of macromolecules

Equipment and reagents

- Dialysis tubing
- Hot-plate
- Cold room
- High salt buffer, e.g. 1 M NaCl
- 1 mM EDTA
- Sample buffer

Method

1. Select dialysis tubing with appropriate pore size and volume.
2. Boil the dialysis tubing in water for ~ 10 min; replace the water and repeat. This procedure removes impurities from the tubing and supports swelling of the tube material.
3. Seal one end of the tube with a double knot; use gloves for handling of the tubes.
4. Fill the solution into the tube and seal the open end of the tube with a double knot.
5. Dialyse first against a large reservoir, e.g. 100 × (volume inside the tube) of buffer with a high salt concentration (e.g. 1 M NaCl; the binding constants of multivalent ions is reduced considerably at high salt and, thus, these ions can be removed more rapidly) and with EDTA (e.g. 1 mM), in order to bind ions like Mg^{2+}, Ca^{2+} etc. The time for equilibration depends on the pore size: several hours for large pores, up to a day for small pores. The dialysis should be performed at low temperatures, e.g. in a refrigerator or a cold room.

   If multivalent ions are required for maintaining the native structure of a protein, dialysis against EDTA may not be appropriate and even may be dangerous; in such cases steps 5 and 6 should be omitted.

6. Repeat step 5. The number of repetitions depends on the residual concentration of contaminants that can be tolerated in the sample.
7. Dialyse against the buffer used for the measurements.

> **Protocol 1** continued
>
> 8 Repeat step 7. The number of repetitions depends again on the amount of salt to be removed and the required degree of purity; several repetitions are recommended.
>
> If the sample should be free of metal ions like Mg^{2+} or Ca^{2+}, the dialysis should not be performed in glass vessels; either vessels from, e.g. polypropylene or, for particularly high purity, quartz should be used.

Application of electric field pulses to solutions often induces formation of air bubbles, which may lead to serious perturbation of the measured signals and in the case of high voltage pulses even may lead to destruction of the measuring cell. These problems may be avoided by careful degassing of solutions by the following standard procedure: the solution, if possible within the measuring cell, is introduced into an appropriate vessel (e.g. desiccator) and then the pressure is slowly reduced by an oil or water-jet pump. Upon reduction of the pressure, small air bubbles appear and grow. Detachment of bubbles from surfaces may be supported by gentle mechanical shocks. Complete degassing is possible by going slightly below the boiling pressure of water. This involves some risk of losing part of the solution by delayed boiling (*Protocol 2*).

Due to the special properties of proteins, degassing of protein solutions may lead to precipitation. In this case it is recommended to degas the protein solution in some vessel and the cell filled with buffer separately. This treatment of the cell is usually necessary in order to remove small bubbles from surfaces and corners. Subsequently, the buffer is removed from the cell—without removing a thin layer of buffer from surfaces or corners. The protein precipitate may be separated by filtration; it is recommended to filter solutions directly into the measuring cell.

Protocol 2
Degassing of solutions

Equipment and reagents

- Desiccator
- Water-jet pump
- Measuring cell
- Solutions

Method

1 Fill the solution into the measuring cell.

2 Put the measuring cell into a desiccator and connect with a water-jet pump or any other pump with moderate pumping capacity (pumps with high capacity, e.g. standard two-step oil pumps used for high vacuum, in general reduce the pressure too quickly, such that solutions are lost due to 'bumping' or 'delayed boiling'—jumping of solutions out of the cell).

Protocol 2 continued

3   Carefully watch the development of bubbles; wait for one or two 'eruptions', corresponding to delayed boiling.

4   Pull off the connecting tube from the desiccator immediately after an eruption, such that the pressure is increased quickly (before new bubbles start to be formed).

5   Check whether air bubbles remained in the cell; if some bubbles still exist, repeat steps 2–5.

4.2 Measurements

Obviously, experimental data should be collected under well defined conditions, which is essential both for comparison with other data in the literature and for comparison with calculations based on models. A particularly important parameter for electro-optical measurements is the temperature because of the strong temperature dependence of the water viscosity, which directly affects the time constants of the electro-optical transients. A change of the temperature from 20 °C to 19 °C leads to a change of the water viscosity by 2.5%—a rather large effect in view of an accuracy of ~ 1%, which is accessible for dichroism decay time constants.

For various reasons it is very useful to collect data over a broad range of electric field strengths. First of all, this is important for evaluation of stationary values of the electric dichroism by orientation functions: the parameters obtained by data fitting are clearly more reliable, if obtained over a broad range of field strengths. Secondly, it is recommended to check for any dependence of dichroism decay time constants on the electric field strength. If a dependence is detected, this is usually an indication for a field induced change of the structure. Under these conditions, the time constants should be extrapolated to zero electric field strength, in order to derive data, which are not affected by any field induced structure change.

Usually, electric dichroism is measured at parallel orientation of the plane of the polarized light with respect to the field vector because the signals are maximal for this orientation. However, it should be checked, by measurements at perpendicular orientation of the polarization plane with respect to the field vector, whether *Equation 4* is fulfilled. If deviations are detected, a further control is possible by measurements at the magic angle orientation, i.e. at an angle of 54.7° between the plane of the polarized light and the field vector. Under 'magic angle' conditions, absorbance changes are not due to orientation, but due to absorbance changes caused by some reaction, e.g. field induced conformation changes.

The wavelength used for the measurements may be selected according to various criteria. In most cases, a wavelength at or close to the maximum of absorbance will be chosen, in order to be able to use a low protein concentration, where aggregation effects can be avoided. A second criterion will be a high light intensity, in order to maximize the signal to noise ratio. As noted above, particularly high light intensities are emitted by high pressure Hg/Xe arc lamps at mercury emission lines. These lines are at 248.2, 265.2, 280.4, 289.4, 302.2, 313.2, 334.2, 366.3, 404.7, 435.8, and 546.1 nm. Because of the relatively close spacing in the spectral range, a suitable line can be found for most applications.

The amplitude of electric dichroism is not only influenced by the magnitude of the absorbance, but also by the orientation of the transition dipole moment with respect to the electric dipole vector. Thus, it is useful to do some test measurements in different ranges of the absorption spectrum. Usually ($\pi^* \leftarrow \pi$) transitions are polarized in the plane of aromatic chromophores, ($\pi^* \leftarrow n$) transitions are usually polarized in a direction perpendicular to the aromatic plane. As represented by *Equation 10*, these differences in orientation may lead to large changes in the observed dichroism amplitudes.

Finally, the signal to noise ratio may be increased by averaging of many transients. Because the signal to noise ratio increases with the square root of the number of shots included in the average, an increase of the ratio by a factor of 5 requires 25 shots and a factor 10 requires 100 shots. Thus, there are practical limits for the increase of the signal to noise ratio by this procedure. Although most samples survive large numbers of shots, it is certainly necessary to do some control measurements on the quality of the sample after many shots. A simple control is possible by repetition of initial experiments at the end of a series, in order to check reproducibility. A comparison of absorbance spectra before and after the measurements may also be useful. A stringent test in the case of proteins may be based on function, e.g. catalytic activity of enzymes.

5 Data evaluation

5.1 Amplitudes: the stationary dichroism

Electro-optical transients are usually composed of several exponential relaxation processes, which are associated with individual amplitudes. The determination of these individual amplitudes requires combined fitting of time constants and amplitudes, which may not be trivial. The sum of these individual amplitudes can be determined much more easily by direct reading of the levels of stationary light intensities. When rectangular electric field pulses are applied, the total amplitude can be read both from the 'on'- and the 'off'-transient: the difference of the light intensity I_t^0 before application of the pulse and the intensity I_t^f at the end of the pulse should be identical with the difference between the intensities I_t^f at the end of the pulse and I_t^∞ at 'infinite' time after the end of the pulse. If these differences are not identical, some special effect contributes to the signal; in this case further experiments are required to test for, e.g. field induced conformation changes or relaxation processes induced by some increase of the temperature.

When electro-optical signals are recorded by transient storage devices, the data do not usually represent the absolute levels of the light intensity but only differences. The offset in the data obtained from the transient storage device cancels, when differences are obtained (e.g. $\Delta I = I_t^f - I_t^0$). The absolute light intensity I^0 before application of the pulse has to be read separately by some voltmeter. Combination of these data provides the change of the absorbance; for example:

$$\Delta A_{\parallel} = -\log[(I_{\parallel}^0 + \Delta I_{\parallel})/I_{\parallel}^0] \qquad [14]$$

where the index $\parallel$ is used for measurements with light polarized parallel to the field vector. According to the standard definition represented by *Equation 5*, the electric dichroism is then given by $\xi = (1.5 \cdot \Delta A_{\parallel})/\bar{A}$. These values are evaluated at different electric field strengths E, where E is the voltage of the pulse applied to the cell divided by its electrode distance.

Protocol 3
Measure dichroism amplitude

Equipment
- Instrument for measuring electric dichroism

Method
1. Measure the total light intensity $I_{\parallel}^0$ for light polarized parallel to the field vector, i.e. the signal recorded by the detector in the absence of an external electric field; do not use more than 90% of the maximal signal. The maximal signal is determined by the upper end of the linear range of the detector.
2. Apply an electric field pulse and record a transient.
3. Read the difference $\Delta I_{\parallel}$ in the light intensities between the levels:
   (a) Before application of the field pulse.
   (b) The final level approached towards the end of the field pulse, corresponding to the stationary state of the signal in the presence of the field.
4. Calculate the change of the absorbance according to $\Delta A_{\parallel} = -\log[(I_{\parallel}^0 + \Delta I_{\parallel})/I_{\parallel}^0]$.
5. Calculate the reduced dichroism amplitude according to $\xi = 1.5 \Delta A_{\parallel}/\bar{A}$. = where $\bar{A}$ is the isotropic absorbance at the given wavelength.

Protocol 4
Check dichroism optics

Equipment and reagents
- See *Protocols 2* and *3*
- Solution with a high electric dichroism

Method
1. Check the measuring cell for strain in the optical windows by viewing it between crossed polarizers against a bright light source (e.g. 100 W bulb); the optical windows should remain dark.

Protocol 4 continued

2. Fill a solution with a high electric dichroism into your measuring cell, e.g. DNA double helices: (sonicated) standard calf thymus DNA in a low salt buffer containing Mg^{2+} ions (e.g. 100 µM; reduces the risk of field induced DNA denaturation). Follow *Protocol 2* for degassing.

3. Follow *Protocol 3*, steps 1–4, with light polarized parallel to the field vector; get by this procedure $\Delta A_\parallel$.

4. Follow *Protocol 3*, steps 1–4, with light polarized perpendicular to the field vector; get by this procedure $\Delta A_\perp$.

5. Check that $\Delta A_\parallel = -2 \times \Delta A_\perp$ (*Equation 4*); the quality of the agreement determines the quality of the measurements; for a reasonably well adjusted instrument the agreement should be within a few per cent.

6. Extension: adjust the polarizer to the magic angle: 55.7° with respect to the field vector. Measure the dichroism amplitude; for optimal adjustment of the optics the dichroism amplitude should be zero.

7. If the test shows significant deviations from theoretical expectation, this may be due to one of the following standard problems:

   (a) If the dichroism effect is too large, i.e. > 10%, *Equation 4* is usually not fulfilled. In that case the test should be repeated either at a lower electric field strength or at a lower concentration of the polymer.

   (b) The windows of the measuring cell may be under strain.

Protocol 5

Check for reaction effects/field induced conformation

Equipment and reagents
- See *Protocols 2, 3*, and *4*

Method

1. Follow *Protocol 4* to ensure a sufficient quality of the polarization optics.

2. Follow *Protocol 3* using the orientations of the polarization planes 0, 55.7, and 90° with respect to the field vector and get the absorbance changes $\Delta A_\parallel$, $\Delta A_{55.7}$, and $\Delta A_\perp$.

3. If there is an amplitude $\Delta A_{55.7}$, check whether $[\Delta A_\parallel - \Delta A_{55.7}] = -2 \times [\Delta A_\perp - \Delta A_{55.7}]$, i.e. check for internal consistency.

5.2 Time constants and individual amplitudes

Transients of the electric dichroism are represented by sums of exponentials:

$$I(t) = \sum_{1}^{n} [\Delta I_n \cdot \exp(-t/\tau_n)] + I_\infty \qquad [15]$$

where $I(t)$ is the light intensity measured in mV or V at time t, n is the number of relaxation processes required to fit the data, I_∞ is the light intensity at time $t = $ infinity, ΔI_n is the change of the light intensity associated with the nth relaxation process, and τ_n the relaxation time constant associated with the nth relaxation process. The parameters n, ΔI_n, τ_n, and I_∞ of *Equation 15* are fitted to the experimental data $I(t) = f(t)$ by least squares minimization routines. Usually the decision on the number of relaxation processes n is not trivial and requires some experience. The decision may be based on direct visual inspection of the quality of the fit(s). A useful indication for the quality of a fit is the sum of squared residuals:

$$S = \sum_{i=1}^{n} (I_i^{exp}(t_i)] - I_i^{fit}(t_i))^2 \qquad [16]$$

where $I_i^{exp}(t_i)$ and $I_i^{fit}(t_i)$ are the measured and the fitted values of the light intensities at different times t, respectively. In general S decreases for fits with an increasing number of relaxation processes n. The decision on the number of 'significant' relaxation processes is usually operational: if S does not decrease 'significantly' upon going from n to n + 1, the number n is taken as valid. The decision on significance requires some experience. Usually a decrease of S by 50% should be accepted as significant, at least if it is reproducible. If the decrease of S is smaller, the available experimental data may not be sufficient to define an additional relaxation process at a satisfactory accuracy and then fitting of the additional process does not make sense. An automatic decision on the number of relaxation processes is taken by the program *'DISCRETE'* (written and distributed by Provencher; cf. above).

5.3 Interpretation

When electro-optical transients can be described by single exponentials to a sufficient accuracy, there is usually no problem with interpretation. Of course, it should checked that there are no hidden contributions; such tests can be performed, for example, by measurements at the magic angle or by examination whether $\Delta A_\parallel$ and $\Delta A_\perp$ values are consistent with *Equation 4*. If these tests are passed without indication of additional processes, the data may be analysed in terms of simple models.

Obviously, the assignment of two or more relaxation processes is more difficult. Again, it should be checked whether there is any contribution by some chemical process. According to theory, electro-optical transients of rigid molecules may contain up to five exponentials. However, numerical simulation of electro-optical transients demonstrate that in most cases the majority of these five exponentials is associated with undetectable small amplitudes. Simulations in the authors laboratory on a wide variety of rigid macromolecules of different shape never revealed transients which required more than two exponentials for satisfactory fitting.

Most biological macromolecules show at least some degree of flexibility. The flexibility can be clearly detected for long linear polymers. Electro-optical experiments may provide quantitative information on the flexibility according to various procedures. Flexibility of linear polymers leads to a reduction of their effective hydrodynamic dimensions. Thus, a very sensitive standard procedure for the assignment of flexibilities is based on a comparison of the dichroism decay time constant measured for a linear polymer with the time constant expected for a rigid rod of the same dimension, i.e. same contour length and diameter. Usually the dipolar forces induced by electric fields tend to elongate the distribution of various bent configurations of flexible polymers (18). Evidence for this type of stretching can be obtained by measurements of both amplitudes and time constants over a broad range of electric field strengths. Stretching is expected to be maximal at high field strengths and may be reflected by a separate relaxation process.

5.4 Simulations of electro-optical data from macromolecular structures

Electro-optical measurements provide information on the hydrodynamic, optical, and electrical parameters of the molecules under investigation. The hydrodynamic parameters reflect the global structure of the complex, whereas the optical and electrical parameters reflect both global and local elements of the structure. The information included in the experimental data may be used up to an optimal level by simulations starting from crystal structures or from molecular models. In the present context it is not possible to describe the procedures of 'quantitative molecular electro-optics' in detail and, thus, the following description is restricted to a short account of major features.

In all cases, the hydrodynamic, optical, and electrical parameters have to be described in terms of the appropriate tensors of the diffusion coefficients, extinction coefficients, and polarizabilities and/or the dipole moment vector.

(a) The diffusion coefficient tensor for objects of complex shape can be calculated by bead model simulations. The global structure is represented by an assembly of spherical beads; the hydrodynamic resistance of each bead is described according to Stokes' law, and the hydrodynamic coupling between the beads is considered by using hydrodynamic interaction tensors, which have been developed and tested (19, 20). In most cases of macromolecules, bead assemblies with overlapping beads of uniform radius have been used. Based on the diffusion coefficient tensor, the 'centre of diffusion' (cf. below) may be calculated according to equations presented by Harvey and Garcia de la Torre (21).

(b) The tensors of extinction coefficients may be calculated by simple tensorial addition of the tensors for the individual molecular components contributing to the absorbance at the wavelength of investigation.

(c) The dipole moment vectors must be calculated with respect to the centre of diffusion, unless the molecule under investigation is without net charge; the positions of all charged residues have to be included in the calculation. Most protein molecules show a considerable permanent dipole moment, which usually dominates any contributions from polarizabilities.

(d) The dichroism decay curves, i.e. the relaxation time constants and the corresponding amplitudes, are calculated according to Wegener *et al.* (22).

(e) The limiting values of the dichroism may be calculated as described by Porschke and Antosiewicz (23).

Short descriptions of a set of procedures and programs used for automatic calculation of electro-optical parameters from macromolecular structures have been published (20, 24–26). A detailed compilation and discussion is in preparation (Antosiewicz and Porschke, in preparation). Programs for calculation/ evaluation of hydrodynamic parameters: *SOLPRO* (using bead models, 27) and *ELLIPS4* (using ellipsoids, 28).

6 Various spectroscopic techniques

6.1 Birefringence

Electric birefringence is complementary to electric dichroism. As described above for the electric dichroism, the molecules are oriented by electric field pulses, but in the case of electric birefringence the orientation is recorded by measurements of the difference in the refractive index for light polarized parallel and perpendicular to the field vector. An advantage of the birefringence approach is the fact that measurements are possible at virtually any wavelength; usually a wavelength outside the range of the absorbance is selected. Lasers are the most convenient light sources for birefringence measurements: major advantages being the very high light intensity and the narrow profile of the light beam.

The standard components of the optical detection system are (cf. *Figure 8*): a laser light source; a crystal polarizer with a polarization plane oriented at 45° with respect to the field vector; a sample cell with an electrode distance in the range of 2–3 mm and a relatively high path length, e.g. 30 mm; a quarter wave plate with its slow axis parallel to the plane of the polarizer—for enhancement

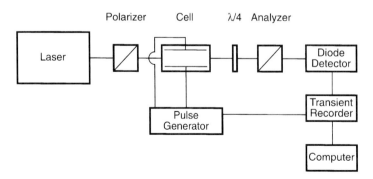

Figure 8 Block diagram with the components of an instrument for measurement of electric birefringence.

of the sensitivity; an analyser, which is rotated by an angle α from the cross position towards the plane of the polarizer; signals (changes of light intensity) are usually recorded by diode detectors.

Under these conditions the change of the signal ΔI_δ induced by an electric field pulse is described by:

$$\Delta I_\delta/I_\alpha = [\sin^2(\alpha + \delta/2) - \sin^2(\alpha)]/\sin^2(\alpha) \qquad [17]$$

where I_α is the total signal measured in the absence of an external electric field at the analyser position rotated at the angle α from the cross position. δ is the phase shift or optical retardation:

$$\delta = 2\pi l \Delta n/\lambda \qquad [18]$$

Δn is the difference in the refractive index, l the optical path length, and λ the wavelength of light used for the measurement.

A comparison of the advantages of electric birefringence versus those of electric dichroism:

- ⊕ The electric birefringence may be measured at any wavelength—usually outside of the spectral range of the absorbance.
- ⊕ Because electric birefringence is usually measured outside the absorbance range, photoreactions are avoided.
- ⊕ Lasers of high power are available as light sources for birefringence measurements. Because of the narrow light beam of lasers, cells with short electrode distances and long optical path lengths can be used.
- ⊖ Electric dichroism can be calculated from molecular structures more easily than the electric birefringence.
- ⊖ Transients of electric dichroism are not perturbed by contributions of the solvent. Water, for example, contributes to birefringence.
- ⊖ Controls for field induced reactions are more easily possible by the optical set-up used for electric dichroism. Signals reflecting field induced reactions can be obtained selectively by measurements at the 'magic' angle.

Protocol 6

Adjustment of the optics for measurements of birefringence

Equipment

- Instrument for measuring electric birefringence

Method

1. Adjust the polarizer, located between light source and measuring cell, to an angular position of 45° in clockwise direction viewed from the direction of the light source.

> **Protocol 6 continued**
>
> 2 Adjust the analyser, located between the measuring cell and the detector, to the cross position, i.e. to the minimum of the light intensity. During this operation the λ/4-plate should not be in the light path.
>
> 3 Insert the λ/4-plate between the measuring cell and the analyser; rotate the λ/4-plate until the light intensity is minimal again. Under these conditions the slow axes of λ/4-plate is parallel to the polarization plane of the polarizer.
>
> 4 The analyser is then rotated by an angle α from the crossed position towards to polarizer. Under these conditions negative birefringence of samples results in a decrease of the light intensity, whereas positive dichroism leads to an increase of the light intensity (1). The magnitude of α depends on the sensitivity of the detector and the light intensity; e.g. for measurements of negative birefringence signals, α may be adjusted such that the total light intensity is at the upper end of the linear range of the detector.

Protocol 7

Measure/calculate birefringence

Equipment

- Instrument for measuring electric birefringence

Method

1 Measure the total light intensity I_α at the angle α from the crossed position.

2 Record transient induced by electric field pulse and read the stationary change of the light intensity ΔI_δ induced by the pulse.

3 Calculate the phase shift or optical retardation δ according to:

$$\delta = 2 \cdot (arcsin[(\Delta I_\delta/I_\alpha) \cdot \sin^2(\alpha) + \sin^2(\alpha)]^{1/2} - \alpha).$$

4 Calculate the birefringence, i.e. the difference in refractive index of the solution parallel and perpendicular to the applied field, $\Delta n = \delta \lambda/(2\pi l)$ where λ is the wavelength used for the measurement and l the optical path length.

6.2 Fluorescence

The main advantage of fluorescence spectroscopy is its high sensitivity and selectivity. The high sensitivity is useful for measurements at very low concentrations; the emission of fluorescence by single or relatively few residues provides selectivity. These advantages may also be used for electro-optical applications. Because both excitation and emission are anisotropic, the dependence of fluorescence intensities on the molecular orientation may be relatively complex. However, the analysis may be simplified considerably by using appropriate boundary conditions.

As described in Section 4.2, the absorbance of light is independent of the molecular orientation, when the plane of the polarization is oriented at the 'magic angle' of 55.7° with respect to the electric field vector. A corresponding magic angle condition is valid for emission of light. This may be used for measurements of fluorescence detected electric dichroism by the following experimental procedure: the light used for excitation is polarized parallel to the field vector, whereas the emitted light is measured at the magic angle orientation. Under these conditions the anisotropy of emission does not affect the measured fluorescence intensity and, thus, changes of this intensity reflect the electric dichroism of light absorbance. This procedure has been used for the characterization of Na/K-ATPase membranes (11); its main advantages are high sensitivity and selectivity.

6.3 Light scattering

Light scattering is also quite sensitive in the detection of field induced orientation, in particular for large molecules or large molecular aggregates. As described in standard textbooks, there are different domains of light scattering, determined by the size of the molecules with respect to the wavelength of light. When the molecules are much smaller than the wavelength of light, the analysis of measured data is relatively simple, but the intensity of scattered light remains rather low. For molecules comparable in size to the incident wavelength the analysis is already relatively complex and, finally, for molecules larger than the wavelength the effects are too complex for any standard analysis. A discussion of the different ranges and phenomena has been presented by Stoylov (4).

7 Kinetics

Electro-optical measurements can be very useful for investigations of kinetics, because electro-optical signals contain information on structures and because these signals can be measured at a very high time resolution. However, filling of standard cells for electro-optical measurements takes some time—usually some minute(s). Thus, the analysis of reactions by electro-optical techniques has been restricted to relatively slow processes. Because many reactions of macromolecules and, in particular, most reactions of biological interest are much faster, their analysis requires a technique with a higher time resolution.

Recently a stopped flow electric field jump instrument (*Figure 9*) has been developed for the analysis of fast reactions by electro-optical measurements (29). The time resolution, defined by the 'dead time' of the stopped flow, is 0.5 msec. The detection of the electro-optical signals, defined by the pulse generator and the time resolution of the detector, can be up to the ns time range. The volume of solution required per shot is ~ 160 μl and, thus, the instrument can be applied to a large variety of reactions of biochemical interest and not only to systems, where the components are available in unlimited amounts.

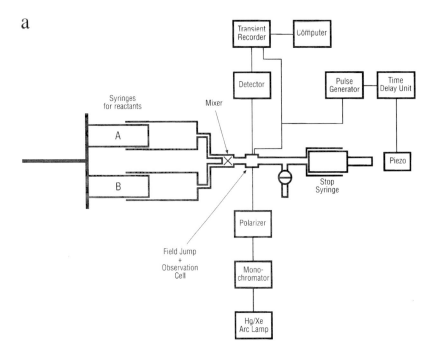
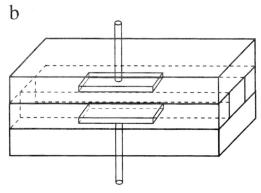

Figure 9 (a) Block diagram of a stopped flow electric field jump instrument. The reactants in the syringes A and B are pushed together through a mixer into the field jump cell; the flow stops, when the piston of the stop-syringe hits the piezo; the signal of the piezo triggers via an adjustable time delay the pulse generator and recording of the optical signal by a transient digitizer (adapted from ref. 29). (b) Construction of the stopped flow field jump cell with Pt electrodes built into a quartz cuvette (from ref. 29).

References

1. Fredericq, E. and Houssier, C. (1973). *Electric dichroism and electric birefringence.* Clarendon Press, Oxford.
2. O'Konski, C. T. (ed.) (1976). *Molecular electro-optics. Part I. Theory and methods.* Marcel Dekker, Inc., New York.

3. O'Konski, C. T. (ed.) (1978). *Molecular electro-optics. Part II. Applications to biopolymers*. Marcel Dekker, Inc., New York.
4. Stoylov, S. P. (1991). *Colloid electro-optics*. Academic Press, London.
5. Porschke, D. and Jung, M. (1985). *J. Biomol. Struct. Dyn.*, **2**, 1173.
6. Porschke, D. (1997). *Biophys. Chem.*, **66**, 241.
7. Cantor, C. R. and Schimmel, P. R. (1980). *Biophysical chemistry. Part II: Techniques for the study of biological structure and function*. Freeman and Co., San Francisco.
8. Tirado, M. M. and Garcia de la Torre, J. (1980). *J. Chem. Phys.*, **73**, 1986.
9. Tirado, M. M. and Garcia de la Torre, J. (1984). *J. Chem. Phys.*, **81**, 2047.
10. Porschke, D. (1996). *Biophys. J.*, **71**, 3381.
11. Porschke, D. and Grell, E. (1995). *Biochim. Biophys. Acta*, **1231**, 181.
12. Perrin, F. (1934). *J. Phys.*, **5**, 497.
13. Antosiewicz, J. and Porschke, D. (1989). *Biochemistry*, **28**, 10072.
14. Schönknecht, T. and Porschke, D. (1996). *Biophys. Chem.*, **58**, 21.
15. Hoffmann, G. W. (1971). *Rev. Sci. Instrum.*, **42**, 1643.
16. Grünhagen, H. H. (1974). *Entwicklung einer E-Feldsprung-Apparatur mit optischer Detektion und ihre Anwendung auf die Assoziation amphiphiler Elektrolyte*. Dissertation, Technische Universität Braunschweig.
17. Porschke, D. and Obst, A. (1991). *Rev. Sci. Instrum.*, **62**, 818.
18. Porschke, D. (1989). *Biopolymers*, **28**, 1383.
19. Garcia de la Torre, J. and Bloomfield, V. (1981). *Q. Rev. Biophys.*, **14**, 81.
20. Antosiewicz, J. and Porschke, D. (1989). *J. Phys. Chem.*, **93**, 5301.
21. Harvey, S. C. and Garcia de la Torre, J. (1980). *Macromolecules*, **13**, 960.
22. Wegener, W. A., Dowben, R. M., and Koester, V. J. (1979). *J. Chem. Phys.*, **70**, 622.
23. Porschke, D. and Antosiewicz, J. (1990). *Biophys. J.*, **58**, 403.
24. Porschke, D. and Antosiewicz, J. (1989). In *Dynamic properties of biomolecular assemblies* (ed. S. E. Harding and A. J. Rowe), p. 103. The Royal Society of Chemistry, Special Publication No. 74.
25. Antosiewicz, J. and Porschke, D. (1995). *Biophys. J.*, **68**, 655.
26. Meyer-Almes, F. J. and Porschke, D. (1997). *J. Mol. Biol.*, **269**, 842.
27. Garcia de la Torre, J. G., Carrasco, B., and Harding, S. E. (1997). *Eur. Biophys. J.*, **25**, 361.
28. Harding, S. E., Horton, J. C., and Colfen, H. (1997). *Eur. Biophys. J.*, **25**, 347.
29. Porschke, D. (1998). *Biophys. J.*, **75**, 528.

Chapter 9
High-flux X-ray and neutron solution scattering

Stephen J. Perkins

Department of Biochemistry and Molecular Biology, Royal Free Campus, Royal Free and University College School of Medicine, University College London, Rowland Hill St., London NW3 2PF, UK.

1 Introduction

Solution scattering is a diffraction technique that is used to study the overall structure of biological macromolecules in random orientations (1–4). A sample is irradiated with a collimated, monochromatic beam of X-rays or neutrons, as the result of which a circularly-symmetric diffraction pattern is observed on a two-dimensional area detector at low scattering vectors Q. The intensities $I(Q)$ are measured as a function of Q, where $Q = 4 \pi \sin \theta / \lambda$ (2θ = scattering angle; λ = wavelength). While X-rays are diffracted by electrons, and neutrons are diffracted by nuclei, the physical principles are the same in both types of scattering experiments. Classical solution scattering views structures at a low structural resolution of about 2–4 nm from data obtained in a Q range of about 0.05–3 nm^{-1} (*Figure 1*). Analyses of $I(Q)$ by Guinier plots lead to the overall molecular weight and the radius of gyration R_G (and in certain cases those of the cross-section and the thickness). The Fourier transform of $I(Q)$ gives the distance distribution function $P(r)$, from which the maximum dimension of the macromolecule and its shape can be deduced. In recent years, a powerful method of constrained solution scattering has been developed (4), in which automated curve fit procedures based on known crystal structures are able to extract structural information to a precision of 0.5–1.0 nm, corresponding to medium resolution structures.

High flux sources make possible many applications of solution scattering in biology. For X-ray scattering, a high flux means that X-ray scattering cameras avoid instrumental scattering curve distortions through the use of a highly monochromatized beam and ideal point collimation geometries (i.e. based on pin-hole optics). As signal–noise ratios are dramatically improved, these make possible the use of:

(a) The study within one experimental beamtime session (one to two days) of a large number of samples.

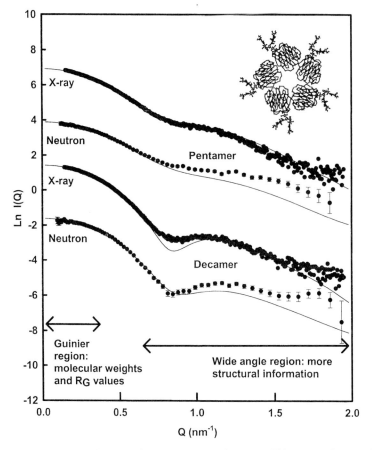

Figure 1 General features of a solution scattering curve $I(Q)$ measured over a Q range. The X-ray and neutron scattering curves of SAP is analysed in two regions, that at low Q which gives the Guinier plot, from which the radius of gyration R_G and the forward scattered intensity $I(0)$ values are calculated, and that at larger Q from which more structural information is obtained. At the lowest Q values, the scattering curve is truncated for reason of the beamstop. The glycosylated SAP model used for the pentamer curve fits (solid lines) is shown as an inset at the top right.

(b) The need for non-physiological high sample concentrations is minimized and macromolecules of low solubility can be studied.

(c) Time-resolved scattering can follow the rate of conformational changes or oligomerization/dissociation processes.

For neutron scattering, the principle utility of a high flux is to make possible the use of contrast variation experiments using mixtures of H_2O and ^2H_2O, and in addition kinetic experiments can be performed in ^2H_2O buffers.

Contrast variation experiments study the internal structure of the macromolecule if this is inhomogeneous in its scattering properties. This is made possible as the scattering powers for different nuclei in neutron scattering (termed 'scattering lengths') are positive and similar to each other with the important exception of that for ^1H nuclei which is negative (Table 1). X-ray scattering differs in that the scattering length increases with atomic number. The scattering density is the total of scattering lengths within the molecule divided by the molecular volume. Since the proportion of non-exchangeable ^1H atoms varies for different biological macromolecules, lipids, proteins, carbohydrates, and nucleic acids each possess distinct neutron scattering densities (Table 2). These fall between the very different neutron scattering densities of H_2O and ^2H_2O. Variations of the ratio of H_2O and ^2H_2O in neutron scattering experiments alters the solvent–macromolecule contrast, and this reveals the structure of these components within the macromolecule. X-ray data can be used as a substitute for neutron data in 0% ^2H_2O, although X-ray data corresponds to hydrated macromolecules while neutron data corresponds to unhydrated structures.

The applicability of solution scattering to study protein–ligand complexes is broad. The ligand can be a small molecular weight species whose size is too small to be detected by scattering (mass < 1000 Da), so this permits study of only the macromolecular structure in the presence and absence of ligand. Alternatively, the ligand can be another macromolecule (mass > 10 000 Da), in which case the

Table 1 Scattering lengths of biologically important nucleii

		Atomic number	f (2θ = 0°) (fm)	b (fm)
Hydrogen	^1H	1	2.81	−3.742
	^2H	1	2.81	6.671
Carbon	^{12}C	6	16.9	6.651
Nitrogen	^{14}N	7	19.7	9.40
Oxygen	^{16}N	8	22.5	5.804
Phosphorus	^{31}P	15	42.3	5.1

Table 2 Scattering densities of solvents and biological macromolecules[a]

	X-rays (e.nm^{-3})	Neutrons (% ^2H_2O)
H_2O	334	0
^2H_2O	334	100
50% (w/w) sucrose in H_2O	402	13
Lipids	310–340	10–14
Detergents	300–430	6–23
Proteins	410–440	40–45
(Hydrophobic residues	410	38)
(Hydrophilic residues	440	54)
Carbohydrates	490	47
DNA	590	65
RNA	600	72

[a] Adapted from ref. 3 (Perkins, 1988), in which a detailed compilation is given for all classes of biological macromolecules.

free and complexed structures of both macromolecules can be studied. The latter can include studies of two interacting proteins (X-rays and neutrons), or a protein in association with lipid, detergent, nucleic acid, or carbohydrate (neutrons). A special case of neutron scattering involves the specific deuteration of a given protein in order that its complex with a protonated protein can be studied.

2 Applications of scattering

2.1 Complementary structural approaches

Scattering fits in well with other methods of biological structure determinations:

(a) X-ray crystallography gives electron density maps at atomic resolution (0.1–0.3 nm), from which atomic structures can be derived, but the results can be affected by conformational or associative artefacts caused by intermolecular contacts in the crystal lattice packing or the high salt concentration used for crystallization. In contrast, scattering is able to identify large-scale conformational changes on ligand binding and the state of oligomerization in near-physiological solution conditions, and has the potential to lead to medium-resolution structures for proteins that cannot be crystallized (see Volume 2, Chapter 1).

(b) 2D-NMR spectroscopy gives atomic structures in solution, but is limited by considerations of macromolecular size where about 25 000 Da is the upper limit. Scattering leads to the solution structure of macromolecules of mass 10 000 Da upwards (see Volume 2, Chapter 10).

(c) Electron microscopy provides a direct visualization of macromolecular structures, but the measurement conditions using stains under vacuum can be harsh. The macromolecular dimensions determined by solution scattering can be compared with those from electron microscopy to confirm the outcome of electron microscopy.

(d) Analytical ultracentrifugation (see Chapters 4 and 5) and light scattering provides structural information through determinations of sedimentation or diffusion coefficients, however this is a single-parameter method analogous to a radius of gyration R_G measurement, while scattering is a multi-parameter method.

Overall, the two main strengths of solution scattering are the provision of a multi-parameter description of a protein structure in near-physiological conditions, and the ability to model the scattering data quantitatively using known atomic structures.

It should be noted that an alternative name for the scattering vector Q is the momentum transfer, and other symbols for this include h and q. It is sometimes redefined as s, the reciprocal of the Bragg spacing d, where $s = Q/2\pi$. Dimensions are commonly reported in nm to reflect the low and medium structural resolutions of the method, but Ångström units (Å) (where 1 nm = 10 Å, and 1 Å is the approximate diameter of a H atom) are also used as their usage is widespread in crystallography.

2.2 Properties of X-ray scattering

The main features of X-ray scattering experiments are summarized as follows:

(a) Most biological macromolecules are studied in high positive solute–solvent contrasts; only lipids have an electron density less than that of water (*Table 2*). As this positive contrast commonly corresponds to the situation when the scattering density of the whole macromolecule is significantly higher than that of the solvent, the effect of this is to minimize systematic errors in the curve modelling of proteins and glycoproteins as internal density fluctuations can be neglected.

(b) Good counting statistics are obtained in this positive contrast, provided that high background levels from the instrument are minimized, which is not always possible at low Q values due to high instrumental background scattering levels. In distinction, the positive contrasts obtained using neutron data for samples in H_2O buffers suffer from a high incoherent scattering background, as a result of which neutron signal–noise ratios can be poor for dilute samples.

(c) Instrumental errors caused by wavelength polychromicity and beam divergence are minimal for synchrotron X-ray scattering experiments, thus Guinier and wide-angle analyses are not affected by systematic errors caused by the instrument geometry, and curve corrections do not have to be applied.

(d) X-ray scattering reveals the hydrated dimensions of the macromolecule. A monolayer of water molecules is hydrogen bonded to the protein surface, meaning that the electron density of bound water is increased compared to that of bulk water, and the macromolecule is made to appear larger by the thickness of this water monolayer, which is readily detected by X-rays.

2.3 Properties of neutron scattering

The main features of neutron scattering experiments are summarized as follows:

(a) Contrast variation experiments using mixtures of H_2O and ^2H_2O buffers permit the analysis of hydrophobic and hydrophilic regions within proteins and glycoproteins, the elucidation of the structure of detergents or lipids in complexation with solubilized membrane proteins, or that of DNA or RNA in complexes with proteins. Deuteration of components in a multicomponent system extends these methods.

(b) No radiation damage effects are encountered in neutron scattering. This can be a severe problem with synchrotron X-rays, where samples may aggregate rapidly after short irradiation times. Sample aggregation in neutron scattering sometimes occurs for a protein or glycoprotein in heavy water buffers if protein solvation is important for its solubility, in which case the use of gel filtration and low protein concentrations may overcome this. Generally neutron samples can be recovered for other structural studies.

(c) The hydration shell is not visible in neutron scattering for reason of the rapid exchange of H and ^{2}H atoms with those in bulk solvent, so the unhydrated dimensions of the macromolecule are studied. As the neutron structure corresponds directly to the water-free protein coordinates obtained by crystallography or NMR, this considerably simplifies the modelling of the neutron curves from atomic coordinates.

(d) Provided that the sample concentrations are known, the absolute molecular weight can be calculated to within ± 5% from the neutron data using H$_2$O buffers, although only Instruments D11 and D22 at the Institut Laue Langevin (ILL: see below) offer sufficient flux to do this for all macromolecules. H$_2$O buffers are advantageous in that the neutron molecular weight calculation is insensitive to the partial specific volume. If ^{2}H$_2$O buffers are used, relative molecular weights can be calculated using calibration graphs for the instrument in question (5) (*Figure 2*), and absolute molecular weights are possible if the partial specific volume is accurately known (6). In distinction, if a suitable standard is available, synchrotron X-ray scattering leads to relative molecular weights if the concentration is known, but the high background at low Q can make this difficult.

(e) The buffer background is very low in ^{2}H$_2$O solvents, even in the presence of high salt concentrations, and this permits studies of macromolecules at very low concentrations (0.5 mg/ml) in a range of salt concentrations.

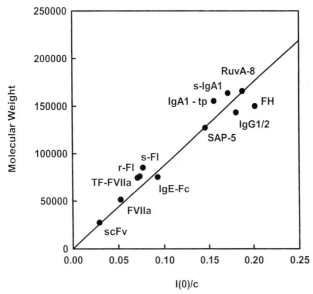

Figure 2 Linear relationship between the molecular weight and the neutron $I(0)/c$ values for glycoproteins in 100% ^{2}H$_2$O buffer measured on LOQ. In order of increasing molecular weight on the vertical axis, the data correspond to the following proteins: scFv antibody, factor VIIa, the factor VIIa–tissue factor complex, the IgE–Fc fragment, recombinant and serum factor I, pentameric serum amyloid P component, bovine IgG1/2, factor H, tailpiece-deleted and serum IgA1, and octameric RuvA. Adapted from ref. 5.

(f) Guinier analyses at low scattering angles are not significantly affected by beam divergence or wavelength polychromicity (9–10% on D11 and D22 at the ILL). However intensities at large Q are noticeably affected by both these factors (*Figure 1*) and this requires consideration in neutron curve modelling from coordinates. Curve modellings for proteins in ^2H_2O also show the effect of a small residual incoherent scattering background which also needs attention.

3 X-ray and neutron facilities

3.1 High flux sources

The probability of a diffraction event when a X-ray photon (or a neutron) approaches an electron or nucleus is similar and very low at 10^{-25} (or 10^{-23}) respectively. In distinction to the use of scattering in other disciplines such as chemistry or metallurgy, a drawback of biological experiments lies in the available signal–noise ratios. The use of high flux sources overcomes this limitation (*Protocol 1*).

Protocol 1

Contact details for synchrotron X-ray and neutron facilities (September 1999)

1 **Daresbury X-ray Synchrotron Radiation Source: Stations 2.1, 8.2, and 16.1, belonging to the Non Crystalline Diffraction group**
   SRS User Liaison Office
   CLRC Daresbury Laboratory
   Warrington, Cheshire WA4 4AD, UK.
   Tel: +44 (0)1925 603223/603636 or 603486
   Fax: +44 (0)1925 603174
   Email: srs-ulo@dl.ac.uk
   Web: http://srs.dl.ac.uk/ULO

2 **European X-ray synchrotron source: Beamlines BM2 and BM26**
   User Office
   European Synchrotron Radiation Facility
   6 rue Jules Horowitz, BP 220, F-38043 Grenoble Cedex, France.
   Tel: +33 (0) 476.88.25.52
   Fax: +33 (0) 476.88.20.20
   Email: useroff@esrf.fr
   Web: http://www.esrf.fr

Protocol 1 continued

3   ISIS spallation neutron source: Instrument LOQ, belonging to the Large Scale Structures group
    User Liaison Office
    ISIS Facility
    CLRC Rutherford Appleton Laboratory, Chilton, Didcot OX11 0QX, UK.
    Tel: +44 (0)1235 445592
    Fax: +44 (0)1235 445103
    Email: uls@rl.ac.uk
    Web: http://www.isis.rl.ac.uk

4   ILL high-flux neutron reactor: Instruments D11 and D22, belonging to the Large Scale Structures group
    Scientific Coordination Office
    Institut-Laue Langevin
    6 rue Jules Horowitz, BP 156, F-38042 Grenoble Cedex 9, France.
    Tel: +33 (0) 476.20.70.82
    Fax: +33 (0) 476.48.39.06
    Email: sco@ill.fr
    Web: http://www.ill.fr/SCO

Protocol 2
Details of other synchrotron X-ray and neutron facilities world-wide

1   The SRS maintains a Synchrotron Radiation WWW Database which attempts to include all synchrotron facilities world-wide, both present and planned, at the Web address http://srs.dl.ac.uk/SRWORLD/index.html. There is another listing at SPring8 at http://www.spring8.or.jp/ENGLISH/other_sr. A X-ray flux of 10^{10} photons/sec is adequate for solution scattering, provided that the camera background is low.

2   The ILL in Grenoble has a Web page devoted to all neutron facilities world-wide at http://www.ill.fr/Info/links.html. Additionally, there is a directory of neutron scattering cameras at http://www.physik.uni-osnabrueck.de/ibel/sansdir.htm. As some biological samples are weak scatterers, it should be confirmed that a given neutron facility will provide adequate signal–noise ratios suitable for biologists. A flux of 10^5 neutrons cm^{-2} sec^{-1} is satisfactory for most uses.

3   The Synchrotron Radiation Newsletter and Neutron News are distributed free of charge to registered users of synchrotron and neutron sources respectively:
    Synchrotron Radiation Newsletter, or Neutron News
    Gordon and Breach Science Publishers
    Marketing Department
    PO Box 197
    London WC2E 9PX, UK.

There are about 50 X-ray synchrotron sources in the world (*Protocol 2*), of which the most powerful ones are currently (September 1999) the 'third-generation' machines such as the 8 GeV Super Photon Ring (SPring8) near Kyoto, Japan, the 6 GeV European Synchrotron Radiation Facility (ESRF) in France, and the 7 GeV Advanced Photon Source (Chicago, Illinois) in the USA. The 'first-generation' sources took X-ray beams parasitically from particle physics experiments. The 'second-generation' machines are dedicated to the production of synchrotron radiation, and sources such as that of the 2 GeV Synchrotron Radiation Source (SRS) Daresbury facility (the first one of this type) provides sufficient flux for scattering data collection. Ironically, while higher fluxes are better for many experiments, too much X-ray flux can occasionally cause undesirable radiation damage effects in biological scattering work. There are about 35–40 neutron sources in the world (*Protocol 2*), mostly reactor sources, but including five spallation sources. Presently, the most powerful reactor source is the 58 MW High Flux Reactor at the ILL in Grenoble, France (thermal neutron flux of 1.3×10^{15} neutrons cm^{-2} s^{-1}), and the most powerful spallation source is that of the ISIS facility near Oxford in the UK (flux of 2.5×10^{16} fast neutrons s^{-1}; approximately 10^{13} to 10^{14} neutrons cm^{-2} s^{-1}). State-of-the art multiuser X-ray and neutron facilities are expensive, but the cost of an individual experiment is much reduced if there is a large international base of users. The full economic cost of beamtime on a scattering camera is of the order £5000 (X-rays) to £10 000 (neutrons) per day. It is very much in a user's interests to list the total annual cost of his/her beamtime awards in any summary of grants that is requested by a University or a funding body. Usually two to four days of beamtime taken in two sessions is sufficient for a given project. On a per user basis, beamtime represents good value for money when compared to the capital cost of an analytical ultracentrifuge costing over £150 000 or an NMR spectrometer costing over £750 000, for which there are hidden extra costs for salaries, instrument consumables and maintenance, and building infrastructure.

A scattering camera is designed to irradiate a solution of path thickness 1–2 mm with a collimated, monochromatized beam of X-rays or neutrons, and to record the resulting scattering (or diffraction) pattern using a two-dimensional area detector interfaced to a computer. X-ray or neutron facilities usually provide at least one scattering camera for multiuser usage. Scattering cameras are shared between workers in metallurgy, polymers, and other fields as well as in biology, and are generally oversubscribed instruments run by ever-patient, overworked instrument scientists. This means that beamtime schedules are tight, and there is a large turnover of users at the facility. Users are not only responsible for providing their samples, but also for providing minor apparatus, and complying with a set of procedures which is ultimately reduced to the application of common sense (see below). It is the job of the User Liaison office to help the user. Contact with the instrument scientist responsible for the scattering camera is usually beneficial. The Web pages for each facility can provide much local information (*Protocols 1* and *2*).

3.2 X-ray instrumentation

Typical synchrotron X-ray cameras include those at Stations 2.1, 8.2, and 16.1 at SRS (*Figure 3*) (7, 8). Electrons circulating at relativistic speeds around the storage ring of the synchrotron emit a 'white' beam of X-rays tangentially from the ring.

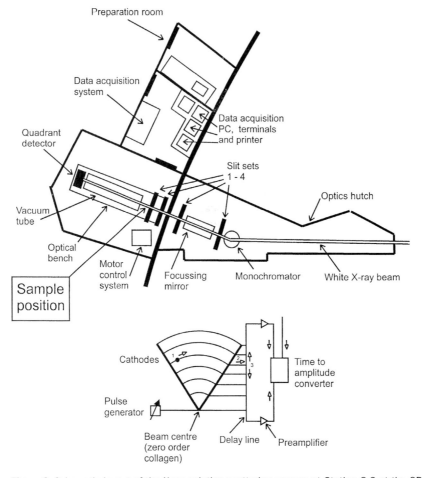

Figure 3 Schematic layout of the X-ray solution scattering camera at Station 8.2 at the SRS Daresbury, not drawn to scale (adapted from http:/srs.dl.ac.uk). (Upper) The camera operates at a wavelength of 0.154 nm using a monochromator-mirror optical system. The beam cross-sectional size is 1 mm × 5 mm at the sample position. The optics hutch is in vacuum and built on a vibration-isolation system. Between the sample and the detector is vacuum tubing of length between 0.5 m to 5 m mounted on an optical bench. The scattering pattern is measured with a quadrant detector which is interfaced to a computer. (Lower) In a quadrant detector, incoming scattered X-rays create ion-electron pairs in the detector gas. The high voltages between the detector planes are responsible for generating an avalanche of electrons which is collected on the cathode (1). This charge moves towards the delay line (2) and splits up into two parts moving in opposite directions (3). The detector electronics measures the time difference between the two pulses in order to calculate the location of the event on the delay line. The increase in size of the cathodes at higher scattering angles compensates for the decrease in scattering intensity at large Q.

The electron beam lifetime is of the order of 12–24 hours, so daily or twice-daily refills are performed. In the camera, the X-ray beam is horizontally focused and monochromated to a wavelength close to 0.15 nm, typically using a Ge or Si perfect single crystal for this, then it is vertically focused by a curved mirror, and collimated by slits to minimize background scatter (8). For the ring operating at 2 GeV and a current of 100 mA, the X-ray flux at the sample position at Stations 2.1, 8.2, and 16.1 is estimated in their routine use to be 2.5×10^{11} photons sec^{-1} (λ = 0.15 nm), 4×10^{10} photons sec^{-1} (λ = 0.15 nm), and about 5×10^{11} photons sec^{-1} (λ = 0.14 nm) respectively, all of which are adequate for biological X-ray scattering work. The optimum sample thickness is a trade-off between thinner samples which absorb less, and thicker samples which scatter more, and is about 1 mm for water and dilute protein solutions. Samples (1 mm path length; surface area 2 mm × 8 mm; total volume 25 μl) are held in Perspex water-cooled cells with 10–20 μm thick mica windows within a brass cell holder (*Figure 4*) or in quartz capillaries. If brass cells are used, the possibility that the samples react with brass (a Cu-Zn alloy) needs consideration; if this is significant, the brass holder can be gold-plated. Other cells with much larger sample volumes and windows permit sample mixing to enable time-resolved experiments to be performed, or the sample to be translated in the beam to avoid radiation damage effects. The sample holder is aligned in the beam with the help of 'green paper' which turns red when exposed to X-rays. The set-up of the camera (detector positioning, beam alignment and focusing, sample and detector alignment) is usually completed in 1–4 hours.

In X-ray scattering, sample–detector distances (0.5–8 m) depend on the desired Q range. Note however that the scattered intensity will decrease according to the inverse of the sample–detector distance squared. A typical two-dimensional detector design is that of the quadrant detector, which corresponds to a multi-wire linear detector that is ideal for isotropic (circularly symmetric) scattering patterns. The scattered intensities are recorded using 512 channels in a two-dimensional angular sector of a circle (70°), at the centre of which is the position of the main beam (*Figure 3*). The maximum count rate is 2×10^4 mm^2/sec. This design leads to much improved counting statistics at large scattering angles, where the hardware itself performs a radial integration of the intensities. On Station 2.1, for a sample–detector distance of 2.5 m, the resulting Q range is 0.1–3.1 nm^{-1}, and for a distance of 7.5 m, this is 0.04–1.1 nm^{-1}. If the main beam intensity is monitored by the use of a 'back' ion chamber positioned after the sample and before the detector, this takes both the sample transmission and the incident flux into account and facilitates data reduction. A lead beam stop in front of the detector protects the latter from the direct main beam; this can be semi-transparent in order to act as another main beam monitor. The detector is interfaced with a computer for data accumulation, which is in turn networked to a Workstation for data storage and processing. On-line data reduction using automated user-customized script files has the major advantage of enabling the experimental data to be assessed as it is being recorded. The whole camera is inside a radiation-shielded hutch, protected by safety interlocks to avoid accidental lethal X-ray exposure to the users.

(a)

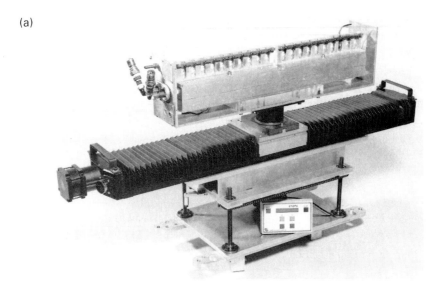

(b)

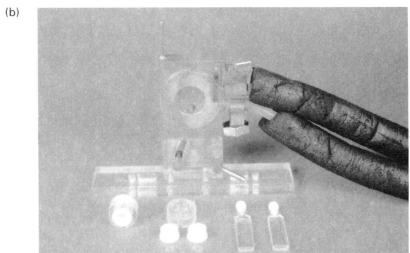

Figure 4 Sample holders and cells used for X-ray and neutron scattering data collection. (Top) The rack is that used on LOQ and takes twenty 1 mm and 2 mm quartz cells; a similar design is also found on D11 and D22. Note the two connections used for coolant on the rack itself, and the motor assembly at the base that enables the rack to be translated (12) (Photograph kindly provided courtesy of ISIS). (Bottom) The X-ray sample holder is shown, together with two Perspex cells (intact and disassembled) in which the mica windows are held in position by Teflon inserts and a Parafilm seal. Two 1 mm- and 2 mm-thick Hellma quartz cells used for neutron scattering are also shown.

3.3 Neutron reactor instrumentation

A neutron camera at a high-flux reactor such as D11 or D22 at ILL Grenoble (9, 10) is designed to maximize the incident flux at a sample by the use of physically large designs and large samples (*Figure 5*). At D11, neutrons enter a beam guide from the reactor after moderation by a cold source (liquid $^2H_2$ at 25 K) positioned 140 m away from the sample position. The cold source maximizes the number of neutrons in a wavelength range of 0.1–1 nm (*Figure 4*). Beam monochromatization is achieved by use of a velocity selector based on a rotating drum with a helical slit in it. The speed at which this rotates allows only the neutrons of the desired mean wavelength (velocity) to pass through, but at the price of a modest wavelength spread of ± 10%. The maximum flux at the sample position is 3×10^7 neutrons cm^{-2} s^{-1} (0.6 nm wavelength; 2.5 m collimation; Dornier selector). Beam collimation employs a series of eight movable straight beam guides of total lengths between 2.5 m to 40 m. The beam size is defined by a diaphragm. A two-dimensional BF$_3$ area detector (now replaced by a ^{3}He detector) is based on 10 mm × 10 mm cells (total of 64 × 64 = 4096 cells) and gives a maximum accessible Q range of 0.006 to 3 nm^{-1}. Sample–detector distances are continuously variable between 1.1 m to 36 m, and wavelengths range from 0.45 nm to 2.0 nm. For a given configuration, the optimal collimation distance is set to be the same as the sample–detector distance. Typically at least two sample–detector distances are used in a biology experiment. Note that sample transmissions are wavelength-dependent. If the same wavelength is used at these different detector positions, and spectra are normalized using H$_2$O runs at the same wavelength at both positions, the merger of data at both configurations is straightforward. For biology work on D11, a typical configuration is based on 2.5 m/2.5 m and 10 m/10 m sample–detector/collimation distances and a wavelength of 1.0 nm, although the flux is much reduced at 10 m distances. This configuration gives a useful Q range of 0.04 to 1.6 nm^{-1} with good spectral overlap. Alignment of the beamstop is performed by the use of a 0.5 mm thick Teflon strip as the sample; this is a strong isotropic scatterer that reveals the

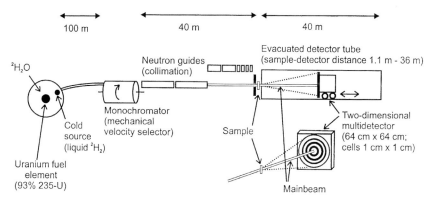

Figure 5 Schematic view of the neutron solution scattering camera D11 at the ILL Grenoble. The polychromatic beam from the reactor cold source is monochromated using a helical velocity selector to give a wavelength ± 10% whose value is determined by the rotational speed of the selector. Eight neutron guides give rise to collimation distances between 2 m and 40 m. The beam size at the sample position is defined by a diaphragm. The 64 × 64 element BF$_3$ detector is housed within a 40 m vacuum tunnel and is moved by computer remote control. The detector is interfaced with a computer for data collection and storage.

shadow of the beamstop on the detector. Up to 22 samples (1 or 2 mm path lengths; surface area 7 mm × 10 mm; total volume 150 μl or 300 μl) in rectangular Hellma quartz cells are loaded onto an automatic sample changer under temperature control, which is programmed to expose them to the neutron beam in sequence (Figure 4). Since the neutron mainbeam intensity is monitored using a small sensor in the incident beam, data normalization requires a separate set of transmission measurements. No radiation hutches are required, however users are prevented from approaching the sample area when the beam is on. Data acquisition, storage, and instrument control are performed by computer, and on-line data analyses can be automated using UNIX script files on a networked Workstation to quickly monitor the progress of the experiment.

The other scattering instrument at the ILL, D22, differs from D11 in having a higher maximum neutron flux of 1.2×10^8 neutrons cm^{-2} s^{-1} and a larger ^{3}He detector of size 96 × 96 cm in which the cell size is 0.75 cm, totalling 128 × 128 = 16384 cells. On D22, the available sample–detector distances are 1.35 m to 18 m, and wavelengths range from 0.3 nm to 4 nm, which gives a maximum Q range of 0.01 to 10 nm^{-1}. A typical biology experiment can involve sample–detector/collimation distances of 5.6 m/5.6 m and 1.4 m/8.0 m, a beam aperture of 7 × 10 mm, and a wavelength of 1.0 nm. This configuration avoids the saturation of the detector and the need for deadtime corrections that arises with the very high flux obtained with short collimation distances, and gives a useful Q range of 0.07 to 2.5 nm^{-1}.

3.4 Spallation neutron instrumentation

A neutron camera at a spallation source such as LOQ at ISIS, Rutherford Appleton Laboratory (11, 12) differs from a reactor-source camera in that its design is based on 50 pulses/sec of neutrons that are emitted from a uranium or tantalum target after proton bombardment from a synchrotron, not a continuous beam as with a reactor source (Figure 6). The neutrons are slowed by a liquid ^{2}H$_2$ moderator positioned 11.0 m from the sample rack, so that each pulse contains neutrons of wavelengths in the range 0.2–1.0 nm. A chopper removes every other pulse to prevent frame overlap between pulses on LOQ, although other chopper configurations can give access to different spectral ranges if needed. Up to 20 samples in rectangular quartz Hellma cells are loaded onto an automatic sample changer which is under temperature control (Figure 4). Unlike the other X-ray and neutron cameras described above, the collimation on LOQ is fixed in length at 4.5 m, and the sample–detector distance is fixed at 4.1 m. The area detector is a ^{3}He-CF$_4$ ORDELA type with an active surface of size 64 × 64 cm and individual cells of side 0.53 cm each. Monochromatization is achieved by time-of-flight techniques based on the total distance of 15.5 m that each pulse travels from the uranium target to the ORDELA detector. The neutrons in each pulse will reach the detector at different times depending on their wavelength (velocity), and this enables a wavelength to be assigned to each neutron. The

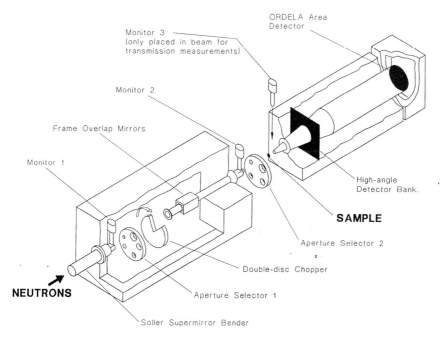

Figure 6 Schematic 3D view of the neutron solution scattering camera LOQ at ISIS. The chopper removes every other pulse of neutrons. After passing through the sample position, the diffracted neutrons are detected on the ORDELA area detector and on a high-angle detector bank. The detectors are interfaced with a minicomputer for data collection and storage. (Figure kindly provided courtesy of ISIS.)

standard LOQ configuration employs 102 time frames for the assignment of wavelengths in each pulse. This is an efficient arrangement as this means that all the neutrons in each pulse are used for data acquisition. A typical time-averaged flux on LOQ is 2×10^5 neutrons cm^{-2} s^{-1}. This flux is adequate for working with 5–10 mg/ml proteins in ^2H_2O buffers or with strong scatterers such as lipoproteins which have a high lipid content. Transmissions at each wavelength are measured either during scattering data acquisition using a semi-transparent beamstop, or separately from the scattering data using an attenuated beam. The major advantage of the camera design is that the entire scattering curve in the Q range of 0.06 to 2.2 nm^{-1} can be measured simultaneously, and this is ideal for kinetic experiments. A side-benefit of this camera is that neutrons of shorter wavelengths are the most abundant in every pulse, and these contribute the most to the scattered curve at large Q, which is where the scattered intensities are lower and signal–noise ratios are poorer. LOQ was upgraded in 1996, when an annular high-angle area detector was added at 11.6 m from the moderator to provide a second Q range of 1.5–14 nm^{-1}. The neutron data is acquired by a DEC VAXstation 3200, and then made available for on-line data processing and hard copies using user-defined automated VMS script files within the COLETTE software package (12) on networked computers.

4 Experimental approach

4.1 Applications for X-ray and neutron beamtime

Forward planning is implicit in applying for beamtime. The Web pages and the User Liaison Office at the synchrotron or neutron facility should be contacted six months in advance of the intended experiment for application forms and the names of suitable instrument scientists (*Protocols 1* and *2*). Sometimes 'Director's discretionary beamtime' can be made available at short notice for exceptional cases. Beamtime applications are competitive. Since over twice as many days is usually applied for than is available, a well-written application is needed (*Protocol 3*).

Protocol 3

Applications for beamtime: experimental reports

A Applications for beamtime

1. Each X-ray or neutron facility has application forms that can be downloaded either from the Web or by anonymous ftp. A beamtime application usually allocates two pages for a scientific proposal, and is a miniaturized version of a project grant application. The key points of a beamtime application are summarized as follows:
   (a) Write a brief abstract of your proposal.
   (b) Give a brief statement of the background and the general importance of the research.
   (c) State the aims of the proposed experiment and describe the experiment for evaluation by scientific reviewers who are not necessarily experts in the field.
   (d) Report preliminary work carried out that is relevant to the proposed experiment.
   (e) Provide and explain the measuring time estimated for the samples.
   (f) State why the X-ray or neutron facility is necessary for the experiment.
   (g) Provide a short bibliography of relevant published literature.
   (h) Specify an experienced instrument scientist at the institution and discuss the proposal with him.

2. Applications are requested at six month intervals, and the beamtime is taken about three to six months after each deadline if the application is successful. Deadlines can vary but are as follows:
   (a) At ISIS, deadlines for receipt are April 16 and October 16.
   (b) At ILL, deadlines for receipt are February 15 and August 31.
   (c) At SRS, the deadlines are at the end of April and October.
   (d) At ESRF, the deadlines are September 1 and March 1.

3. User Liaison Offices are usually overwhelmed by proposals arriving at the last minute, and efforts to minimize inconvenience are usually appreciated.

Protocol 3 continued

B Experimental reports and publications

1. The experimental reports are important as these are published in the Annual Report of the facility, and are used as essential evidence of progress by scientific reviewers when assessing subsequent beamtime applications. A failure or delay in providing these may affect future work. Of course, peer-reviewed publications in good journals are also important.

4.2 Preparation of protein–ligand complexes for scattering

Users are responsible for ensuring that their samples are ready in time for what is a short but intensive one or two day experiment (*Protocol 4*). The basic requirement is for pure, monodisperse solutions at a high enough sample concentration for a scattering curve to be observable in the required solute–solvent contrast. Large-scale preparations involving up to 50 mg of material and concentrations in the range of 5–10 mg/ml will be required for a complete project in two or three beam sessions, although it is possible to work with less material. For a single beam session, 0.5 ml of material at 10 mg/ml is ideal for synchrotron X-ray work, and 1.5 ml at 10 mg/ml for neutron work in three contrasts. If stopped-flow methods are to be used, these amounts are considerably larger.

Protocol 4
Pre-experimental organization

1. Most importantly, notify biochemical collaborators of beamtime dates in good time, organize a sample preparation timetable, and notify the beamtime institution of the names of the actual people coming.
2. Book accommodation and travel.
3. Establish any visa requirements to enter the country.
4. Establish correct safety procedures to be followed (assignment of a safety hazard level according to the local safety procedures, especially if samples from a pathogenic organism or from humans are to be used, or if genetic engineering is involved), and determine safe areas for working with hazardous materials if needed.
5. Determine members of team and allocate tasks (on-line data processing [for which a basic knowledge of UNIX is needed], sample manipulations, and station operation).
6. Determine the available Q range on the camera, choose the sample–detector distance and wavelength to be used, and specify the detection system if required.
7. Construct and bring a suitable sample holder, or arrange for the use of a suitable one.
8. Obtain water-bath or oil-bath for temperature control of the sample holder between 0–100 °C.

Protocol 4 continued

9 Confirm availability of cold room and fridge.

10 Cancel experiment in good time if the samples will unfortunately not be ready.

In terms of sample preparations, biochemical standards of purity are adequate, but physical monodispersity is more important. Since scattered intensities are proportional to the square of the molecular weight at low Q, scattering curves are affected by non-specific aggregation. Aggregates are revealed in a scattering curve by performing Guinier analyses of ln $I(Q)$ vs. Q^2, which normally gives a linear plot of intensities at low Q values, except when aggregates are present, in which case the plots are curved upwards at the lowest Q values and are unusable. If the sample is prone to aggregation, it is essential to remove all traces of aggregates prior to measurement by gel filtration and re-concentration of the samples. Microfiltration is not sufficient. In fact, it is a good precautionary practice to subject all scattering samples to gel filtration immediately prior to experimental work. This is advantageous in that the column eluate after the sample has eluted can be used as the buffer background for X-ray scattering experiments without the need for further dialysis. If the equipment is available, laser light scattering can be used to test for the presence of aggregates before beamtime is taken.

Protocol 5

General experimental procedures

A. Immediately before start-up of scattering experiment

1 Most importantly, read the instrument manuals in either paper or Web format, or at least look through them before data acquisition is started.

2 Before travelling, check the beam status on a Web page (see *Protocols 1* or *2*).

3 Bring accessories that are not usually available at the facility (log book, pager or mobile, laptop or palmtop, hand calculator, Allen keys, screwdriver set, two alarm timers, temperature thermocouple, dental mirror, tweezers, scalpel knife, scissors, Stanley knife, adhesive tape, 5 m tape measure, stapler, Eppendorf vials, Gilson or Finn pipettes and tips, Parafilm, syringes, needles, 1 mm and 2 mm thick quartz Hellma cells, Hellma cleaning fluid, dialysis bags and clamps, magnetic stirrers).

4 Complete user registration procedure at the reception desk or control room.

5 Obtain computer account for local processing and determine a method for data archival (FTP, CD-ROM, or Exabyte) and for passage through the local firewall.

6 Locate or obtain on-site copies of the instrument and safety manuals (hardcopy or Web pages), especially the trouble-shooting guide and the 'frequently-asked questions' pages!

Protocol 5 continued

7 Edit a script file for automated data collection and another one to initiate on-line data processing.

8 Maintain an indexed notebook to summarize the most frequently-used procedures and commands.

9 Often overlooked: Measure protein concentrations on a UV spectrometer at 280 nm (these can be measured directly in the quartz cells used for neutron scattering), and arrange for other concentrations to be measured.

10 Agree an acceptable method for contacting the instrument scientist during the experiment.

B. End of experiment

1 Finish data collection at the agreed time.

2 Clean up the experimental area, and remove your equipment.

3 Return any borrowed equipment (e.g. the Fe^{55} source at Daresbury) and any keys or swipe cards.

4 Check neutron-irradiated samples for residual radioactivity.

5 Perform a full data reduction before leaving (XOTOKO at SRS; COLETTE at ISIS; DETEC, RNILS, SPOLLY, RGUIM, and RPLOT at ILL). Script files prepared during data collection make this very easy.

6 Most importantly, back up the raw and reduced data at least twice and verify this (if using FTP, be aware of the difference between binary and ASCII transfers to avoid file corruption).

7 Thank the instrument scientist, report suggestions for improvements, and never complain.

Prior to measurements (*Protocol 5*), sample concentrations c must be accurately known for the molecular weight and neutron matchpoint calculations. A neutron matchpoint corresponds to that percentage ^2H_2O in which the scattering density of the macromolecule is the same as that of the solvent (*Table 2*). For proteins, concentrations are measured from the optical density at 280 nm in a UV spectrophotometer, and converting this reading using the absorption coefficient calculated from the sequence composition (6). Protein concentrations can usually be measured in the same quartz cell used for neutron data collection. Lowry or Bradford assays can be performed if optical densities cannot be measured, and commercial kits for these are widely available. For DNA, concentrations are measured from the absorbance at 260 nm, and converting this using an absorption coefficient of 1 in a pathlength of 1 cm for 50 µg of DNA as in fully paired duplex DNA (13). For lipids, concentrations are measured in individual colour reagent assays for each type of lipid that is present (phospholipid, triglyceride, free and esterified cholesterol), for which commercial kits are available (14). In addition to concentration measurements, biochemical assays of the samples for purity, functional activity, and absence of radiation damage are required to demonstrate the validity of the scattering data, as these details are needed for publication purposes.

In terms of the buffer, dialysis procedures and its X-ray or neutron absorbance are important parameters in scattering experiments. Sample dialysis into the buffer used for data collection is essential, as the buffer data is used for an accurate subtraction of the scattering from the solvent, as well as from the mica or quartz cell windows and the camera background. Usually the final dialysate buffer is used for this purpose (but see above), and the sample and buffer must be well stirred during dialysis. In X-ray work, slight differences in the electron density of the buffer caused by variation in the salt concentration can prevent an accurate buffer subtraction. In neutron work, differences in the exchangeable proton content of the buffer can likewise invalidate the buffer subtraction. For this reason, neutron samples and buffers must be sealed against atmospheric exchange with moisture. The absorbance of the buffer will also affect the scattering experiment. In X-ray work, the closer a protein buffer is to pure water, the higher the sample transmission becomes, and the better the counting statistics. Phosphate buffer saline (12 mM phosphate, 140 mM NaCl pH 7.4) is commonly used, where its X-ray transmission is about 40% for a path length of 1 mm, while the use of 1 M NaCl reduces the transmission to about 10%. In neutron work, the strong incoherent scattering of ^1H nuclei reduces the sample transmission, in which H_2O buffers have a neutron transmission of about 45% for a path length of 1 mm and a wavelength λ of 1.0 nm, while ^2H_2O buffers have a neutron transmission of about 88% for a path length of 2 mm at λ of 1.0 nm.

4.3 Data collection strategies

Data collection using any synchrotron or neutron source (*Protocols* 6 and 7) is best designed on the assumption that the main beam or the camera will unexpectedly stop at some point during data collection. Beam may not be available again until many months later. The chances for a failure of some kind are not small. The beam can be lost due to the weather (e.g. thunderstorms, snowstorms, tornadoes); a major accident; electricity power surge; coolant leaks; vacuum or electronics failures; operator error; etc. Alternatively the scattering camera can fail due to computer malfunction; jamming of the port shutter, beam chopper or sample holder; electronics breakdown; vacuum leaks; etc. This means that all the basic data essential for subsequent analysis should be collected before the sample and buffer runs are measured, and that the samples should be measured in order of priority with the easiest ones first. As the beam may be off for 12–24 hour periods, sessions should be booked for at least 48 hours in order to be reasonably sure of collecting some data. Two-day experiments also make better use of the time spent in setting up the camera and recording the basic data and buffer backgrounds, while one-day experiments are better for test experiments or the final data collection for a project.

Protocol 6
X-ray data collection

A. Prior to X-ray scattering

1. Order ruby mica windows (15–20 μm thick) two months in advance (Attwater and Sons Ltd., PO Box 39, Hopwood St Mills, Preston PR1 1TA, UK).
2. Obtain rat tails and prepare wet rat tail collagen fibrils for mounting in a sample cell.
3. Prepare green paper to mount in sample cells to check the beam position.
4. Prepare a data acquisition plan to allow a given dilution series to be completed before a beam dump and refill.
5. Dialyse samples into their buffers and bring the final dialysate for use as the buffer background.

B. During X-ray scattering

1. Determine beam centre.
2. Measure attenuated collagen peaks at each new beam refill, and confirm that the 1st order peak is visible.
3. Measure detector responses during beam refills and note how long each one lasted.
4. Measure samples and buffer in alternation, using time-frames to acquire the data in order to check for radiation damage.
5. Identify the maximum allowed count rate. At all times, check that the detector is not being saturated, especially just after a refill, and that air bubbles have not appeared in the cell.
6. Process and print scattering curves using script files to check the progress of the experiment.

Protocol 7
Neutron data collection

A. Prior to neutron scattering

1. Purchase or borrow rectangular quartz Hellma cells (1 mm and 2 mm thicknesses). Clean the cells using Hellma cleaning fluid, rinsing at least five times, and dry thoroughly.
2. Obtain strips of Teflon polymer and cadmium cut to the same rectangular dimensions as the Hellma cells (0.5 mm and 1 mm thicknesses).
3. Purchase and bring 99.8% heavy water, NaO^2H, and ^2HCl.

Protocol 7 continued

4   Dialyse samples into their heavy water buffers (four changes over 36 hours) and use the final dialysate for use as the buffer background, keeping all sample and buffer cells sealed with Parafilm against atmospheric exchange with H_2O.

5   Check the temperature of the fridge used for storage, as heavy water freezes at 4 °C. Freezing may cause sample aggregation or break the container through expansion. The same applies when travelling by car in icy winter conditions. Never store heavy water samples directly on wet ice.

6   Prepare a prioritized list of samples and buffer runs.

B. During neutron scattering data collection

1   Align beam stop using Teflon (D11, D22).

2   Measure scattering data for the standards first (H_2O, empty cell, and cadmium backgrounds at D11 and D22; standard polymer and empty position at LOQ). The H_2O run should be repeated subsequently as a check of the detector response.

3   Next measure a basic set of transmissions. Check the basic set of transmissions immediately to confirm the amount of heavy water present. Complete a full set of these when the data collection is going well (empty cell, empty position, pure H_2O, and ^2H_2O; samples and buffers).

4   Measure the samples and buffers for short periods in order of priority. Heavy water samples are usually measured early as the acquisition times are short. Save the long acquisitions in H_2O buffer for overnight runs.

5   Identify the maximum allowed count rate. At all times, check that the detector is not being saturated, especially for H_2O samples which have a high incoherent scattering background, or for known strong scatterers.

6   Process and print scattering curves using script files to check the progress of the experiment, in particular the signal–noise ratios, and abandon aggregated samples at once.

7   Long runs should be stored at hourly intervals.

8   Neutron curves can be summed, so rerun samples if the statistics are weak until sufficient counts are obtained. To halve the errors, the sample has to be run for four times longer. Good buffer statistics as well as sample statistics are required. Abandon samples for which adequate counts will not be possible in the available beamtime.

Modern scattering cameras routinely provide access to the low Q values required for Guinier analyses. These provide a good starting point for the interpretation of the scattering data as they provide both the radius of gyration R_G and the molecular weight from $I(0)$ and c. The criterion for a satisfactory Guinier analysis is that data is measurable in the Q range below $Q.R_G$ of 1. The smallest Q that has to be measured to give an R_G value is given by the relationship $Q_{min} D_{max} \leq \pi$, in which D_{max} is the maximum dimension of the protein (1). The choice of

Q involves the selection of the sample–detector distance and the wavelength λ for cameras where these can be varied (see above). If the protein is expected to possess a compact globular structure, the R_G value can be estimated from a plot of 47 experimental X-ray R_G values versus molecular weight (3):

$$\log R_G = 0.365 \log M_r - 1.342,$$

where M_r is the molecular weight. If the protein is expected to be elongated, the R_G value will become much increased. In the latter case, the R_G value can be estimated from the overall length D using the approximation $R_G = D/\sqrt{12}$, assuming that the two cross-sectional axes (leading to the R_{XS} value: see below) are low and can be neglected. Estimates of D can come from electron microscopy or from hydrodynamic calculations of sedimentation coefficients assuming a rigid rod model. When the R_G value is large, the sample–detector distance is set to greater distances in order to access a low enough Q value, however at the cost of decreased counting rates. If in doubt, it is preferable to err on the side of lower Q. For neutron scattering on Instruments D11 or D22, a choice of wavelength λ is required. The advantage of λ at 1.0 nm or greater is that no wavelength corrections for recoil are required for the calculation of the absolute molecular weight (15).

The distance distribution function $P(r)$ is most useful to have (see below). For a $P(r)$ curve, sufficient data is needed over a wide enough Q range (about two orders of magnitude) to be able to calculate this, including both the Guinier region at low Q and the wide-angle region at large Q. If the instrumental geometry on an X-ray camera does not permit access to the full Q range needed for the $P(r)$ curve, it will be necessary to reposition the X-ray detector, refocus the beam and rerun the samples. For the neutron instruments D11 and D22, even though short sample–detector distances are needed for a $P(r)$ curve, it is a matter of minutes to move the detector inside its vacuum tunnel to obtain the desired distance and to rerun the samples.

X-ray data collection (*Protocol 6*) involves determinations of the Q range on the detector, the detector response, and the scattering from the samples and buffers.

(a) The Q range is defined using the strong diffraction peaks from wet slightly stretched collagen which have a spacing of 67 nm (16). Rat tail tendons are a ready source of collagen fibrils, requiring only a scalpel, two pairs of tweezers, and a Petri dish to remove the skin, followed by incisions to remove the tendons from the bone. The fibrils are readily separated from the tendons and are stored in aqueous buffer prior to use. Reflections are indexed based on the stronger intensities of the 1st, 3rd, 5th, and 9th order peaks (*Figure 7*). The peaks should be re-measured after every beam refill, and the mainbeam may require attenuation to avoid damage to the detector or saturation of the detector channel.

(b) The efficiency of the 512 channels on the multiwire detector is not uniform. A detector response is measured by exposure of the detector for several hours to a radioactive ^{55}Fe source during periods when there is no beam for data collection (*Figure 7*).

(c) Data collection from the samples and buffers needs to allow for reduction of the mainbeam intensity during the beam session and for possible radiation damage. An ion chamber placed after the sample but before the detector or the use of a semi-transparent beamstop will provide normalization data to allow for the changes in mainbeam intensity. Since the background intensity can be high at low Q, small sample–buffer subtraction errors in the Guinier region can occur as the mainbeam intensity decreases (*Figure 7*). These are minimized by measuring samples in duplicate with a buffer run in between

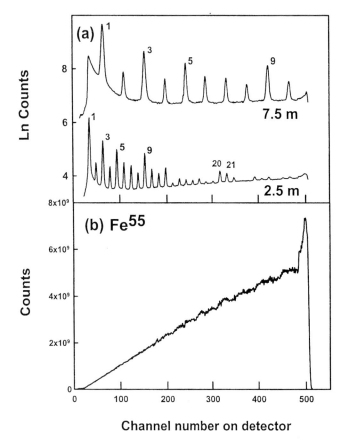

Figure 7 A basic set of X-ray scattering data. (a) Collagen diffraction patterns. A total of 10 and 31 peaks are seen at the 7.5 m and 2.5 m detector positions, the peak separation being $2\pi/67$ nm^{-1}. For examples, this means that the 9th order peak occurs at a Q value of 0.84 nm^{-1} and the 20th order peak at one of 1.88 nm^{-1}. The collagen peaks are used to calibrate the Q values of the channels on the multiwire quadrant detector. Note the higher camera background level at low channel numbers. (b) A detector response recorded using radiation uniformly emitted by a ^{55}Fe radioactive source. The stability of the detector is checked during an experiment by dividing a series of short responses obtained during beam refills by a long one obtained during the small hours.

them (all in the same cell with the same mica windows) for equal periods of time each (usually 10 min, sometimes 20 min). Radiation damage is monitored by recording the data in 10 equal time slices of 1 or 2 min during the acquisition, and examining the 10 subcurves for time-dependent effects. Should changes occur, possible options are to use only the first time frame only, or to process all 10 time frames and extrapolate back to zero time. The use of additives such as 100 mM formate or an antioxidant such as 1 mM dithioerythritol has been proposed as a means of retarding attack by X-ray induced free radicals (17, 18). The final X-ray scattering curve is calculated by subtracting the buffer curve from the sample curve (normalizing each on the basis of the ion chamber counts), and dividing the result by the detector response.

In neutron data collection (*Protocol 7*), the Q range is defined by the dimensions of the instrument, and does not require measurement. A neutron experiment thus involves (a) background runs, (b) transmissions, and (c) the sample and buffer runs.

(a) For D11 or D22, the basic background runs (9) are cadmium (to monitor the ambient neutron and electronic noise background in the detector), Teflon (for defining the beam centre on the detector and the detector area masked by the beamstop), a 1 mm thick H_2O sample (for determination of the detector response and the absolute scale), an empty cell (the background for the H_2O sample), and an empty cell holder (a check for stray reflections or scattering). The H_2O sample can usually be replaced by the buffer in 0% ^2H_2O if this is a solution in dilute salts such as phosphate-buffer saline. The H_2O standard should be re-measured several times during a beam session to check for detector reproducibility. A special case arises at large sample–detector distances. For reason of the low counts rates in this position, the H_2O standard in this case is in fact measured at a short sample–detector distance at which the counting rates are much higher, and the run is adapted for beamstop position and intensity ratios before use for data reduction involving the large distance. For LOQ, the two basic background runs are those for a standard deuterated polymer used to determine absolute intensities and the empty position as the background for the polymer run.

(b) Neutron transmission measurements (with an attenuated beam) are required for all the samples and buffers in order to perform molecular weight and matchpoint calculations, and to confirm that the dialysis into ^2H_2O buffers has proceeded to completion. They are also required for the time frame analysis of the LOQ data reduction. Transmissions vary from about 0.5 for 1 mm thick H_2O samples to about 0.9 for 2 mm thick ^2H_2O samples, depending on the wavelength and temperature. These values mean that solutions are usually measured in 1 mm thick cells for 0–30% ^2H_2O buffers and 2 mm thick cells for 40–100% ^2H_2O buffers.

(c) Sample counting times depend on the amount of ^2H_2O in the buffer. On D11 and D22, protein samples and buffers in H_2O each require about 1-2 hours of acquisition, while those involving ^2H_2O require 5-20 minutes each. DNA samples are more difficult to measure as the matchpoint of DNA is high, and about 1 hour is needed for both H_2O and ^2H_2O buffers. On LOQ, protein samples and buffers in ^2H_2O buffer require about 1-3 hours each. Buffer background subtractions are usually straightforward for reason of instrumental stability. The lack of radiation damage and the stability of the instrument means that the neutron runs can be added if required to improve the signal-noise ratio. After D11 or D22 data collection is completed, the raw scattering curves are firstly corrected by subtracting the cadmium background from each one. The buffer curve is subtracted from the sample curve, then the final reduced curve is calculated by dividing this by the water background (corrected for the transmission of water) minus the empty cell background (9). For LOQ data, the (sample-buffer) subtraction is also performed, but the water correction is not used, and the scattering intensities are normalized using data from a standard polymer corrected using data for an empty position.

4.4 Guinier analyses of the reduced scattering curve

Guinier fit analyses of the scattering curves ln $I(Q)$ as a function of Q^2 at low Q values (*Figure 8*) give the radius of gyration R_G and the forward scattered intensity $I(0)$ from the slope and intercept of linear plots (1):

$$\ln I(Q) = \ln I(0) - R_G^2 Q^2/3$$

For elongated macromolecules, the corresponding mean cross-sectional radius of gyration R_{XS} and the cross-sectional intensity at zero angle $[I(Q)Q]_{Q \to 0}$ (19) are obtained from curve fit analyses in a different Q range that is larger than and does not overlap with the one used for the R_G determinations (*Figure 8*):

$$\ln [I(Q)Q] = \ln [I(Q)Q]_{Q \to 0} - R_{XS}^2 Q^2/2.$$

Concentration series for the samples should yield linear, reproducible Guinier plots.

After the Guinier plots have been processed, the results should be validated by analysing the $I(0)$ data. Plots of the $I(0)/c$ data as a function of c should show the same molecular weight or a small concentration dependence, thereby confirming the absence of sample association or dissociation. Molecular weights can be deduced from $I(0)/c$ values (c = sample concentration in mg/ml), either as relative values from the X-ray data, or as absolute values from neutron data by referencing $I(0)$ to a standard (15, 20), thereby confirming the integrity of the sample. Matchpoints are obtained from a linear plot of $\sqrt{(I(0)/ctT_s)}$ values (t: sample thickness; T_s: sample transmission) as a function of the volume percentage ^2H_2O in the buffer, remembering that the $\sqrt{I(0)}$ values are negative above the matchpoint. The intercept at zero $I(0)$ gives the observed matchpoint, and this should agree within error of the value calculated from the composition for an unhydrated macromolecule (6) (i.e. the partial specific volume will be higher than that conventionally measured by densitometry).

HIGH-FLUX X-RAY AND NEUTRON SOLUTION SCATTERING

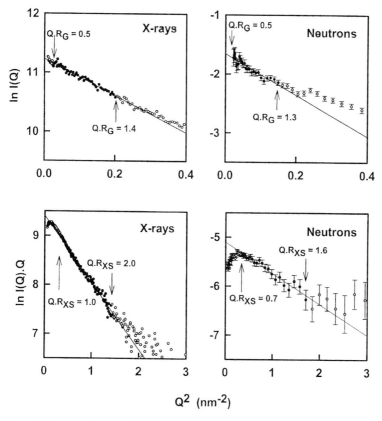

Figure 8 X-ray and neutron Guinier R_G and R_{XS} plots for factor VIIa of blood coagulation at concentrations of 1.9 and 3.9 mg/ml for X-rays and neutrons respectively. Filled circles between the $Q.R_G$ and $Q.R_{XS}$ ranges show the data points used to determine the R_G and R_{XS} values. Note that the R_{XS} value is much higher by X-ray scattering than by neutron scattering, mainly for reason of the hydration shell that is visible by X-rays but not by neutrons, and also because of the use of different solute–solvent contrasts.

The R_G values are a measure of the degree of particle elongation. Once the R_G data have been validated using the $I(0)$ values, the R_G data can be analysed to give their mean ± standard deviation. From the available X-ray and neutron data in several solute–solvent contrasts $\Delta\rho$, the magnitude of any concentration dependence of the R_G data in at least one contrast should be determined. If the R_G increases with dilution, interparticle effects may be important (when the protein molecules are too close together). If it decreases with dilution, dissociation may be occurring. The X-ray R_G value will correspond to a hydrated structure in a high positive solute–solvent contrast. The neutron R_G values correspond to an unhydrated structure and will depend on the solute–solvent contrast. In a full contrast variation study involving at least three contrasts, this dependence is described by the Stuhrmann equation (21):

$$R_G^2 = R_C^2 + \alpha.\Delta\rho^{-1} - \beta.\Delta\rho^{-2}$$

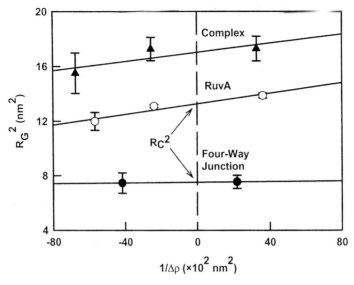

Figure 9 Stuhrmann contrast variation analysis for RuvA, four-way junction and their complex. The R_G^2 values are shown as a function of the reciprocal solute–solvent contrast difference $1/\Delta\rho$. Note that the intercepts are well separated and that the slopes of the graphs are positive and small for both RuvA and the complex, indicating that DNA is buried within the centre of a conformationally-rearranged complex.

where R_C is the R_G at infinite contrast (when $\Delta\rho^{-1}$ is zero), α is the radial distribution of scattering density fluctuations within the macromolecule, and β measures the effect on the R_G if the change of contrast causes a displacement of the centre of gravity of the scattering density (*Figure 9*). A similar analysis can be made using the R_{XS}^2 values.

In order to interpret the X-ray or neutron R_G values, the ratio of the observed R_G or R_C value with that calculated for a sphere of the same hydrated or unhydrated volume will indicate the degree of macromolecular elongation. If the cross-sectional analyses have been made, the combination of the R_G and R_{XS} analyses will give the longest macromolecular dimension D. In neutron contrast variation R_G analyses of protein–protein complexes and two-component centrosymmetric complexes with two very different scattering densities (e.g. protein and DNA), the resulting Stuhrmann analyses should give linear R_G^2 and R_{XS}^2 plots as a function of $\Delta\rho^{-1}$ since β is small and unmeasurable. The value of the Stuhrmann α can be compared with other known values to show whether the radial distribution of internal components is as expected (3), where the comparison is achieved by noting that α is proportional to R_C^2. If the two components in a complex are asymmetrically distributed with respect to their centre of mass, and have very different scattering densities from each other, the Stuhrmann plot is a parabola. Its curvature β provides the distance between the centres of mass of the two components (e.g. protein and lipid). The reader is referred to refs 1–3 for more details.

4.5 Distance distribution function analyses

If sufficient intensity data is obtained over about two orders of magnitude of Q, the scattering curve $I(Q)$, which is measured in reciprocal space with units of nm^{-1}, can be converted into the distance distribution function $P(r)$, which reveals the structure in real space with units of nm.

$$P(r) = \frac{1}{2\pi^2} \int_0^\infty I(Q) \, Qr \, \sin(Qr) \, dQ$$

$P(r)$ corresponds to the distribution of all the distances r between all the volume elements within the macromolecule (*Figure 10*). The transformation step is ordinarily unstable, since data points will be missing at the lowest Q values due to the beamstop, and are truncated at the highest Q values at the limit of the active detector region. In the classical Indirect Transformation Procedure of Glatter (1), this problem is dealt with by the use of 10–20 B-spline mathematical functions that are optimally fitted to the scattering curve in order to represent the data as a continuous analytical function, after which the B-splines are transformed to give $P(r)$ (*Figure 5*). An alternative and more automated procedure is the GNOM program of Svergun (22), in which a regularization procedure is employed with an automatic choice of the transformation parameter α to stabilize the $P(r)$ calculation.

In practise, the calculation of $P(r)$ curves involves testing a range of maximum assumed dimensions D_{max} for the macromolecule. For proteins, the final choice of this dimension is based on the knowledge that the $P(r)$ curve should

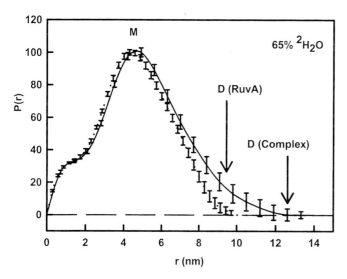

Figure 10 Distance distribution function curves $P(r)$ for RuvA and its complex with four-way junction. The $P(r)$ curves in 65% ^2H_2O correspond to the DNA matchpoint at which the four-way junction becomes invisible. It is clearly seen that the maximum dimension D is increased in the complex compared to unbound RuvA, and shows that a large conformational change has occurred in octameric RuvA when this binds DNA.

exhibit positive intensities and that it should be stable as this dimension D_{max} is increased beyond the macromolecular length D. From the P(r) calculation, an alternative calculation of the R_G and I(0) values is obtained that is based on the full Q range of the observed scattering curve, which complements the Guinier analysis made at low Q values. Clearly the R_G value from the P(r) curve should agree with the R_G from Guinier analyses. Another result from the P(r) curve is that, at the point at which P(r) becomes zero at large r, the value of r gives the length of the protein D. However, as this region of the P(r) curve contains the lowest intensities, the determination of D can be subject to error. The P(r) curve also gives the r value of one or more maxima M which corresponds to the most common scattering vector(s) within the macromolecule, and this can be useful. For example, if the r value at M is half that of D, the macromolecule is deduced to be spherical. If a protein possesses a large globular domain attached to an elongated rod-like domain, this will lead to a biphasic P(r) curve with narrow and broad profiles and giving two M values.

5 Structural modelling of scattering data

Modelling extends the interpretation of the scattering analyses by producing a three-dimensional model of the structure giving rise to the observed scattering curve. This work is best left until the scattering experiments have been fully processed. Many approaches have been developed over the years. Even though the information content of a scattering curve is limited (see Section 5.2), and unique structure determinations are not possible for reason of the random molecular orientations in solution and the spherical averaging of scattering data, modelling is nonetheless able to rule out classes of structures that are incompatible with the scattering curves. A most promising development in recent years involves the use of known atomic structures to strongly constrain the modelling of the scattering curve. On applying this constraint, it has been shown that only a small fraction of a large number of trial atomic models will fit the scattering curve, and the resulting structures can provide biologically useful information.

5.1 Sphere modelling of scattering curves: use of atomic structures

The use of small spheres (*Figure 11*) remains the most flexible and powerful approach to model scattering curves (1). The scattering curve I(Q) is calculated using Debye's Law adapted to spheres, essentially by computing all the distances r from each sphere to the remaining spheres and summing the results. For a single-density macromolecule, the X-ray and neutron scattering curve I(Q) can be calculated assuming a uniform scattering density for the spheres using the Debye equation adapted to spheres (23):

$$\frac{I(Q)}{I(0)} = g(Q)\left(n^{-1} + 2n^{-2}\sum_{j=1}^{m} A_j \frac{\sin Qr_j}{Qr_j}\right)$$

$$g(Q) = \frac{(3(\sin QR - QR \cos QR))^2}{Q^6 R^6}$$

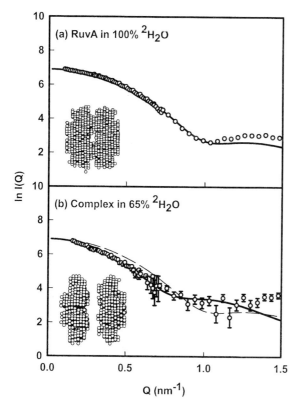

Figure 11 D22 neutron scattering curve fits for the RuvA octamer and its complex with four-way junction. The insets show the small sphere models used for the curve fits. (a) The best-fit model curve for the octamer (continuous line) is compared with experimental data for RuvA in 100% ^2H_2O. (b) The best curve fit for the octamer in the complex with four-way junction in 65% ^2H_2O, in which the DNA component is invisible. The dashed curve illustrates the large difference between complexed and unbound RuvA.

where $g(Q)$ is the squared form factor for the sphere of radius R, n is the number of spheres filling the body, A_j is the number of distances r_j for that value of j, r_j is the distance between the spheres, and m is the number of different distances r_j. If it is necessary to incorporate two different scattering densities 1 and 2 in the model, the above expression is modified as follows (23):

$$[I(Q)/I(0)] = g(Q) \, [n_1\rho_1^2 + n_2\rho_2^2 + 2\rho_1^2 \sum A_j^{11} (\sin Qr_j/Qr_j) + 2\rho_2^2 \sum A_j^{22} (\sin Qr_j/Qr_j)$$
$$+ 2\rho_1\rho_2 \sum A_j^{12} (\sin Qr_j/Qr_j)] \, (n_1\rho_1 + n_2\rho_2)^{-2}$$

The model is constructed from n_1 and n_2 spheres of two different densities ρ_1 and ρ_2. A_j^{11}, A_j^{22}, and A_j^{12} are the number of distances r_j for that increment of j between the spheres 1 and 1, 2 and 2, and 1 and 2 in that order. The summations $\sum$ are performed for j = 1 to m, where m is the number of different distances r_j.

For modelling an atomic structure, the diameter of the small spheres has to be significantly less than the resolution of the scattering experiment (*Figure 11*), so that the form factor $g(Q)$ becomes almost independent of Q. The coordinates of an atomic structure or model are converted to spheres by placing all the atoms within a three-dimensional grid of cubes of side about 0.6 nm. A cube is included as a sphere in the model if it contains sufficient atoms above a specified cut-off level such that the total volume of all the cubes that results from this cut-off equals that of the unhydrated protein calculated from the sequence in the case of neutron modelling (6). If residues are missing in the crystal structure or a homologous crystal structure is used, the ensuing volume discrepancy can be compensated by adjustment of the cut-off value for sphere generation in order to reach the correct full volume. X-ray curve modelling requires hydrated structures. A hydration shell is well-represented by 0.3 g of water/g glycoprotein and an electrostricted volume of 0.0245 nm^3 per bound water molecule, and corresponds to a water monolayer surrounding the protein surface. In comparison, the volume of a free unbound water molecule is 0.0299 nm^3, and is therefore less electron-dense. The simplest way to hydrate the cube models is to increase the size of the cubes so that their total volume matches the hydrated volume, however this can significantly distort the macromolecular structure, for example, if there is a void space at its centre (*Figure 1*). An improved approach is to add a layer of additional spheres evenly over the surface of the dry sphere model, where the total number of spheres gives the hydrated volume (24).

For modelling a multidomain structure, for which separate crystal structures are available for the domains within the structure, but the overall domain structure is unknown, the modelling analyses are powerfully constrained by the known steric connections between the domains (4). In this approach, a large number of complete multidomain structures are made with connections that are randomly generated using molecular graphics macros. The models are then compared using automated script files against the scattering data by a trial-and-error process in order to identify the structures that are consistent with the scattering curves. The essential stage of this procedure is the use of three filters to eliminate unsatisfactory models:

(a) The process of model creation can result in steric overlap between the domains, so the number of spheres in each model produced from the grid transformation is compared to that expected from the dry volume. The model is retained only if it possesses at least 95% of the expected number of spheres.

(b) Models are retained if the modelled R_G and R_{XS} values are within 5% or $\pm$ 0.3 nm from the experimental values.

(c) All the models that pass the first two filters are next classified using a goodness-of-fit R-factor $= 100 * \Sigma \ |I(Q)_{exp} - I(Q)_{cal}| \ / \ \Sigma \ |I(Q)_{exp}|$ to measure the agreement between the calculated and observed scattering curve.

A large spreadsheet is prepared with the geometrical steps used to define each model, the number of spheres in it, the R_G and R_{XS} values, and the R-factor values. This is used to optimize the best cut-off filters, sort the models, and identify the best ones that identify the unknown structure.

For both X-ray and neutron cameras, R_G analyses at low Q are not affected by instrumental corrections for wavelength spread or beam divergence as these generally employ an ideal pin-hole configuration that does not lead to geometrical distortion of the beam. At large Q on synchrotron X-ray cameras, no instrumental corrections are needed as well (7, 8). At large Q on neutron cameras, instrumental corrections are now required for reason of the physically large dimensions of the camera (9–12). A Gaussian function to allow for wavelength spread and beam divergence has been used for data from both D11/D22 and LOQ. In addition, it is noticeable that at large Q the neutron fits deteriorate (*Figure 1*), and this indicates a small residual flat background from incoherent scatter from the protons in the sample that needs correction.

5.2 Spherical harmonics modelling

Molecular shapes can be determined directly from a scattering curve in a model-independent manner if a full scattering curve has been acquired. If it is assumed that the protein can be represented by a compact structure of uniform scattering density, its shape can be defined by a surface, i.e. by a molecular envelope function $F(\theta,\phi)$ which has a scattering density of 1 inside it and 0 outside it. $F(\theta,\phi)$ can be expanded into a series of spherical harmonics $Y_{lm}(\theta,\phi)$ (25):

$$F(\theta,\phi) = R_0 \sum_{l=0}^{L} \sum_{m=-1}^{l} f_{lm} Y_{lm}(\theta,\phi)$$

where R_0 is a scale factor and f_{lm} are complex multipole coefficients that are related to coefficients of the power series describing the experimental scattering curve. The f_{lm} coefficients can be determined by the automated minimization of a R-factor between the calculated and observed data points in the scattering curve (26). The number of harmonics terms L that can be used in the above summation is determined by the number of degrees of freedom (or the minimum number of independent parameters) that describes the experimental scattering curve. This is given by Shannon's sampling theorem as $D_{max} Q_{max} / \pi$ (1), e.g. this is 20 for a protein of length $D = 25$ nm for which a scattering curve was measured out to $Q = 2.5$ nm^{-1}. If L is 4, which is typical for the determination of a molecular envelope for a protein, there are $(L + 1)^2$ coefficients to be determined, of which 6 are arbitrary rotations and translations that do not affect the scattering curve, leaving 19 coefficients. If molecular symmetry is present, the number of coefficients to be determined is twofold or threefold less if a dimer or trimer is considered, and higher values of L are permitted. It is possible to detect conformational changes on the binding of a small ligand by inspection of the molecular surfaces before and after ligation. If crystal structures are available, reasonable agreement between the molecular surface and the coordinates can usually be obtained by superimposition by eye using molecular graphics.

5.3 Other modelling using spheres, shells, or ellipsoids

In the absence of a crystal structure to constrain the curve modelling, sphere modelling can be used. This proceeds by determining the dimensions of the simplest triaxial object constructed from assemblies of spheres that are varied in their dimensions and arrangement until these account for the scattering curve. Typically curve fitting starts from the R_G value of the sphere model which is compared with the scattering curve at low Q (i.e. the Guinier region), and progressively extending the fits to large Q by a trial-and-error adaptation of the sphere model. A new development in this area involves the use of genetic algorithms, in which an assembly of an arbitrarily-chosen array of spheres is refined to fit the observed scattering curve (27, 28). A large number of sphere arrangements are generated, in which spheres at individual positions in the array are added or deleted during the optimisation procedure in order to fit the experimentally-observed curve. The best-fit arrangement of spheres may provide a useful approximation of the protein structure.

Depending on the known features of the macromolecule to be modelled, other modelling approaches employ spherical multi-shell models or prolate or oblate ellipsoids. In all these approaches, the incorporation of other structural information lessens the impact of the ambiguity caused by the lack of a unique structure determination by scattering. For example, subunit structures within the macromolecule may be known from sequences, or electron microscopy images can be incorporated as constraints.

5.4 Comparison with sedimentation and diffusion coefficients

Analytical ultracentrifugation will provide sedimentation and diffusion coefficients and dynamic light scattering will also provide diffusion coefficients (see Chapters 4 and 5). These are analogous to R_G values in being single-parameter values that provide a measure of macromolecular elongation. Sphere models (commonly called 'bead' models in this field) can be used to fit these hydrodynamic parameters (29). A survey of methods for the construction of 'bead' models has been recently reported (30), in which these models fall into one of three types. The earliest models were constructed from a small number of large spheres of different sizes. This approach is nowadays replaced by models based either on a hollow shell of equally-sized spheres that represents the macromolecular surface or filling the whole protein volume by small spheres. The computing overheads of hydrodynamic programs such as *GENDIA*, *HYDRO*, and *SOLPRO* are more substantial than those used for calculating an R_G value. Nonetheless, starting from 1992, the same hydrated 'filled' sphere models used for X-ray fits have been used to calculate sedimentation and diffusion coefficients (24, 31). This is an important development as the use of identical models unifies both the scattering and ultracentrifuge analyses. The consistency of the modelling from both disciplines constitutes strong support for the outcome of modelling.

6 Conclusions and worked examples

Solution scattering is a low- to medium-resolution structural discipline that is applicable to a broad range of protein–ligand complexes. As described above, modern solution scattering methods benefit from the availability of an extensive range of powerful cameras with large Q ranges and good counting statistics at high-flux X-ray and neutron sources. Data processing using UNIX script files or EXCEL macros to handle many experimental data files considerably eases the Guinier and P(r) analyses, and provides a detailed characterization of the system of interest. Data interpretation is much improved by the ability to perform molecular modelling using tight constraints based on available atomic structures in the Protein Data Bank. By this technology, significant new insights on structure–function relationships are frequently obtained, even in cases where a crystal structure determination has already provided much information. Once the data collection is completed and interpreted, the time-scale for project completion is a matter of several weeks. Some worked examples are summarized below.

6.1 Small ligand binding to a protein (AmiC)

An interesting case of scattering applied to a protein–small ligand interaction is provided by the cytoplasmic protein AmiC, which is the amide sensor protein and negative regulator of the five-gene amidase operon in an important opportunistic pathogen *Pseudomonas aeruginosa*. It is a structural analogue of the periplasmic binding proteins that mediate the process of small molecule transport into bacteria (32). Its ligand substrate is acetamide, while butyramide is an anti-inducer of AmiC that binds 100-fold more weakly. As crystal structures show that large domain rotations of 18° to 52° occur between the ligand-free and ligand-bound forms of the periplasmic proteins, and these are reflected in predicted R_G decreases of 0.08–0.21 nm on ligand binding, it was of interest to know whether the same changes could be observed in the cytoplasmic analogue AmiC. Another interest of this work is that periplasmic binding proteins are generally monomeric in solution, while it had been proposed that AmiC formed an antiparallel back-to-back dimer as viewed in its crystal structure (33). The ensuing scattering study of AmiC produced unexpected results (34). Dilution series in the presence of each of the two ligands showed pronounced concentration dependencies, with interparticle interference effects at the highest concentrations, and sample dissociation at the lowest ones. It was deduced that AmiC existed as a monomer–trimer equilibrium in solution, and not as a stable dimer, and that the presence of acetamide or butyramide ligands affected the amounts of monomer or trimer formed. No conformational change could be detected between the acetamide- or butyramide-bound forms. Modelling based on the AmiC crystal structure showed that good curve fits could be obtained with a monomeric and a symmetric trimeric structure for data obtained at 1.0 mg/ml and 6.8 mg/ml respectively. The outcome of this study gave the useful result that the solution properties of AmiC are different from those of typical periplasmic binding proteins.

Another example of the application of scattering was made for iron binding to a two-lobed protein transferrin, where extensive use was made of spherical harmonics. This work showed that a clear conformational change occurred between the surfaces of the iron-free and iron-bound states, similar to the large changes seen in the crystal structures of periplasmic binding proteins (35).

6.2 Ligand binding to a homodimeric complex (serum amyloid P component)

Human serum amyloid P component (SAP) is a normal plasma glycoprotein and the precursor of amyloid P component which is a universal constituent of the abnormal tissue deposits in amyloidosis (Figure 1). The pentamer is formed from five identical glycosylated subunits that are non-covalently associated in a disc-like configuration with cyclic symmetry, for which a crystal structure is available (36). One face of the disk contains an α-helix (the 'A-face') and contains five glycosylation sites, while the other contains the Ca^{2+} binding sites (the 'B-face'). In the presence of Ca^{2+} and a monosaccharide (methyl 4,6-O-(1-carboxyethylidene)-β-D-galactopyranoside: abbreviated MOβDG), SAP forms pentamers in solution. In the presence of EDTA, SAP forms a very stable decamer composed of two pentameric discs interacting face-to-face. X-ray and neutron scattering data showed that pentameric or decameric ring structures for SAP in solution were readily distinguished from each other (Figure 1) (37). This is difficult to demonstrate by gel filtration as the latter is a qualitative method.

Interestingly, given that the SAP decamer resisted extensive efforts to crystallize it, scattering was able to produce the first molecular structure for this (24). It was found that the two pentamers associate through contacts at their A-faces, and not their B-faces. This result was derived from the $I(0)/c$ values which had resulted in an abnormally low molecular weight calculation for the decamer. As the crystal structure of the pentamer showed that 20 Trp residues were close to the A-face, this meant that, if the two A-faces formed the contacts in the decamer, the proximity of 40 Trp residues close to each other in the decamer would alter the 280 nm absorption coefficient of the decamer relative to the pentamer, and in turn the determination of the $I(0)/c$ value. This analysis was confirmed by ultra-violet difference and fluorescence spectroscopy. It was also supported by modelling calculations in which structures for the decamer were determined using the pentamer crystal structure. Slightly better curve fits were obtained for the decamer with the A-faces in contact with each other (Figure 1). In this modelling, the two alternative decamer structures were distinguished from each other by the presence of 10 oligosaccharide chains on the two A-faces but not on the B-face. The two models accordingly resulted in different R_G values for the decamer. Thus the scattering data on SAP provided novel insights on its decameric association, and the addition of ligand caused it to dissociate into pentamers. In fact, since the use of other ligands are able to cause the reversed B-face-associated decamer to be formed (38), scattering has great potential for rapid serial studies of SAP–drug interactions, in which the medium-resolution structures of many SAP–ligand complexes can be identified for more detailed subsequent investigation.

6.3 A heterodimeric complex (factor VIIa–tissue factor complex)

Factor VIIa (FVIIa) is a four-domain serine protease that forms a tightly-bound complex with a two-domain membrane-bound tissue factor (TF) in order to trigger blood coagulation after tissue factor is exposed to plasma. As the FVIIa-TF complex catalyses the activation of other multidomain serine proteases, its solution structure is of great interest. In distinction to the preceding examples, the TF ligand is large enough to be detected by scattering (FVIIa mass of 51 400 Da; TF mass of 24 800 Da). Accordingly Guinier and $P(r)$ analyses were reported for the two proteins and their complex, confirming the analyses by molecular weight calculations (*Figures 2* and *8*). Knowledge of homologous crystal structures for the domains of FVIIa was used to show that these domains formed an extended arrangement in solution. The length calculations from scattering showed that no significant increases were seen in the R_G or length D values on formation of the complex. Thus the observed lengths of FVIIa (10.3 nm), TF (7.7 nm), and the complex (10.2 nm) were readily explained by a compact structure for the complex, in which FVIIa and TF formed extensive side-by-side contacts with each other (39). These results were confirmed by the subsequent crystal structure of the complex (40). The automated modelling of the scattering curve verified the extended domain structure of FVIIa in solution (41), which is useful as only a three-domain form of FVIIa has been crystallized to date (42). The application of automated modelling to determine a structure for the heterodimeric FVIIa-TF complex produced a large number of compatible structures. Here, additional information is needed to reduce the number of structures for the complex to give a useful outcome, and this is readily incorporated into a well-designed modelling search.

The specific deuteration of one protein in a protein–protein complex permits the study of the arrangement of the two proteins in the complex to be studied by neutron contrast variation. This is a particularly useful approach for large uncrystallizable protein complexes whose shape can be determined by electron microscopy, but for which information is needed on the arrangement of individual proteins within the complex. The validity of this approach is well justified by the recent crystal structure for the 30S ribosomal subunit, in which the positions of the ribosomal proteins were in good agreement with those for deuterated proteins determined by neutron scattering in 1981–1987 (43).

6.4 A protein–DNA complex (RuvA–Holliday junction complex)

Holliday junctions are crossover DNA structures with four arms that enable the pairing and exchange of strands between two homologous DNA molecules. In *E. coli*, the main pathway for the processing of Holliday junctions is the RuvABC system, in which RuvA is a disk-like tetrameric protein (mass 82 800 Da) that forms a face-to-face complex with the four-way Holliday junction. This protein–ligand complex was studied by neutron contrast variation, where the DNA ligand (64 base pairs; mass 39 400 Da) is large enough to be seen by scattering except when the buffer contains 65% ^2H_2O, the matchpoint of DNA (44). The free protein is an octamer formed by the face-to-face association of RuvA. Its oligomeric structure was readily identified by molecular weight calculations (*Figure 2*) and curve fits based on the crystal structure of the tetramer (below). Thus in the complex, the R_G^2 values showed that DNA binds between the two tetramers which had moved apart to accommodate the DNA (*Figure 9*). The slopes α were positive and small compared to other complexes such as nucleosome core particles where DNA is located outside a central core of protein. This structural change was clearly seen in *P(r)* curves recorded in 65% ^2H_2O when only the protein component is visible (*Figure 10*). Unlike the case of the heterodimeric FVIIa-TF complex above, scattering curve modelling is much facilitated by the four-fold symmetry of RuvA, and the detail was precise enough to show that salt bridge formation was the likely reason for octamer formation. Hence, by modelling of the 65% ^2H_2O data using small spheres derived from the RuvA crystal structure, it was shown that the RuvA tetramers had separated by about 1–2 nm in the complex (*Figure 11*). This solution scattering work was important to complement two different tetramer–DNA and octamer–DNA crystal structures for the RuvA–Holliday junction complex (45, 46) by showing that an octameric complex can exist in solution.

Acknowledgements

Scattering software is available from the author on request (Email: steve@rfhsm.ac.uk). The beamtime work has been supported by the BBSRC and the Wellcome Trust. Dr W. Bras (SRS), Dr J. G. Grossman (SRS), Dr R. K. Heenan (ISIS), Dr S. M. King (ISIS), Mrs S. L. Slawson (SRS), and Dr P. A. Timmins (ILL) are thanked for support and advice during many scattering experiments during the past decade.

References

1. Glatter, O. and Kratky, O. (ed.) (1982). *Small angle X-ray scattering*. Academic Press, New York.
2. Perkins, S. J. (1988). *Biochem. J.*, **254**, 313.
3. Perkins, S. J. (1988). *New comprehensive biochemistry*, Volume 18B (Part II), pp. 143–264.
4. Perkins, S. J., Ashton, A. W., Boehm, M. K., and Chamberlain, D. (1998). *Int. J. Biol. Macromolecules*, **22**, 1.
5. Boehm, M. K., Woof, J. M., Kerr, M. A., and Perkins, S. J. (1999). *J. Mol. Biol.*, **286**, 1421.
6. Perkins, S. J. (1986). *Eur. J. Biochem.*, **157**, 169.
7. Nave, C., Helliwell, J. R., Moore, P. R., Thompson, A. W., Worgan, J. S., Greenall, R. J., et al. (1985). *J. Appl. Crystallogr.*, **18**, 396.
8. Towns-Andrews, E., Berry, A., Bordas, J., Mant, G. R., Murray, P. K., Roberts, K., et al. (1989). *Rev. Sci. Instrum.*, **60**, 2346.
9. Lindner, P., May, R. P., and Timmins, P. A. (1992). *Physica B*, **180**, 967.
10. Ghosh, R. E., Egelhaaf, S. U., and Rennie, A. R. (1998). *A computing guide for small-angle scattering experiments*. Institut Laue Langevin Publication 98GH14T.

11. Heenan, R. K., Penfold, J., and King, S. M. (1997). *J. Appl. Crystallogr.*, **30**, 1140.
12. King, S. M. and Heenan, R. K. (1996). *The LOQ instrument handbook*. Rutherford-Appleton Laboratory Publication RAL-TR-96-036.
13. Sambrook, E. F., Fritsch, E. F., and Maniatis, T. (ed.) (1989). *Molecular cloning: a laboratory manual*, 2nd edn. Cold Spring Harbor Laboratory Press, New York.
14. Converse, C. A. and Skinner, E. R. (1992). *Lipoprotein analysis: a practical approach*. IRL Press, Oxford.
15. Jacrot, B. and Zaccai, G. (1981). *Biopolymers*, **20**, 2413.
16. Folkhard, W., Geercken, W., Knorzer, E., Mosler, E., Nemetschekgansler, H., Nemetschek, T., *et al.* (1987). *J. Mol. Biol.*, **193**, 405.
17. Durchschlag, H. and Zipper, P. (1988). *FEBS Lett.*, **237**, 208.
18. König, S., Svergun, D., Koch, M. H. J., Hübner, G., and Schellenberger, A. (1993). *Eur. Biophys. J.*, **22**, 185.
19. Hjelm, R. P. (1985). *J. Appl. Crystallogr.*, **18,** 452.
20. Kratky, O. (1963). *Prog. Biophys. Chem.*, **13**, 105.
21. Ibel, K. and Stuhrmann, H. B. (1975). *J. Mol. Biol.*, **93**, 255.
22. Semenyuk, A. V. and Svergun, D. I. (1991). *J. Appl. Crystallogr.*, **24**, 537.
23. Perkins, S. J. and Weiss, H. (1983). *J. Mol. Biol.*, **168**, 847.
24. Ashton, A. W., Boehm, M. K., Gallimore, J. R., Pepys, M. B., and Perkins, S. J. (1997). *J. Mol. Biol.*, **272**, 408.
25. Stuhrmann, H. B. (1970). *Acta Crystallogr.*, **A26**, 297.
26. Svergun, D. I. and Stuhrmann, H. B. (1991). *Acta Crystallogr.*, **A47**, 736.
27. Chacon, P., Moran, F., Diaz, J. F., Pantos, E., and Andreu, J. M. (1998). *Biophys. J.*, **74**, 2760.
28. Svergun, D. I. (1999). *Biophys. J.*, **76**, 2879.
29. Perkins, S. J. (1989). In *Dynamic properties of biomolecular assemblies* (ed. S. E. Harding and A. J. Rowe), pp. 226–45. Royal Society of Chemistry, London.
30. Carrasco, B. and de la Torre, J. G. (1999). *Biophys. J.*, **76**, 3044.
31. Smith, K. F., Harrison, R. A., and Perkins, S. J. (1992). *Biochemistry*, **31**, 754.
32. Tam, R. and Saier, M. H. Jr. (1993). *Microbiol. Rev.*, **57**, 320.
33. Pearl, L. H., O'Hara, B., Drew, R. E., and Wilson, S. A. (1994). *EMBO J.*, **13**, 5810.
34. Chamberlain, D., O'Hara, B. P., Wilson, S. A., Pearl, L. H., and Perkins, S. J. (1997). *Biochemistry*, **36**, 8020.
35. Grossman, J. G., Crawley, J. B., Strange, R. W., Patel, K. J., Murphy, L. M., Neu, M., *et al.* (1998). *J. Mol. Biol.*, **279**, 461.
36. Emsley, J., White, H. E., O'Hara, B. P., Oliva, G., Srinivasan, N., Tickle, I. J., *et al.* (1994). *Nature*, **367**, 338.
37. Wood, S. P., Oliva, G., O'Hara, B. P., White, H. E., Blundell, T. L., Perkins, S. J., *et al.* (1988). *J. Mol. Biol.*, **202**, 169.
38. Hohenester, E., Hutchinson, W. L., Pepys, M. B., and Wood, S. P. (1997). *J. Mol. Biol.*, **269**, 570.
39. Ashton, A. W., Kemball-Cook, G., Johnson, D. J. D., Martin, D. M. A., O'Brien, D. P., Tuddenham, E. D. G., *et al.* (1995). *FEBS Lett.*, **374**, 141.
40. Banner, D. W., D'Arcy, A., Chene, C., Winkler, F. W., Guha, A., Konigsberg, W. H., *et al.* (1996) *Nature*, **380**, 41.
41. Ashton, A. W., Boehm, M. K., Johnson, D. J. D., Kemball-Cook, G., and Perkins, S. J. (1998). *Biochemistry*, **37**, 8208.
42. Pike, A. C. W., Brzozowski, A. M., Roberts, S. M., Olsen, O. H., and Persson, E. (1999). *Proc. Natl. Acad. Sci. USA*, **96**, 8925.
43. Clemons Jr., W. M., May, J. L. C., Wimberly, B. T., McCutcheon, J. P., Capel, M. S., and Ramakrishnan, V. (1999). *Nature (London)*, **400**, 833.

44. Chamberlain, D., Keeley, A., Aslam, M., Arenas-Licea, J., Brown, T., Tsaneva, I. R., *et al.* (1998). *J. Mol. Biol.*, **285**, 385.
45. Roe, S. M., Barlow, T., Brown, T., Oram, M., Keeley, A., Tsaneva, I. R., *et al.* (1998). *Mol. Cell*, **2**, 361.
46. Hargreaves, D., Rice, D. W., Sedelnikova, S. E., Artymuik, P. J., Lloyd, R. G., and Rafferty, J. B. (1998). *Nature Struct. Biol.*, **5**, 441.

Chapter 10

Isothermal titration calorimetry of biomolecules

Ronan O'Brien and John E. Ladbury
Department of Biochemistry and Molecular Biology, University College London, Gower Street, London WC1E 6BT, UK.

Babur Z. Chowdhry
School of Chemical and Life Sciences, University of Greenwich, Wellington Street, Woolwich, London SE18 6PF, UK.

1 Introduction

One of the primary goals of contemporary molecular biology is to gain an understanding of the molecular basis of specificity and recognition phenomena, between proteins and ligands. Proteins can specifically recognize, and reversibly bind to small molecular weight target molecules, other macromolecules (e.g. proteins, nucleic acids) and to macromolecular assemblies. Characterization of these interactions necessarily involves investigation of the inter-relationship between function, structure (including dynamics), kinetics, and energetics (i.e. thermodynamics) of a system under defined physico-chemical conditions. In this chapter we focus on the determination of the latter of these quantities. The measurement of thermodynamic parameters is important because all reversible biomolecular interactions involve a redistribution of non-covalent bonds. The most experimentally accessible of the thermodynamic quantities occurring on a protein going from the free (unbound) to the bound state is the heat (enthalpy) uptake or release.

A wide variety of experimental methods can be, and still are, employed for the *indirect* (non-calorimetric) determination of thermodynamic quantities for the interaction of biomolecules. These inevitably involve the calculation of the thermodynamic quantities from theoretical relationships. For example, the enthalpy changes can be determined from the temperature dependence of the equilibrium binding (or dissociation) constant (van't Hoff analysis). There are severe limitations associated with indirect approaches of this nature. The recent development of high sensitivity calorimetric instrumentation allows precise values for the change in enthalpy ($\Delta H°$) to be directly obtained.

All chemical reactions, by definition, involve changes in enthalpy. Theoretically, therefore, calorimetric techniques are of universal applicability, provided that enthalpy changes for the reaction can be measured precisely and accurately. Isothermal titration calorimetry (ITC) is the only technique currently available that can be used to measure directly the interaction enthalpy for almost any bimolecular process. Within certain limits, and depending upon experimental design, this technique can also be used to obtain the equilibrium constant (K_B) and the molar ratio of interacting species (n) for an inter (bi-) molecular binding interaction. In principle, it is therefore possible to obtain a comprehensive thermodynamic profile for a protein–ligand binding reaction (including change in free energy and entropy, $\Delta G°$ and $\Delta S°$ respectively) under defined solution conditions. If experiments are conducted as a function of temperature, using standardized experimental conditions the value of the change in heat capacity, $\Delta C_P°$ for the system may also be obtained.

2 The principle of operation

2.1 Instrumentation

Until the 1970s the majority of bio-calorimetric analyses were performed in laboratories equipped to build their own instrumentation (1). Commercial batch and flow calorimeters then became available, e.g. from LKB (Sweden); now Thermometric AB (Spjutvägen 5A, S-175 61 Järfälla, Sweden; E-mail: mail@thermometric.se; http://www.thermometric.com). These types of instruments were widely used, for example, in the field of enzyme kinetics as well as for the investigation of the kinetics and energetics of cellular metabolism and cell–ligand (drug) interactions. The present generation of such instruments, e.g. the TAM (manufactured by Thermomeric) continue to be used for investigations of this nature.

Since the early-eighties commercially available microcalorimeters (both DSC and ITC) have been developed. The term 'isothermal microcalorimeter' is not well defined, but is now commonly used for calorimeters designed for work in the microwatt range conducted under (essentially) isothermal conditions. 'Nanocalorimeters' (indicating a detection limit approaching one nanowatt) are included in this group. Substantial developments in ITC equipment have occurred over the last decade and several easy-to-use instruments are now commercially available.

From the point of view of heat measurement it is possible to divide calorimeters into three main groups: adiabatic, heat conduction, and power compensation calorimeters (2–5). In an ideal adiabatic calorimeter there is no heat exchange between the calorimetric vessel and its surroundings. Adiabatic conditions are usually obtained by placing an 'adiabatic shield' between the vessel and the surroundings. During a measurement the temperature difference between the vessel and the shield is kept at zero. The heat which is evolved, or absorbed, during an experiment with an ideal adiabatic calorimeter is equal to the product between the temperature change and the heat capacity of the calorimetric vessel. In a heat conduction calorimeter heat released (or absorbed) in the reaction vessel is allowed to flow to (or from) a surrounding heat sink, e.g. an aluminium block. Normally, a thermopile positioned between the sample container and the heat sink is used as a sensor for the heat flow. In a power compensation calorimeter the thermal power from an exothermic process is balanced by a cooling power, normally by use of water or Peltier effect cooling. For endothermic processes, compensation can be achieved by reversing the Peltier effect current or by use of an electrical heater.

Most isothermal microcalorimeters in current use are of the heat conduction type, for example the microcalorimetric systems marketed by CSC (earlier Hart Scientific; USA), Setaram (France), and Thermometric (Sweden). In contrast, Microcal's (USA) titration microcalorimeters uses the same principle as an adiabatic shield DSC. The temperature is allowed to increase, very slowly, during an experiment and the heat evolution from a reaction is balanced by a corresponding change in the heating rate (power compensation, see below).

Isothermal microcalorimeters form a heterogeneous group of instruments and many different designs have been described. Regardless of the calorimetric principle used, most microcalorimeters are designed as twin or differential instruments. The sample and reference cells, which are manufactured with an inert material, should preferably be identical with respect to geometry and physical properties, particularly with respect to heat capacity and thermal conductance. In some cases microcalorimetric reaction vessels are taken out from the calorimeter for cleaning and charging operations ('insertion vessels'); alternatively vessels are permanently mounted in the heat sensitive zone of the calorimeter.

To our knowledge, at the time of writing, the VP-ITC (Microcal Inc., 22 Industrial Drive E., Northampton, MA 01060-2327, USA; http://www.microcalorimetry) and the CSC nano-ITC calorimeter (Calorimetry Sciences Corp., 155 West 2050 South, Spanish Fork, Utah 84660, USA; email: CalSCorp@aol.com) are currently the most sensitive ITC instruments commercially available, for the study of macromolecule–ligand interactions. Experiments can be carried out relatively quickly, typically in about 1–2 hours, due to rapid equilibration, allowing a respectable sample throughput rate compared to other techniques for K_D determination. Baselines are stable, typically $\leq 0.04\ \mu J.s^{-1}$ constancy over a 15 minute period, with stirring. The VP-ITC has a useful temperature range of 2–80 °C. Computer control and automation makes this instrument very user friendly, and instrument operation is straightforward. The ORIGIN software used for data acquisition and analysis is very powerful and versatile.

In instruments of this type the calorimeter measures the extent of binding of one molecule (component A) to another (component B) using the heat of interaction as the probe. ITC is a differential technique having both a reference and a sample cell which are maintained at a constant and identical temperatures at least 5 °C above the surroundings. The reference cell is usually filled with water or buffer and plays no part in the titration. Component A is placed in a device from which small injections can be made. Component B is in the calorimetric cell. The temperature of this sample cell is accurately measured and

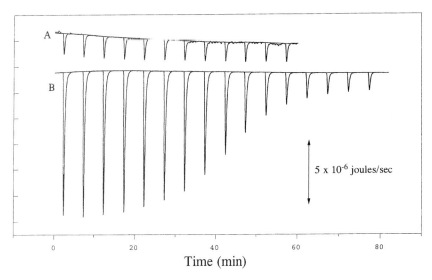

Figure 1 Raw data output from the MCS (MicroCal Inc.) instrument for a titration (B) and the heats of dilution (A). As can be seen the heats of dilution into buffer are equal to the heats of the last few injections indicating that the binding sites are fully saturated at the end of the titration.

continuously compared to that of the reference cell. A feedback mechanism via a thermopile/thermocouple system is used to ensure that whatever temperature change occurs in the sample cell heat is supplied (or restricted) to retain thermal equilibrium between the cells, i.e. return the sample cell temperature to that of the isothermal reference cell. Thus, when an injection of a small volume of A is made into B if, for example, the interaction is endothermic more heat is required to maintain isothermal conditions. The electrical energy required to maintain this parity upon injection of ligand is measured and converted into a heat of interaction.

In most cases the experiment is designed such that the initial injection of A (from a syringe) into B (in the calorimeter reaction cell) results in the binding of most of the added ligand, generating the maximal heat associated with the system (i.e. the total enthalpy of the interaction). On subsequent injections less of B is available for binding resulting in a reduction in the formation of the complex and the associated heats of interaction. Eventually with the continued injection of A, all of B is found in the complex and no further heat of binding is observed. Raw data for a typical titration experiment is shown in *Figure 1*.

The heats of each individual injection, shown in *Figure 1* are then integrated with respect to time and plotted against molar ratio of components. These are then fitted to the appropriate binding model enabling the calculation of the enthalpy, the equilibrium dissociation or association (binding) constant (K_D or K_B respectively; where $K_D = 1/K_B$) and stoichiometry (see *Figure 2*). If the K_B and ΔH of an interaction are known then the free energy (ΔG) and the entropy (ΔS) can be determined (see below).

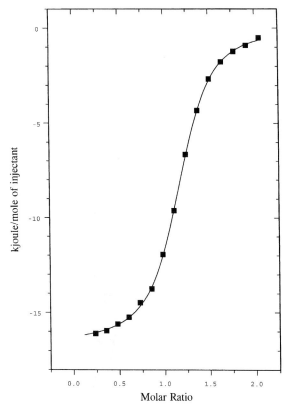

Figure 2 ITC isotherm of the integrated raw data from *Figure 1*. The data is fit using the ORIGIN software to a model for one independent binding event. The three variables are best fit to the following values: stoichiometry, n = 1.14; equilibrium binding constant, $K_B = 3.6 \times 10^5$ M^{-1}; enthalpy of interaction, ΔH_{obs} = 16.585 kJ.mol^{-1}.

The thermodynamic parameters determined from an ITC experiment for any equilibrium binding event contain the sum of all the individual changes in non-covalent interactions occurring on changing the state of the system. For example, the ΔH measured is directly related to the number and strength of bonds formed, or broken on going from the free to the bound state including those associated with the solvent. As a result the thermodynamic parameters derived directly from the ITC experiment are often more accurately described as the observed or apparent parameter and given the subscript 'obs', e.g. ΔH_{obs}. Some of the contributions to and the method of deconvolution of the ΔH_{obs} will be described later in this chapter.

2.2 Instrumental calibration

Calorimetric techniques are, by their very nature, uniquely vulnerable to systematic experimental errors. These errors are usually caused by contributions from processes accompanying the interaction being investigated. Moreover such errors are difficult to control and include, e.g. evaporation, condensation, incomplete mixing, non-specific adsorption, as well those due to spurious (side-reaction) chemical processes. Errors in the calibration (and time constant of the instrument in relation to the kinetics of the process) are another possible contribution to systematic errors in calorimetric studies. Although often difficult, attempts can be made to identify, control and minimize such errors.

Unfortunately there is no universally accepted method of calibration in use for ITC. There are several methods available for calibration. One commonly used method is that of electrical calibration. Here the calorimetric signal is standardized by a pulse of an accurately known quantity of heat (Q; or thermal power, i.e. dQ/dt) released by an electrical heater positioned in or close to the calorimetric vessel (i.e. cell) used in the ITC experiment. Other calibration methods used include acid–base interactions, e.g. the interaction of HCl with NaOH (contamination with CO_2 should be avoided). Two ligand reactions have been proposed for use as calibration reactions. The interaction of Ba^{2+} and 18-crown-6 (1,4,7,10, 13,26,-hexaoxacyclooctadecane) is one such reaction for which the pertinent parameters at 298.15 K are: $\Delta H = -31.42 \pm 0.20$ kJ.mol^{-1}, $K_B = 5900 \pm 200$ M^{-1}, $\Delta C_p = 126$ JK^{-1}.mol^{-1} (288–310 K) using 0.001–0.01 mol.l^{-1} for the crown ether in the calorimetric cell (vessel) and 0.01–0.1 mol.l^{-1} for barium chloride in the injection syringe. The binding of 2' CMP to bovine pancreatic ribonuclease has also been proposed as a calibrant (6).

3 Full thermodynamic characterization of biological interactions

This section describes how a full thermodynamic characterization of a biological interaction can be obtained from ITC data and allows the researcher to gain some insight into the information which can be obtained from such analysis. Clearly an interpretation on an atomic level of what is occurring in an ITC experiment is complex. The combination of ITC data and high resolution structural detail provide a powerful tool to be able to begin this process.

3.1 Derivation of thermodynamic parameters

As described in Section 2.1 the observed enthalpy, ΔH_{obs}, includes contributions not only from the binding, ΔH_{bind}, but also all associated events. These may include heats from conformational changes, ΔH_{conf}, and the combined heats of ionization, ΔH_{ion}, of the buffer and the interactants. This can be expressed as:

$$\Delta H_{obs} = \Delta H_{bind} + \Delta H_{conf} + \Delta H_{ion} \quad [1]$$

This is often written as:

$$\Delta H_{obs} = \Delta H_{int} + \Delta H_{ion} \quad [2]$$

if $\Delta H_{int} = \Delta H_{bind} + \Delta H_{conf}$; where ΔH_{int} is the intrinsic enthalpy.

It can be seen from this that before any interpretation of the thermodynamic data, in terms of molecular mechanisms can be conducted it is necessary to quantitate the various contributions. For some comparative studies (e.g. comparing protein binding data for a range of mutants) this may not be important. However, one should always be aware that changes in ΔH_{obs} are not necessarily due only to the formation, or removal of non-covalent bonds between the protein and the ligand at the binding interface, but also involve the effects of solvent and co-solute interactions in the free and bound states.

3.2 Heats of ionization and pH dependence

Here we will attempt to explain how to quantitate ΔH_{ion}. Many protein–peptide interactions show some pH dependence indicating the linkage of the binding of ligand and the binding of protons. This is termed proton linkage (7). Any change in ΔH_{obs} and K_{obs} with pH indicates that there is a change in the ionization state of one or more groups upon binding of protein and ligand. This will contribute to the 'observed' data, and has to be taken into account before the intrinsic values of enthalpy, ΔH_{int}, and binding constant, K_{int}, can be elucidated. The equations used here have been derived for a protein interaction with a single proton linkage (for more complicated models see ref. 8).

The change in K_{obs} with pH can be used to estimate the number of protons, n, (not to be confused with the previously mentioned value, n, for the stoichiometry of an interaction measured by ITC) linked to the ligand binding using the expression (8):

$$-d(\log K_B)/ d(pH) = n \qquad [3]$$

A good example is the complex formation of endothiapepsin and pepstatin A which has been studied in this way (9). Care should be taken to ensure that the buffers used have identical ionic strengths to ensure that this is not the reason for changes in the measured parameters. The use of a three buffer system has been used to circumvent any such problems (10). A mixture of buffers with a range of pK$_a$'s can be employed over the entire pH range.

In cases where the binding is too tight to measure readily at all pH's an alternative approach exploits the different heats of ionization of the ions in the buffer solution. ΔH_{ion} can itself be subdivided into contributions from the heats of ionization of the buffer, ΔH_{ion}^{buff}, and the protein, ΔH_{ion}^{prot}:

$$\Delta H_{ion} = n\,(\Delta H_{ion}^{buff} + \Delta H_{ion}^{prot}) \qquad [4]$$

To obtain ΔH_{ion} we need to determine independently the individual components of equation [4]. For the purposes of this discussion ΔH_{ion}^{prot} incorporates all the enthalpic changes associated with the ionization of the protein in the free and bound state. A list of ionization enthalpies of some commonly used buffers is shown in *Table 1*. As:

$$\Delta H_{obs} = \Delta H_{int} + n\,(\Delta H_{ion}^{prot} + \Delta H_{ion}^{buff}) \qquad [5]$$

and

$$\Delta H_{int} + \Delta H_{ion}^{prot} = \Delta H_R \qquad [6]$$

by performing titrations in a range of buffers at constant pH, a plot of ΔH_{obs} against ΔH_{ion}^{buff} will have a slope equal to n and an intercept of ΔH_R (7, 9, 11–15) (*Figure 3*).

Table 1 The heats of ionization of some commonly used buffer systems (adapted from ref. 11)

Buffer[a]	pK	ΔH (kJ mol^{-1})
MES	6.07	15.53 ± 0.03
Cacodylate	6.14	−1.96 ± 0.02
Pipes	6.71	11.45 ± 0.04
ACES	6.75	31.41 ± 0.05
BES	7.06	25.17 ± 0.07
MOPS	7.09	36.59 ± 0.06
TES	7.42	32.74 ± 0.03
Hepes	7.45	21.01 ± 0.07

[a] Abbreviations. MES: 2-(N-morpholino) ethanesulfonic acid. Pipes: piperazine-N,N'bis(2-ethanesulfonic acid). ACES: N-(2-acetamido)-2-aminoethanesulfonic acid. BES: N,N-bis(2-hydroxyethyl)-2-amino-ethanesulfonic acid. MOPS: 3-(N-morpholino) propanesulfonic acid). TES: N-Tris(hydroxymetnyl)methyl-2-aminoethanesulfonic acid. Hepes: N-2-hydroxyethylpiperazine-N'-2-ethanesulfonic acid.

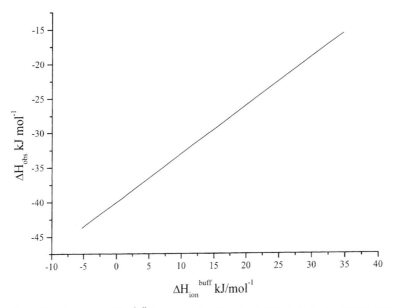

Figure 3 ΔH_{obs} versus ΔH_{ion}^{buff} for a model system involving binding of a protein and a ligand coupled to a protonation event on the protein. The line is calculated using equation ($\Delta H_{obs} = \Delta H_{int} + n\Delta H_{ion}^{buff}$) using values of n = 0.7, ΔH_R of −40 kJ.mol^{-1}, and ΔH_{ion}^{buff} for Pipes, ACES, cacodylate, and MOPS, shown in Table 1.

It is clear that this gives an independent method of calculating n and determines the enthalpy in the absence of effects from buffer. It is however important to note that $\Delta H_{obs} = \Delta H_{int}$ and $K_{obs} = K_{int}$ at a pH where there is no change in ionization states (7). Depending on the nature of the study it may be advisable

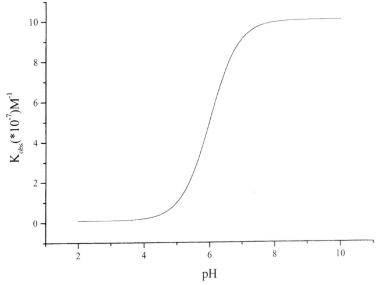

Figure 4 The pH dependency of a model system involving binding of a protein and a ligand coupled to a single protonation event on the protein. The pK_a's of the free and complexed form of the protein are 4 and 6, respectively with K_{int} of 1×10^8 M^{-1}. $K_{int} = K_{obs}$ above pH 8.

to work under these conditions where it is possible to determine ΔH_{int} and K_{int} directly using calorimetry (above pH8 in *Figure 4*).

Determination of the pK_a's of the free and bound states and the enthalpy associated with the ionization of the protein group can lead to the identification of amino acids which are crucial to binding by comparison with literature values. This is extremely useful in the absence of structural data and can only be done readily using ITC. Methods for describing these types of analysis have been outlined in detail (7, 16) but a brief description will be given here.

The change in the number of protons bound to the protein in the free and bound states depends on the difference in the pK_a's of these states and is described by:

$$n = p^b - p^f = (K_p^b \cdot 10^{-pH} / 1 + K_p^b \cdot 10^{-pH}) - (K_p^f \cdot 10^{-pH} / 1 + K_p^f \cdot 10^{-pH}) \quad [6]$$

where p^f and p^b are the fractional saturation of protons in the free and bound protein. K_p^f and K_p^b are the proton binding constants for the bound and free forms, respectively, and are equal to 10^{pKa} of the ionizing group. It can be seen from this that by obtaining n from equation [5] at two pH values and solving with simultaenous equations allows for the determination of the pK_a's in the free and bound state of the protein. In practice this involves performing titrations using at least two pH values, in a range of buffers.

Determination of $n.\Delta H_{ion}^{buff}$ at a third pH also allows for the estimation of the final missing variable parameter from equation [5] ΔH_{int} as well as the parameters which contribute to ΔH_{ion}^{prot}. As:

$$\Delta H_R = \Delta H_{int} - p^f \Delta H_p^f + p^b(\Delta H_p^f + \delta \Delta H_p) \quad [7]$$

where ΔH_p^f is the enthalpy of protonation of the free protein and $\delta \Delta H_p$ is the change in this quantity in the bound state, which is the difference in ΔH_{obs} well below and above the pK_a (7), the information obtained at all three conditions can be used to determine all these quantities, including ΔH_{int}.

Estimation of the binding constant at a pH where it is too tight to measure directly can also be achieved by performing experiments at a range of pH's in at least two types of buffers. These buffers should be chosen to maximize the difference in ΔH_{obs} (16) (see *Table 1*).

3.3 Determination of ΔG and ΔS

Once you have obtained the binding constant and the change in enthalpy of the interaction the change in free energy, ΔG, and entropy, ΔS, associated with going from the free to the bound state, can be derived using simple and well known expressions. ΔG is related to the binding constant by:

$$\Delta G = -RT.\ln K_B \qquad [8]$$

where R is the gas constant and T is the absolute temperature expressed in Kelvin. Hence, the independent determination of the K_B in the ITC experiment allows calculation of ΔG. Since the ΔH is also experimentally determined the ΔS can be calculated using the relationship derived from Second Law of thermodynamics:

$$\Delta G = \Delta H - T \Delta S \qquad [9]$$

For an interaction to occur spontaneously the change in ΔG must be negative. It is clear then from the above equation that a negative (exothermic) ΔH and a positive ΔS term are favourable for binding. In cases where the exothermic enthalpy term is dominant, the reaction is said to be enthalpically driven and vice versa.

3.4 Determination of the change in heat capacity

An additional thermodynamic parameter that can readily be gained from ITC experiments is the change in constant pressure heat capacity, ΔC_p. This parameter is determined from the temperature dependence of the ΔH as described for *Equation 10*.

$$\Delta C_p = d(\Delta H)/d(T) \qquad [10]$$

For a rigid body, bi-molecular interaction the ΔH should change linearly with temperature, i.e. the ΔC_p is temperature independent. Any observed non-linearity is an indication of coupled equilibria occurring on complex formation. Determination of ΔH at a range of temperatures, therefore, provides a method of confirming that the thermodynamic parameters obtained are associated solely with the binding event. Non-linearity is often seen in ΔC_p determinations for protein–ligand interactions where at a given temperature the protein starts to thermally denature. This effect can actually be used to determine the thermodynamic parameters associated with protein thermal unfolding (17).

4 Preparation for an ITC experiment

4.1 Concentration requirements

One of the major perceived disadvantages of ITC determinations of equilibrium binding constants (K_B) compared to spectroscopic methods is the amount of material required. Since the calorimeter measures the heat given out (or taken in) to effect a change in the system (i.e. for the compounds of interest to go from an unbound to a bound state) the technique is limited only by the amount of heat associated with a given interaction. Instrumentation currently available can measure down to the microjoules (μJ) to nanojoule (nJ) range. To produce data that enables the accurate determination of the K_B, the total amount of heat derived from the saturation of the binding sites available (component B) in the calorimeter cell has to be of the order of 200 μJ. Thus for a typical interaction that has an enthalpy of 40 kJ.mol^{-1}, between proteins of molecular weights of approximately 20 kDa and being performed in a calorimeter cell with an active volume of about 1 ml, using a total injection volume of 250 μl, something like 0.2 mg of component A and 0.1 mg of B is required (in other words 50 μM of A and 5 μM B). With respect to many methods for determination of the K_B these amounts represent a lot of material. However, considering that recombinant proteins can now be expressed in milligram quantities it makes ITC not only accessible but the method of choice.

The affinity of the interaction also needs to be considered in choosing the required concentration. The ability of the technique to obtain good estimates of the binding constant depends partly on the unitless 'c' value (18, 19):

$$c = K_B.c.n \qquad [11]$$

where c is the concentration of the component in the calorimetric cell and n is the stoichiometry of the interaction. ITC data can be used when the c value is between 1 and 1000 but optimally between 10 and 100 (for demonstration of this see ref. 18). At low c values the shape of the titrations are shallow and nondescript whereas at high values the transition region between ligand binding and saturation is devoid of data points. In both cases data fitting is not feasible. An ideal 'c' value of 10 requires that the concentration is 10 fold higher than the dissociation constant K_D (= $1/K_B$). This is a very different scenario to that used to determine binding constants by spectroscopic methods which utilise a 'c' value at least 2 orders of magnitude lower.

Based on the ΔH values usually associated with equilibrium biomolecular interactions and the limitations of the required c values, the current sensitivity of the technique limits measuring binding constants, routinely, to between 10^3 and 10^8 M^{-1}. There are however methods for extending this range which will be discussed later in this chapter.

4.2 Sample preparation

In choosing the buffer it is helpful to have some information about your particular system. Several effects can give rise to problems in obtaining useful ITC data. The sort of questions one must ask before starting any binding or biophysical experiments are the following:

(a) Do your components have a propensity to oligomerize at the concentrations required?

(b) Are your samples sensitive to high or low salt concentrations?

(c) Does your system adopt different conformations at different pHs?

(d) Do you require a reducing agent such as dithiothreitol (DTT) to prevent non-native disulfide bridge formation?

(e) Do you require detergent or other solubilizer in the buffer solution?

In the absence of detailed knowledge of the behaviour of the components of the interaction in a given buffer, it is best to use solutions in which the components are stable and readily soluble. It might be wise to adopt a buffering system that has been used in other studies on the same or similar system. Both reactants should then, if possible, be extensively dialysed against the buffer in the same container. This is because the exceptional sensitivity of the instrument means that even small differences in co-solute concentrations can generate heats similar in magnitude to that of binding when adding component A to component B. In some cases it will not be possible to dialyse components A and B together. For example if one or both of the components are the small, such as in the case of short peptides or drug compounds. However discrepancies generated in the binding data by buffer mismatch can be compensated for in some cases by floating the control heats of dilution in the data fitting procedure.

4.3 Use of mixed solvent systems

For some interactions it is necessary to use organic solvents or detergents to solubilize one or both of the components. Since most biological molecules are water soluble the requirement for organic solvents is not common. However in studies on the binding of, for example, largely hydrophobic drug compounds to proteins this problem does arise. This problem can be addressed as in the case of the binding of drug compounds, which are soluble in DMSO, to gyrase (20). The drug was serially diluted with buffer from a stock solution in 100% DMSO and was found to remain in solution at DMSO concentrations as low as 0.1% (v/v). Concentrations of organic solvents of less than about 2% (v/v) can be safely added to many proteins and therefore large buffer mismatches can be avoided. It is wise to check the effect of mixed solvent systems on the stability of the protein using an independent method.

One word of caution when using organic solvents, since many biomolecular binding sites contain a significant hydrophobic component, there is the risk that the co-solvent may compete for the binding site of the ligand. 1% (v/v) DMSO does not sound like very much, but is actually about 140 mM which is well above the K_D of even weak interactions. It is therefore advisable to take care when comparing data in different solvent systems containing co-solvent.

4.4 Concentration determination

The accuracy of any thermodynamic parameter obtained by ITC is directly dependent on the accuracy of concentration determination of the interacting components. For proteins and nucleic acids this is usually done by measuring the absorption at 280 nm and 260 nm, respectively. This is routine in many laboratories but should be determined with great care. One major reason for obtaining inaccurate concentrations is the use of single wavelength measurements. If at all possible one should scan over the entire absorption region and beyond to check for scattering problems and slight cell mismatch. These problems are readily observed for proteins and nucleic acids by sloping and non-zero absorption spectra in the region between 320 and 350 nm (where these macromolecules should not absorb). If, in this spectral region, the absorption is non-zero but constant, it is more accurate to remove this value from that measured at 280 nm (for proteins). If it is sloping, there are contributions from scattering requiring less trivial procedures for correction. A plot of the logarithm of absorption, log(Abs), versus the logarithm of wavelength (log λ) between 320 and 350 nm will be linear. The antilogarithm of the log λ value, obtained from the extrapolated line at log(A_{280}) should be subtracted from the absorption at this wavelength to obtain estimates of the true absorption in the absence of scattering (21).

Another problem frequently encountered in spectroscopic concentration determinations is when a co-solute has an absorbance in the spectral range. For example, many proteins require the presence of a reducing agent such as DTT to avoid non-specific disulfide bond formation. This molecule has an absorption spectra not dissimilar to proteins which will change as the DTT oxidizes. As complete prevention of oxidation is near impossible during sample preparation, it is imperative to dialyse the samples in the buffer containing the reducing agent and to measure the absorption, with the dialysis buffer as the reference, immediately after removal from the dialysis membrane. DTT is a better reducing agent to use than 2-mercaptoethanol in ITC studies since the oxidation of the later reagent has a large heat associated with it which can cause large baseline drifts during the calorimetric titration (22).

To determine the molecular concentration spectroscopically it is necessary to have an extinction coefficient. The concentration can then be determined using the well known Beer–Lambert relationship, $A = \varepsilon c l$ where, A, is the absorption, ε, is the extinction coefficient at a given wavelength, c is the molar concentration, and l is the path length of the cuvette in centimetres. In the absence of quoted extinction coefficients one can determine this experimentally and a comparison of the methods for doing this are given in ref. 23. Alternatively, reasonable estimates can be obtained by summing the contributions from the various chromophores if the primary sequence of the protein is known (23). Peptides frequently do not contain chromophores, in such cases independent amino acid determination is advisable as using weight alone is notoriously difficult and inaccurate due to accompanying counter-ions and the hygroscopic nature of some of these compounds. Alternatively it may be wise to incorporate a tryptophan (for example at one of the termini in the sequence) if possible.

The quoted extinction coefficient for the purines and pyrimidines that make up nucleic acids are derived from the absorption of these bases free in water, these values are somewhat larger than is observed when they are incorporated into a secondary structural unit adopted by DNA and RNA. The best way to remove this effect of hypochromicity is to determine the concentration by measuring the absorption before and after enzymatic cleavage by an endonuclease.

If it is only possible to accurately determine the concentration of one of the interacting components (for example in the case of drug compounds that have no spectral absorbance) it is still possible to use the ITC method. In the data fitting procedure the concentrations of the components of the initial interaction are required and the stoichiometry is a dependent variable. However, where the stoichiometry of the interaction in known (for example from structural studies under similar conditions as those employed in the ITC experiment) it may be possible to float the concentration of the unknown component in the fitting procedures.

5 ITC experimental protocol

5.1 Performing a titration

The protocol for performing a typical titration is outlined below (*Protocol 1*). Our description is based on the use of the Microcal MCS and VP calorimeters, however the general principles detailed here are also applicable to other titration calorimetric equipment currently commercially available.

Protocol 1
Setting up a titration

Equipment

- Titration calorimetric equipment, e.g. Microcal MCS and VP calorimeters

Method

1. The temperature of the calorimeter cells should optimally be 5–10 °C below the experimental temperature. If the calorimeter is temperature controlled using a water-bath, which circulates water around the insulating jacket of the calorimeter, this has to be pre-set to a suitable temperature at least 8 h prior to performing an experiment. Although reasonable temperature stability is achieved from the instrument control (water-bath or Peltier) it is advisable to keep the equipment in a room with constant temperature to avoid temperature drifts and baseline instability. Experiments can be routinely performed at temperatures between 5–70 °C.

2. Before loading, the samples should be degassed to reduce the possibility of bubble formation in the calorimeter cell during the titration. This is particularly important

Protocol 1 continued

when performing titrations at low temperatures. If you are working with particularly volatile buffers such as acetic acid and wish to avoid degassing then it is advisable not to use samples immediately after removing them from the refrigerator. Allow them to stand at room temperature or warm the samples by holding in the palm of your hand. The weight of sample should be taken before and after degassing. Any discrepancy in weight after degassing should be made up with water.

3 Fill the calorimeter sample cell with care. Air bubbles can lead to baseline instability, and furthermore, those which force their way out of the calorimeter cell, can result in heat pulses of similar magnitude to that generated by the heat of binding. The active volume of the calorimeter cell is approximately 1.3 ml, however about 2 ml of component B is required to fill the sample chamber and the fill tube.

4 Component A is loaded into a syringe. These come in a variety of volume capacities however, the most commonly used is that of 250 µl. To obtain a complete binding isotherm the solution in the 250 µl syringe should be 10 times more concentrated than the concentration of binding sites in the cell. This results in a final concentration regime in the calorimeter cell of approximately two moles of A to every mole of B (for a 1:1 binding event).

5 Set the rotation speed of the syringe to a suitable level. This is usually approximately 400 r.p.m. However, particularly viscous samples may require adjustment of this (see manufacturer's handbook). In the MicroCal instruments the injection syringe needle is fashioned into a paddle. Rotation of this enhances the mixing of interacting components reducing times required for equilibrium to be reached in the sample cell.

6 Set up menu for titration. In the absence of any prior knowledge of the system a reasonable experimental set-up is 16, 15 µl injections of 12 sec duration, 5 min apart, of 100 µM compound A into 10 µM of compound B.

5.2 Trouble-shooting

1 Sometimes it is observed that the heat pulse is incomplete when the next injection is made (i.e. the cell feedback does not return to the equilibrium value). The cell feedback power should approximate to the same value before and after the injection. This effect can be the result of, for example, a slow reaction occurring on binding, solvent effects, pressure build up in the calorimeter cell due to ill-fitting syringe, or conformational changes occurring which accompany binding. This effect leads to a stepped baseline and inaccurate measurement of the heat for the injections. Care to ensure that the time between injections is appropriate, the solvent system is optimal, that the syringe is free to rotate, and the syringe collar is not obstructing the calorimeter fill-tube should be taken prior to performing a complete titration.

2 A permanent change in the cell feedback indicates that the heat capacity of the solution has changed and may be indicative of aggregation or precipitation. In this case the sample conditions need to be changed.

3 Each heat of injection should be, at the very least, 5-10 μcals over the first two-thirds of the titration. If the heats are too small then either the concentration of the reactants or the size of the injection should be increased. Some advantage may also be gained by increasing the rate of injection. Increasing the sample concentration is preferable if the affinity is not too high because increasing the size of the injections will decrease their number and therefore the number of data points to be used to fit the data. It should also be noted that there will be a maximal injection volume (depending on the size of the syringe) that should be used when working at temperatures far from room temperature. This is because the needle of the syringe contains this volume and is always equilibrated at the experimental temperature, whereas material in the rest of the syringe is not. Using larger volumes will add a heat associated with thermal equilibration.

5.3 Controls

Heats of dilution experiments are a necessary control required to assess the heat energy associated with dilution of the interacting components into the buffer solutions in which they are dissolved. Experiments whereby (1) component A is injected into buffer and (2) buffer is injected into component B are required. Furthermore, a control experiment of (3) buffer into buffer is also strictly necessary. These experiments should use exactly the same solutions of A and B as used in the binding experiments and, ideally, the buffer solutions used for dialysis of the interacting components respectively. These experiments should reveal an approximately equal heat for each injection, and hence in practice it is not usually necessary to set up a full titration schedule since the consistency of the heats per injection can be seen in three or four injections. The heat per injection for the dilution experiments should be determined and these should be removed from the binding data in the following way:

$$\text{The heats from binding data} - (1) - (2) + (3) \qquad [12]$$

The value from (3) is added because the buffer into buffer effects are also an integral part of the other two controls and therefore have been subtracted twice. In practice the latter two components ((2) and (3)) are usually small and of similar size. Particular care should be taken over (1) as changes in the heat of each injection may indicate dissociation of ligand upon dilution (see *Figure 1*) and thus, a competing equilibria which must be taken into account and will be discussed later in this chapter. Here is a good place to reiterate the usefulness of dialysis since buffer mismatching can result in the inherent inaccuracies associated with subtracting large heats from the binding data.

6 ITC data handling

6.1 Data fitting

The data shown in *Figure 1* now has to be processed to obtain the thermodynamic characterization of the interaction. It is not the purpose of this chapter to give details of the algorithms used to fit ITC data. The methods adopted will usually depend on the software which accompanies the instrument. There are a number of suitable software packages available to analyse binding data. Perhaps the most popular (and that supplied with MicroCal instrumentation) is ORIGIN. The raw data output from the calorimeter should consist of a file of the heats per injection (integrated from the power per second required to maintain isothermal conditions between the ITC cells; see above), the volumes and concentrations of interacting components at each injection.

A plot of the ΔH against molar ratio of the interacting components for a simple 1:1 interaction is shown in *Figure 2*. In this case the titration curve is sigmoidal with the equivalence point at a 1:1 ratio of A to B. These data can be fit using a non-linear least squares Marquardt algorithm with three independent variables (as described elsewhere) (18, 19). In the ORIGIN software package these variables are K_B, ΔH, and stoichiometry, n. In other packages the floating value for stoichiometry can be replaced by that of concentration of either A or B, however, this depends on the stoichiometry being characterised *a priori*.

6.2 Data analysis

The model required to fit the data is usually based on one independent binding site, however, using the ORIGIN software (and numerous other software packages) more complex binding events can be fit to give the thermodynamics of binding and the stoichiometry of interaction. More complex models such as multiple co-operative binding events invoke more independent variables and hence although apparent improvement of the fit can be achieved, this does not necessarily reflect the fact that the most appropriate model has been chosen. It should be noted that the only parameter that is model independent is the enthalpy. In the absence of further experimental evidence it is useful only to apply the simplest models.

In some cases there is the possibility for more than one independent interaction to occur. This is normally seen by a bi (or multiphasic) titration isotherm (*Figure 5*).

In this case there are two distinct interactions taking place with two different K_B values. Most software accompanying current instrumentation can accommodate these types of binding and provide the enthalpies and stoichiometries in the same way as for the single site model. One way to confirm the presence of two (or more) independent binding events is to perform the titrations at different temperatures. Assuming that the binding events have different associated ΔC_p values the isotherm will change profile from one temperature to another (24).

6.3 Choosing the appropriate model

Many protein–ligand interactions in isolation are of the 'simple' one type of binding site system. Nature rarely evolves proteins with multiple component binding interactions. From the experimentalists point of view this is the most straightforward to analyse. We can check for the presence of additional equilibria (*Protocol 2*).

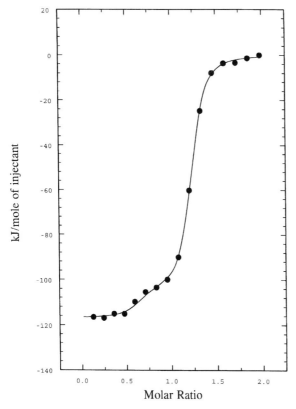

Figure 5 ITC binding isotherm for the interaction of bisphosphorylated peptide to the p85 subunit of PI-3 kinase. The binding shows two distinct binding events. The first binding event occurs with an apparent stoichiometry of approximately 0.5:1, corresponding to the binding of two molecules of p85 to one bisphosphorylated peptide. The second binding event occurs with a stoichiometry of approximately 1:1 (44).

Protocol 2
Checking for additional equilibria

In all titrations there is the potential for additional equilibria to occur simultaneously with the binding event under study. For example, effects of the binding of solvent, the protonation of a charged group, the dissociation/association of a moiety involved in the interaction.

There are a number of ways of observing such effects.

1. Check for asymmetry of the binding isotherm.
2. Check if the determined heats of dilution (see Section 5.3) are constant and equal to the last few points of the titration (see *Figure 1*).
3. Check the linearity of the temperature dependence of the ΔH for the interaction.
4. Perform titration at a range of concentrations.

Protocol 2 continued

5 Perform 'reverse' titrations, i.e. two sets of experiments should be carried out, one with the ligand in the syringe and the macromolecule in the calorimetric cell and another with the position of the two reactants reversed can give an indication of additional enthalpic contributions in addition to the heat of binding. If there are competing equilibria due to oligomerization of one or more of the components this will be observed by changes in the derived thermodynamic parameters. This is probably the easiest and most efficient way of checking a range of models. No matter what their complexity, if the model is correct the same values should be obtained however the experiment is carried out. In cases where one of the components is known to oligomerize then the association constant and stoichiometry of this process need to be quantified and incorporated into the model. If however one can work at concentrations where the degree of association is negligible (well below the K_D) then one can circumvent this problem. This usually implies having that component in the calorimeter cell and not at the higher concentrations required when in the syringe.

7 Structure/thermodynamic relationships

7.1 Heat capacity versus buried surface area

Data derived from the heats of transfer of non-polar hydrocarbons into aqueous solvents and from protein folding/unfolding equilibria are important in identifying the correlation between thermodynamic data and structural detail (25-28). The ΔC_p was shown to correlate with the amount of surface area buried from solvent upon binding. Quite clearly it is of great interest to be able to predict thermodynamic data directly from structural studies. This would have particular relevance in the field of rational drug design, for example, where the financially and temporally expensive processes of synthesis and performing binding studies could be circumvented if thermodynamic data was accessible for a range of potential lead compounds from the structure of the target with its substrate. A significant effort has been put into optimizing the reliability of this correlation in recent years. Several groups have derived equations in which this correlation can be used to predict thermodynamic or structural information (9, 26, 29) and these have stimulated the first sojourns into thermodynamic predictions from X-ray crystallographic and NMR solution structures (9, 30, 31).

In reality the formation of an interface between two molecules involves a large number of contributions which, in most cases, these correlations mask because of their generality. The tolerance of these correlations is low and potentially breaks down when dealing with small compounds where the surface area changes resulting from substitutions of chemical groups give rise to changes in ΔC_p which are lost in the inherent error in determining these values. As such, changing a chemical moiety on a small drug compound may have a dramatic effect on binding, but the thermodynamic effects of this are lost in the errors in ΔC_p.

Additional sources of error in the correlation between the experimentally determined ΔC_p and surface area burial can include the effects of bound water molecules or other components of the solvent system in the biomolecular interface. Furthermore, contributions from multiple protonation events can have a significant effect on the ΔC_p (29).

7.2 Thermodynamic insights into the mechanism of interactions involving DNA

We have shown above the dramatic effects that physico-chemical environment can have (pH and temperature) on the thermodynamic parameters and how their direct determination by calorimetry can lead to useful insights into the mechanism of interaction. In addition manipulation of the salt conditions has proved extremely insightful for the study of the DNA with proteins (32–37) and with drugs (38, 39). It is widely known that increasing the salt concentration to moderate levels (< 1 M) decreases the affinity of protein–DNA interactions. This is thought to be due to the 'polyelectrolyte' effect which is the release of counterions from the polyanionic DNA upon binding (32). The release of these cations becomes less favourable at higher salt concentrations and was inferred to be of entropic origin using van't Hoff analysis of spectroscopically determined binding isotherms (32). The entropic origin of this salt effect has been confirmed more recently using isothermal titration calorimetry (36) and lends weight to the proposed mechanism. However it should be pointed out that salt effects on interactions involving protein and DNA can be far more complex. The enthalpy of binding of *E. coli* SSB protein to single-stranded DNA has been shown to depend on salt type and concentration. This phenomenon is thought to be due to weak anion binding to the protein (35). It should be reiterated that these sorts of effects, from coupled equilibria, observed in this study are extremely difficult to observe, and hence deconvolute using indirect van't Hoff type analysis.

8 Extending the range of binding constants determination

ITC is generally limited to measuring binding constants between approximately 10^3 and 10^8 for reasons already mentioned above. In the event that changing solvent conditions are not sufficient to bring the K_{obs} into the experimentally accessible range then displacement studies can be used. The success of this technique depends on the availability of competitive binders which have measurable affinities differing in several orders of magnitude from the ligand of interest (40, 41).

For high affinity ligands component B is titrated to saturation with a 'weak' inhibitor and the enthalpy, ΔH_{wi}, and affinity, K_B, are determined from the resulting isotherm. The resulting mixture is left in the cell and then titrated with the high affinity ligand of interest. This second experiment is fitted in the normal way to give the observed ΔH_{obs} and K_{obs}. If the difference in affinity is several orders of magnitude the terms ΔH_{obs} and K_{obs} are related to the desired ligand parameters ΔH_{lig} and K_{lig} by:

$$\Delta H_{lig} = \Delta H_{obs} + \Delta H_{wi} \qquad [13]$$

$$K_{lig} = K_{obs} * K_{wi} \qquad [14]$$

For low affinity essentially the same technique as that described for high affinity ligands is used, but in addition the enthalpy and binding constant has to be determined for a stronger inhibitor (42).

9 Advantages and disadvantages

In terms of thermodynamic studies of biomolecular systems, ITC techniques are advantageous for the following reasons.

1 The technique is widely applicable to biologically significant reactions. The interaction of proteins with peptides, other proteins, lipids (including lipid assemblies, i.e. liposomes and proteoliposomes), carbohydrates, receptors (soluble and membrane bound), nucleic acids (including synthetic oligonucleotides), and drugs have, to date, been examined. Such systems are, of course, not only of fundamental interest but also important in relation to the manipulation of biological events within the pharmaceutical and biotechnology areas. Moreover many abnormal protein–protein and other types of protein–ligand interactions play an important role in various disease states. Hence understanding the difference between disease and wild-type states is fundamental to pharmaceutical intervention.

2 For a given protein–ligand interaction the experimental parameters which can be varied include, but are not limited, to reagent concentration (especially protein concentration), temperature, ionic strength, pH, and buffer composition at each pH. The nature and concentration of salts, mutational changes in the protein (site-specific or group mutagenesis), systematic changes in ligand structure, and of course combinations of the foregoing can and should also be varied. This is particularly the case if a systematic and thorough study of a particular protein–ligand interaction is to be interrogated.

3 No chemical modification of the reactants (cf., spectral probes) or immobilization of any of the reactants (cf., SPR methods) is required. Specific chromophores used in many spectroscopic techniques, as a basis for analysis, need not be present in either of the reacting species. In fact the reagents are used in their 'native' form. Furthermore, 'the equilibrium state' of the interaction is not (theoretically) perturbed because it is not necessary to separate bound from free species as is the case in e.g. ELISA (enzyme-linked immunosorbent assays) experiments which are sometimes used to obtain association (affinity) constants. ITC is also non-invasive and non-destructive, such that if required the formed complexes could subsequently be further analysed by other methods. Of course this also allows (*provided the correct controls are instituted*) the ability to investigate the

interaction of a protein–ligand complex with a third reaction component. This is particularly useful for proteins or peptides, which have more than one binding site for different ligands and or exhibit allosteric behaviour.

4 With careful measurements the reproducibility in measuring the molar binding ratios (i.e. n), has been claimed to be as good as 1% (43).

The possible disadvantages of ITC methods include the following:

1 Corrections due to buffer ionization enthalpies resulting from protonation/deprotonation reactions also need to be assessed, ideally, for *each* titration which is conducted in order to obtain the true value of the binding enthalpy as opposed to the observed enthalpy of binding.

2 High affinity reactions cannot simply be measured because of the need to use high concentrations of reactants (> micromolar) which can cause, or be accompanied by attendant problems, e.g. aggregation, non-availability of materials, etc. Using alternative experimental conditions such as a different temperature and/or pH, measuring the ΔH_b and $\Delta C_{p,b}$, and then extrapolating to 'ambient' conditions can sometimes overcome this problem (see above). In the case of temperature the van't Hoff equation is normally used.

3 Although not an intrinsic disadvantage, ITC data obtained from model compound studies (although often invaluable) should be viewed with caution when relating the results of such experiments to protein–ligand systems.

4 Again, although not an intrinsic disadvantage of the ITC method itself, there are problems in parsing the forces that are responsible for non-covalent intermolecular interactions (and relating them to structural factors). The forces that are important in protein folding (hydrophobicity, hydrogen bonding, electrostatic interactions, van der Waals interaction, etc.) are precisely those that are responsible for protein–ligand interactions. Not only are non-covalently controlled phenomena poorly understood but it is possible that the repertoire of non-covalent interactions that underlie complex recognition events are not yet fully defined. The issue is made more complex in relation to molecular recognition, because the ways in which individual non-covalent forces compete with or reinforce one another in complex systems is (extremely) poorly understood. In addition the magnitude and signs of the thermodynamic parameters characterizing each of the (known) non-covalent interactions are, currently, disputed.

References

1. Laidler, K. J. (1995). In *The world of physical chemistry*, Chapter 4, pp. 83–130. Oxford University Press.
2. Wadso, I. (1987). In *Thermal and energetic studies of cellular systems* (ed. J. A. M. Wright), p. 34. Bristol.
3. Wadso, I. (1994). In *Solution calorimetry* (ed. K. N. Marsh and P. A. G. O'Hare), p. 161. Blackwell, Oxford.

4. Ott, J. B. and Wormald, C. J. (1994). In *Solution calorimetry* (ed. K. N. Marsh and P. A. G. O'Hare). Blackwell, Oxford.
5. Stodeman, M. and Wadso, I. (1995). *Pure Appl. Chem.*, **67**, 1059.
6. Wiseman, T., Williston, S., Brandts, J. F., and Lin, L. N. (1989). *Anal. Biochem.*, **179**, 131.
7. Baker, B. M. and Murphy, K. P. (1996). *Biophys. J.*, **71**, 2049.
8. Wyman, J. and Gill, S. J. (1990). *Binding and linkage: functional chemistry of biological macromolecules*. University Science Books, Mill Valley, CA.
9. Gomez, J. and Freire, E. (1995). *J. Mol. Biol.*, **252**, 337.
10. Ellis, K. J. Louie and Morrison, J. F. (1982). In *Methods in enzymology* (ed. D. L. Purich), Vol. 87, p. 405. Academic Press, London.
11. Fukada, H. and Takahashi, K. (1998). *Proteins*, **33**, 159.
12. Murphy, K. P., Xie, D., Garcia, K. C., Amzel, L. M., and Friere, E. (1993). *Proteins*, **15**, 113.
13. Kreshek, G. C., Vitello, L. B., and Erman, J. E. (1995). *Biochemistry*, **34**, 8398.
14. Eftink, M. and Biltonen, R. (1980). In *Biological microcalorimetry* (ed. A. E. Beezer). Academic Press, San Diego.
15. Bradshaw, J. M. and Waksman, G. (1998). *Biochemistry*, **37**, 15400.
16. Doyle, M. L., Louie, G., Dal Monte, P. R., and Sokoloski, T. D. (1995). In *Methods in enzymology* (ed. M. L. Johnson and G. K. Ackers), Vol. 259, p. 183. Academic Press, London.
17. Thomson, J., Ratnaparki, G. S., Varadarajan, R., Sturtevant, J. M., and Richards, F. M. (1994). *Biochemistry*, **33**, 8587.
18. Jelesarov, I. and Bosshard, H. R. (1999). *J. Mol. Recognit.*, **12**, 3.
19. Ladbury, J. E. and Chowdhry, B. Z. (1996). *Chem. Biol.*, **3**, 791.
20. Tsai, F. T. F., Singh, O. M. P., Skarzynski, T., Wonacott, A. J., Weston, A. J., Tucker, A., et al. (1997). *Proteins*, **28**, 41.
21. Perkins, S. J. (1986). *Eur. J. Biochem.*, **157**, 169.
22. Cooper, A. and Johnson, C. (1998). In *Protein protocols on CD-ROM* (ed. J. M. Walker). Humana Press, New Jersey, USA.
23. Pace, C. N., Vasdos, F., Fee, L., Grimsley, G., and Gray, T. (1995). *Protein Sci.*, **4**, 2411.
24. Ladbury, J. E., Wright, J. G., Sturtevant, J. M., and Sigler, P. B. (1994). *J. Mol. Biol.*, **238**, 669.
25. Spolar, R. S., Livingstone, J. R., and Record, M. T. (1992). *Biochemistry*, **31**, 3947.
26. Spolar, R. S. and Record, M. T. (1994). *Science*, **263**, 777.
27. Baker, B. M. and Murphy, K. P. (1997). *J. Mol. Biol.*, **268**, 557.
28. Gomez, J. and Freire, E. (1997). In *Structure-based drug design; thermodynamics, modeling and strategy* (ed. J. E. Ladbury and P. R. Connelly). Springer–Verlag, Berlin.
29. Morton, C. J. and Ladbury, J. E. (1996). *Protein Sci.*, **5**, 2115.
30. Pantoliano, M. W., Horlick, R. A., Springer, B. A., Vandyk, D. E., Tobery, T., Wetmore, D. R., et al. (1994). *Biochemistry*, **33**, 10229.
31. Haq, I., Ladbury, J. E., Chowdhry, B. Z., Jenkins, T. C., and Chaires, J. B. (1997). *J. Mol. Biol.*, **271**, 244.
32. Ha, J.-H., Capp, M. W., Hohenwalter, M. D., Baskerville, M., and Record, M. T. (1992). *J. Mol. Biol.*, **228**, 252.
33. Lohman, T. M. and Mascotti, D. P. (1992). In *Methods in enzymology* (ed. D. M. J. Lilley and J. E. Dahlberg), Academic Press, London. Vol. 212, p. 424.
34. Mascotti, D. P. and Lohman, T. M. (1990). *Proc. Natl. Acad. Sci. USA*, **87**, 3142.
35. Kozlov, A. G. and Lohman, T. M. (1998). *J. Mol. Biol.*, **278**, 999.
36. Lohman, T. M., Overman, L. B., Ferrari, M. E., and Kozlov, A. G. (1996). *Biochemistry*, **35**, 5272.
37. O'Brien, R., DeDecker, B., Fleming, K. G., Sigler, P. B., and Ladbury, J. E. (1998). *J. Mol. Biol.*, **279**, 117.

38. Chaires, J. B., Priebe, W., Graves, D. E., and Burke, T. G. (1993). *J. Am. Chem. Soc.*, **115**, 5360.
39. Haq, I., Lincoln, P., Suh, D., Norden, B., Chowdhry, B. Z., and Chaires, J. B. (1995). *J. Am. Chem. Soc.*, **117**, 4788.
40. Sigurskjold, B. W., Berland, C. R., and Svensson, B. (1994). *Biochemistry*, **33**, 10191.
41. Bains, G. and Freire, E. (1991). *Anal. Biochem.*, **192**, 203.
42. Berland, C. R., Sigurskjold, B. W., Stoffer, B., Frandsen, T. P., and Svensson, B. (1995). *Biochemistry*, **34**, 10153.
43. Liu, Y. and Sturtevant, J. M. (1995). *Protein Sci.*, **4**, 25559.
44. O'Brien, R., Rugman, P., Renzoni, D., Layton, M., Handa, R., Hilyard, K., *et al.* (2000). *Protein Sci.*, **9**, 570.

Chapter 11
Differential scanning microcalorimetry

Alan Cooper, Margaret A. Nutley, and Abdul Wadood
Chemistry Department, Glasgow University, Glasgow G12 8QQ, Scotland.

1 Introduction

Differential scanning calorimetry (DSC) is an experimental technique to measure the heat energy uptake that takes place in a sample during controlled increase (or decrease) in temperature. At the simplest level it may be used to determine thermal transition ('melting') temperatures for samples in solution, solid, or mixed phases (e.g. suspensions). But with more sensitive apparatus and more careful experimentation it may be used to determine absolute thermodynamic data for thermally-induced transitions of various kinds. Formerly this was more the realm of the dedicated specialist, but now with the ready availability of sensitive, stable, user-friendly DSC instruments, microcalorimetry has become part of the standard repertoire of methods available to the biophysical chemist for the study of macromolecular conformation and interactions in solution at reasonable concentrations. And, to the extent that thermal transitions might be affected by ligand binding, DSC can provide useful information about protein–ligand binding. The advantages of calorimetric techniques arise because they are based on direct measurements of intrinsic thermal properties of the samples, and are usually non-invasive and require no chemical modifications or extrinsic probes. Furthermore, with careful analysis and interpretation, calorimetric experiments can directly provide fundamental thermodynamic information about the processes involved.

This chapter concentrates on the basic theory and practical applications of DSC in the field of protein stability and ligand interactions, with practical examples of its use, and details of data analysis and pitfalls. It should be said from the outset, however, that DSC is at best only a rather indirect way of studying protein–ligand interactions, and in most cases other and more direct methods (including isothermal titration calorimetry, ITC) might be better suited to the problem. However, the technique has proved useful in some cases, and can provide preliminary information that might form the basis for more detailed studies by other techniques.

2 DSC basics

A sketch showing the typical layout of a DSC instrument is shown in *Figure 1*. In a DSC experiment a solution of protein (typically 1 mg/ml or less in modern instruments) is heated at constant rate in the calorimeter cell alongside an identical reference cell containing buffer. Differences in heat energy uptake between the sample and reference cells required to maintain equal temperature, correspond to differences in apparent heat capacity, and it is these differences in heat capacity that give direct information about the energetics of thermally-induced processes in the sample. Correct use of such instruments requires careful attention to sample preparation, buffer equilibration, and baseline controls, together with accurate measures of sample concentration if absolute thermodynamic data are required.

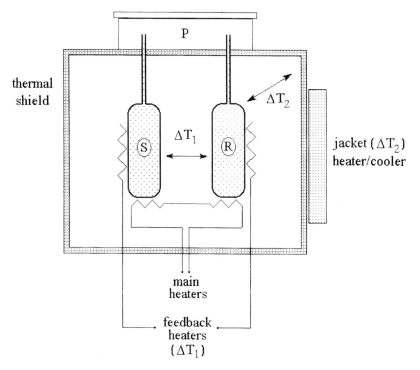

Figure 1 Sketch diagram of a typical DSC used for thermal studies of dilute solutions of biomolecules (adapted from ref. 2). Identical, total-fill sample (S) and reference (R) cells (typically 0.5–2 ml volume) containing protein solution and buffer, respectively, are held under elevated atmospheric or inert gas pressure (P) to inhibit bubble formation during heating. During up-scan operation, power is supplied to the main heaters to raise the temperature of the cells at a steady rate, whilst monitoring the temperature differences between sample and reference cells (ΔT_1) and between cells and the surrounding adiabatic jacket (ΔT_2). Feedback through the jacket heater allows the thermal shield temperature to follow that of the cells, and feedback heaters on the cells compensate for any temperature differences between the cells during the scan.

2.1 Instrumentation

To avoid confusion, it must be emphasized that the differential scanning microcalorimeters described here for work on dilute biomolecular solutions are specialized instruments that differ significantly from the possibly more familiar 'DSC' or 'DTA' instruments commonly used in less demanding thermal analysis measurements. In particular they are designed to accommodate relatively large volumes (0.5–2 ml) of dilute solutions (1 mg/ml or less), in true differential mode, rather than the typically 50 µl pans used for DTA studies of solids or pastes. Currently available instruments are based primarily on pioneering work by the Privalov and Brandts groups (1–3), including the following: Microcal MC2, MCS, and VP-DSC (Microcal Inc., 22 Industrial Drive E., Northampton, MA 01060–2327, USA; http://www.microcalorimetry.com); CSC NANO II DSC (Calorimetry Sciences Corp., 155 West 2050 South, Spanish Fork, Utah 84660, USA; email: CalSCorp@aol.com); DASM-4 (Bureau of Biological Instrumentation, Russian Academy of Sciences, Moscow, Russia). Current versions of these instruments are comparable in sensitivity and stability though they may differ somewhat in cell configuration and control and analysis software options. Slightly less sensitive instruments, but with greater flexibility in sample handling, are available from Setaram (7 rue de l'Oratoire, F-69300 Caluire, France; http://www.setaram.fr). The experiments described here have been done using Microcal equipment, but this does not imply any particular preference.

Protocol 1
DSC of protein unfolding—basic procedure using lysozyme as a model

Equipment and reagents

- DSC instrument (Microcal MCS, VP-DSC, or equivalent)
- Dialysis tubing or cassettes (e.g. Pierce Slide-A-Lyser®)
- Degassing equipment (vacuum desiccator, magnetic stirrer)
- Buffer: 20 mM Na acetate pH 5.2[a]
- Protein: hen egg white lysozyme (e.g. Sigma L-6876), typically 2 ml at a concentration of 1 mg/ml (for MCS) or 1 ml at 0.1 mg/ml (for VP-DSC). For more demanding work, commercial samples of lysozyme may be dialysed against ultrapure water and lyophilized to remove extraneous salts before use.

A. Sample preparation

1 Prepare the protein solution and dialyse several changes of appropriate buffer. Each DSC run will typically require 1–2 ml of protein solution at a concentration of around 1 mg/ml, or less, depending on the DSC instrument.

2 Retain the final dialysis buffer for DSC reference, equilibration, and dilutions.

3 Determine the protein concentration by 280 nm absorbance or other appropriate

Protocol 1 continued

method. For lysozyme the molar extinction coefficient, $\varepsilon_{280} = 37900$ ($A_{280}^{1mg/ml} = 2.65$, $M = 14300$).

4 Immediately prior to the experiment, degas portions of the sample mixture and buffer for 2-3 min under gentle vacuum with gentle stirring. Be careful to avoid excessive degassing or frothing of the mixture at this stage.

B. DSC procedure

1 Load DSC sample and reference cells with degassed buffer and collect baseline scan(s) using appropriate temperature range and scan rate (typically 20-100 °C, 60 °C/h).

2 Allow the DSC cells to cool and refill the sample cell with protein solution.

3 Repeat the DSC scan(s) using the same parameters as in step 1. (Depending on circumstances, it may be useful to do repeat scans with the same sample to establish reversibility and reproducibility. It can also be useful to run a preliminary scan, stopping some way before the unfolding transition begins, before cooling and performing the complete scan. This minimizes baseline artefacts that can be induced by the thermal shock involved in loading the sample or reference cells.)

4 After final cooling, remove the sample and examine for turbidity, aggregation, or other visible changes. (Precaution: traces of aggregated protein or other contaminants in the DSC cell will cause erratic baseline behaviour. Routine vigorous cleaning of the DSC cells, using detergents or strong acids/bases as recommended by the manufacturer, is essential for reliable DSC operation, especially when working with readily aggregating systems.)

5 Process data using instrumental software. This normally involves subtraction of buffer baseline (from step 1), concentration normalization, followed by deconvolution of the resultant thermogram using an appropriate model.

[a] Note: most buffers, including organic solvent mixtures, are compatible with DSC, but mercaptoethanol is best avoided because of adverse thermal effects due to oxidation and thermal degradation. Other reducing agents such as DTT or DTE are usually satisfactory, if needed.

2.2 Quantitative analysis of DSC data—practical considerations

The typical experimental procedure for following the thermal unfolding of a simple globular protein is described in *Protocol 1*. Representative data from such an experiment are shown in *Figure 2*. In this section we shall outline some of the practical aspects related to analysis and interpretation of such data, leaving the theoretical background and justification for some of the points to be described in later sections.

The output from any DSC experiment is a thermogram showing the excess heat capacity (C_p, sample minus reference) as a function of temperature. For a simple globular protein the thermogram comprises three regions: the pre-

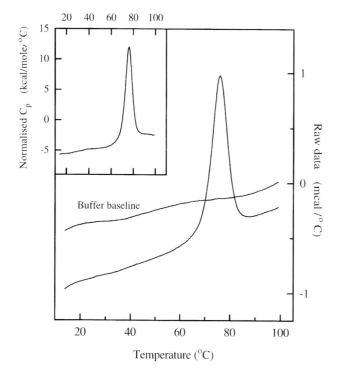

Figure 2 Raw DSC data for thermal unfolding of hen egg white lysozyme (1.2 mg/ml) in 20 mM Na acetate buffer pH 5.2, at a scan rate of 60 °C/h. These data were obtained following *Protocol 1*, using a Microcal-MCS system. (Similar data are obtained with 10-fold lower concentrations using the more recent VP-DSC instrument.) The inset shows the same data after subtraction of the instrumental (buffer) baseline and concentration normalization, illustrating also the pre- and post-transition baseline behaviour typical of such processes.

transition baseline, the endothermic unfolding transition, and the post-transition baseline. At temperatures well below the onset of thermal unfolding, the C_p simply reflects the difference in heat capacity between the protein and the solvent (usually mainly water) it has displaced. Since water has a high heat capacity compared to most organic substances, including proteins, the apparent C_p in this region will normally be negative. For most proteins, this pre-transition baseline also shows a slight positive slope, indicating a gradual increase in heat capacity with temperature—a characteristic also of organic solids. As the protein begins to unfold, the C_p increases as more heat energy is taken up in denaturing the protein, reaching a peak at approximately the mid-point (T_m) temperature of the process (assuming a single cooperative unfolding process), before dropping down to the high temperature baseline. This post-transition baseline, representing the relative heat capacity of the unfolded polypeptide, is usually found at a higher level (positive ΔC_p) and has a lesser slope than the pre-transition baseline. Similar effects are seen for organic liquids also. Consequently, as a first approximation, one might picture the unfolding of a globular protein in water as the 'melting' of an organic microcrystal suspended in an aqueous environment.

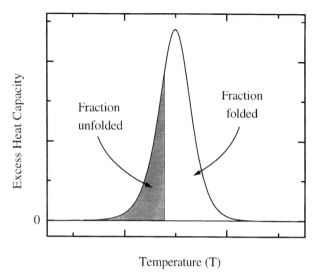

Figure 3 Sketch showing the use of DSC thermograms to determine the van't Hoff enthalpy (ΔH_{VH}) of a two-state transition. (This is the peak as it might appear ideally after baseline correction.) At any particular temperature, the extent of unfolding is represented by the area under the thermogram up to that point (shaded area). Consequently, the ratio of shaded:unshaded areas corresponds to $K\,(=[U]/[N])$ at that temperature, which can be used in the van't Hoff equation to determine ΔH. Note that, since only ratios of areas are used in this calculation, neither the absolute units of C_p nor the protein concentrations are required. The method is, however, dependent on the validity of the two-state (or other) model adopted.

The calorimetric enthalpy (ΔH_{cal}) is the total integrated area under the thermogram peak which, after appropriate baseline correction, represents the total heat energy uptake by the sample undergoing the transition. This heat uptake will depend on the amount of sample present in the active volume of the DSC cell and is, in principle at least, a model-free absolute measure of the absolute enthalpy of the process involved. [It is axiomatic that *equilibrium* transitions observed in a DSC up-scan experiment must be endothermic. Exothermic heat effects can be observed, but when these are encountered it is usually an indication of thermodynamically irreversible, non-equilibrium processes, kinetically activated by elevated temperatures. Aggregation of thermally denatured protein is one such example.]

The van't Hoff enthalpy (ΔH_{VH}) is an independent estimate of the enthalpy of the transition, based on an assumed model for the process. Here one simply uses the area under the C_p peak at any temperature (see *Figure 3*), divided by the total area, as a measure of the fraction or extent of unfolding that has occurred at that temperature. In this way one is using the calorimetric signal in just the same way as any other indirect method for following the unfolding transition, such as CD or fluorescence, for example. Assuming a simple two-state model, one can then relate the temperature variation of the fraction unfolded to the apparent enthalpy of the process using the van't Hoff equation (see below). The advantage of this approach is that, since it relies only on ratios of areas under the experimental curve, it does not require any information about concentration or purity of the sample. Ideally, ΔH_{cal} and ΔH_{VH} should be identical in any calorimetric experiment, and comparison of the two can be quite revealing about factors such as the purity and concentration of the sample, and can also give information about the reversibility and apparent mechanism of the process.

2.3 Concentration measurements

The accuracy of any calorimetric ΔH_{cal} (as opposed to ΔH_{VH}) estimate is critically dependent upon the purity of the sample and on the reliability of the methods used to determine its concentration. For proteins, the most convenient and straightforward method for concentration measurement is usually the UV (280 nm) absorbance, provided a reliable molar extinction coefficient (ε_{280}) is available. This is non-destructive and can frequently be done on the actual sample solution prior to insertion in the DSC. ε_{280} may usually be estimated to reasonable precision (typically ± 5%) from the aromatic amino acid (Trp, Tyr) composition of the protein (4). It goes without saying that such measurements should follow good working practices using reliable instrumentation and clean cuvettes, since the entire DSC analysis may depend on this one measurement. In our experience it is unwise to rely on a simple A280 measurement at fixed wavelength, but better to record a complete UV spectrum (240–400 nm), since this can show up immediately any problems due to incorrect baselines, light scattering by aggregated protein, or other impurities. Colorimetric methods of protein estimation (e.g. 'Bradford' or other dye-binding assays) are generally less reliable unless previously calibrated for the specific protein under investigation.

One must remember, of course, that most methods of protein estimation will also measure contributions arising from misfolded protein and other protein impurities. For example, if some of the protein sample is already misfolded or unfolded prior to the DSC experiment, and does not contribute to the unfolding transition, then the calorimetric enthalpy for that transition will be reduced accordingly, even though one might be unaware of this problem from simple concentration measurements. Interestingly, the van't Hoff enthalpy is not affected by such impurities, provided they don't interfere with the cooperative transition of the correctly folded fraction.

2.4 Units

The SI unit for energy is the joule (J). Consequently, the conventional units are kJ mol^{-1} for molar thermodynamic energies such as enthalpy (H) or free energy (G) and J mol^{-1} K^{-1} for molar entropy (S) or heat capacity (C_p). Despite this, many (particularly in the US) still use the older system of units based on the calorie, and some instruments (e.g. Microcal) are still calibrated in such units. For conversion: 1 calorie = 4.184 J ; the gas constant, R = 1.987 cal mol^{-1} K^{-1} = 8.314 J mol^{-1} K^{-1}.

2.5 Scan rates/reversibility

A scan rate of 60 °C/h is usually adequate for simple, reversible unfolding transitions, and in theory the DSC thermogram should be unaffected by use of different scan rates. However, there are many instances of kinetically-determined irreversible process (such as aggregation or chemical degradation at higher temperatures) that can affect the shape of the thermogram, and which are scan rate-dependent. It is always prudent to repeat experiments at different scan rates to determine whether this is a problem in particular instances. Analysis of DSC data in such cases is beyond the scope of the current chapter, but details may be found in refs 5 and 6.

2.6 The baseline problem

Reliable interpolation of the baseline is crucial to the estimation of both calorimetric and van't Hoff enthalpies of a DSC transition, since both the area under the thermogram and its shape will be affected by this. There are two separate aspects to this that one might refer to as the 'instrumental' baseline and the 'sample' baseline problems, respectively.

The instrumental baseline is quite straightforward and is just the measured DSC response one would get in the absence of sample. This is typically obtained from scans under identical conditions using sample buffer or appropriate solvent in both cells of the DSC. Since instrumental baselines are susceptible to long-term drift and can vary with ambient conditions, such measurements are best made on a well-equilibrated instrument, both before and after the experimental scans. For samples involving irreversible transitions, some workers prefer to use a second sample scan as baseline. 'Annealing' of the sample prior to the transition can also be useful. Since the greatest variation in instrumental baseline usually occurs in the first scan after reloading the DSC cell, due to the relatively large thermal disturbance that this involves, heating the sample in the DSC one or more times to a temperature below the onset of the transition and cooling, prior to execution of the full scan ('annealing'), can give more reliable baseline stability and reproducibility.

Estimation of the sample baseline is a thornier problem. Since the heat capacity baseline of a sample rarely returns to the same level after the transition as it was before, because of ΔC_p effects, one needs to be able to estimate what the sample baseline might have been in the region under the endotherm peak in the absence of the transition. A typical DSC endotherm for a simple globular protein undergoing a cooperative two-state unfolding transition is illustrated in *Figure 2*. At any point under the transition endotherm, the sample comprises a mixture of folded and unfolded proteins, and the problem is how best to estimate what the heat capacity of this mixture should be. Various strategies have been adopted and are illustrated in *Figure 4*.

The 'progress baseline' is obtained by extrapolation of pre- and post-transition baselines and, at any particular temperature, calculating the baseline heat capacity in proportion to the estimated amounts of folded and unfolded

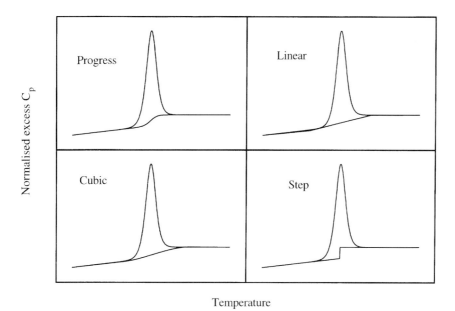

Figure 4 Examples of the different baseline assumptions that may be used in analysis of DSC transitions. The data here correspond to a typical two-state cooperative unfolding transition, calculated for a 25 kDa protein with T_m of 50 °C, $\Delta H_{cal} = \Delta H_{VH} = 100$ kcal mol^{-1}, ΔC_p (at T_m) = 1.5 kcal mol^{-1} K^{-1}, with a pre-transition slope of 0.025 kcal mol^{-1} K^{-2}, and a post-transition slope of zero. Note that this means that the ΔC_p is temperature-dependent in this case.

material present at that temperature from the area under the curve. Simpler approaches use a step baseline at the mid-point of the transition—either at the peak of the thermogram or at a halfway point in terms of area under the curve. Alternatively one may choose linear or quadratic interpolations of pre- and post-transition baselines. The next step in any DSC analysis is usually to subtract the chosen baseline prior to area integration or fitting of the unfolding endotherm, and in practice there seems little to choose between the various methods of baseline correction: each of them produces inevitable distortion in the corrected thermogram that can affect both the apparent shape and area under the transition that translate into possible errors in the ΔH_{cal} and ΔH_{VH} estimates (see *Table 1* for example). However, for good data these differences are relatively small, and usually smaller than errors arising from concentration estimates or instrumental baseline drifts.

Interestingly, and paradoxically, even in cases where the chosen baseline correction is patently wrong, an empirical relationship has been discovered that combines the apparent computed ΔH_{cal} and ΔH_{VH} to give an enthalpy close to reality (but only for transitions that are truly two-state and for which experimental data, including protein concentrations, are otherwise correct). This comes about because errors in choice of baseline correction tend to have opposing effects on the calorimetric and van't Hoff enthalpies. Any baseline error that

Table 1 Effects of different baseline assumptions on deconvoluted DSC data[a]

	Baseline	T_m	ΔH_{cal}	ΔH_{VH}	ΔH_{WA}
Figure 4:	Progress	49.9	99.3	100	99.8
	Linear	50.1	104	97.5	99.8
	Cubic	50.1	105	96.9	99.7
	Step	49.7	99.5	99.7	99.6
Figure 5:	No. 1	50.2	109	94.3	99.4
	No. 2	50.1	96.8	103	100.8
	No. 3	50.2	110	91.3	97.8
	None	50.3	143	73.8	98.0
	(True)	50	100	100	100.0

[a] Hypothetical data were calculated for a two-state transition with T_m = 50 °C, ΔH_{cal} = ΔH_{VH} = 100 kcal mol^{-1}, and ΔC_p = 1.5 kcal K^{-1} mol^{-1} at T_m, with pre- and post-transition baseline slopes of 0.025 and 0 kcal K^{-2} mol^{-1}, respectively. The various baselines were then subtracted, and data fitted using Microcal ORIGIN software to a two-state model that independently estimates apparent ΔH_{cal} and ΔH_{VH} values. These were then combined using the empirical equation (see text, Section 2.6) to give the weighted average ΔH_{WA}.

reduces the apparent area under the peak, thus lowering ΔH_{cal}, will also tend to sharpen the peak, raising the estimate of ΔH_{VH}. Conversely, any baseline correction that broadens the transition endotherm, giving a reduced estimate of ΔH_{VH}, will also increase the area under the curve to give a higher apparent ΔH_{cal}. This is illustrated in *Figure 5* and *Table 1* using ideal calculated data for a representative two-state transition.

Table 1 shows how, even for perfect data, the choice of baseline can lead to distortions that affect subsequent estimates of calorimetric and van't Hoff enthalpies, usually in opposite directions. The differences are small and usually experimentally insignificant between the various conventional progress, step, or interpolation baselines. However, the differences in apparent ΔH_{cal} and ΔH_{VH} are much larger when seriously distorted baselines are involved—the sort of thing that can arise experimentally from baseline fluctuations due to particulate matter in the sample or other instrumental 'glitches'. Empirically we have found that, even in such pathological cases, by combining the enthalpy estimates using the following formula:

$$\Delta H_{WA} = 0.65 \times \Delta H_{VH} + 0.35 \times \Delta H_{cal} \qquad [1]$$

one obtains a weighted average estimate (ΔH_{WA}) remarkably close to the true value. This relationship was first obtained by Haynie (7) as a means of correcting data in special cases, but we have subsequently shown that the formula is more generally applicable to any kind of baseline uncertainty. The reason for the success of this relationship is not fully clear, but probably stems from the mathematical properties of curves of this kind, and the inverse correlation between peak area and peak width during baseline interpolations. Despite this, one must be very cautious in using any such empirical relationship as a substitute for good experimental technique. Differences in ΔH_{cal} and ΔH_{VH} often arise for reasons

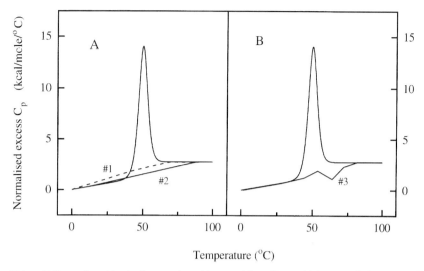

Figure 5 Some (exaggerated) examples of incorrect baselines which, nevertheless, can give reasonable estimates of ΔH when using the empirical equation for ΔH_{WA} (see *Table 1*). The data are calculated as in *Figure 4*. (A) Alternative, incorrect linear baselines. (B) A baseline 'glitch', such as might arise experimentally from electronic interference, or from the formation of a small air bubble in the DSC cell, or from convection artefacts due to particulate matter in the sample or reference solution.

other than poor baseline correction, and application of this empirical formula would be inappropriate in such cases.

3 Thermodynamic background

For many applications it is not necessary to have a full understanding of the theoretical thermodynamic background—and it is perfectly possible to use the calorimeter as a convenient qualitative analytical instrument, just as one might use many other devices, without regard to theory. However, in order to fully appreciate the quantitative limitations on experimental observations and their thermodynamic interpretation, it is preferable to have an understanding of at least some of the basics. This is particularly important if one wishes to avoid some of the more common pitfalls in the (over)interpretation of thermodynamic data.

3.1 Heat capacity, enthalpy, and entropy

Differential scanning calorimetry of the kind used here measures the excess heat capacity of the sample solution with respect to the reference (usually aqueous) solvent. The heat capacity (or specific heat) of any substance (usually designated C_p at constant pressure) reflects the ability of the substance to absorb heat energy without increase in temperature, and this is central to DSC measurements and to the fundamental underlying thermodynamics. Liquid water has a relatively high C_p because of the extensive ice-like hydrogen-bonded network in the liquid that allows heat energy to be used up in breaking bonds between water molecules rather than increasing their kinetic energy (i.e. temperature). Organic matter, including proteins and nucleic acids, has a lower specific heat than water, except possibly when undergoing some process such as unfolding or melting involving breaking of bonds. Consequently the heat capacity of a dilute biomolecule solution is dominated by the water in the system and great care has to be taken to subtract this in any DSC measurements to give the excess differential heat capacity contribution arising from the process of interest.

Heat capacity is the fundamental property from which all thermodynamic quantities may be derived. In particular, the absolute enthalpies (H) and entropies (S) of any substance are related to the total heat energy uptake involved in the (imaginary) process of heating from absolute zero to temperature T, as represented in the following integrals:

$$H = \int_0^T C_p.dT + H_0 \quad [2]$$

where H_0 is the ground state energy (at 0 K) due to chemical bonding and other non-thermal effects, and since classically from the 2nd law of Thermodynamics:

$$dS = dH/T = (C_p/T).dT \quad [3]$$

it follows that:

$$S = \int_0^T (C_p/T).dT \quad [4]$$

The molecular interpretation of H (enthalpy, or heat content) is fairly easy to grasp since it is just the total energy (including pressure/volume work terms) taken up in raising the system to temperature T whilst keeping the pressure constant. This will include the energy associated with all the atomic and molecular motions—translation, rotation, vibration, etc.—together with energy taken up in changes in inter- and intramolecular interactions ('bonds'). By contrast, the absolute entropy (S) is a rather more difficult quantity to comprehend. Usually it is described in terms of 'molecular disorder'—the higher the disorder the higher the entropy—but this obscures the connection with heat capacity evident in the above integral definition. Perhaps a better way of viewing entropy is as the multiplicity of ways in which the molecules in a system can take up energy without increasing temperature.

The *magnitude* of the heat capacity depends on the numbers of ways there are of distributing any added heat energy to the system, and so is related to entropy. Consider the energy required to bring about a one degree rise in temperature. If a particular system has only relatively few ways of distributing the added energy, then relatively little energy will be required to raise the temperature, and such a system would have relatively low C_p. If, however, there are lots of different ways in which the added energy can be spread around amongst the molecules in the system (such as different modes of vibration and rotation, or breaking of bonds), then much more energy will be needed to bring about the same temperature increment. Such a system would have a high C_p. In this way, adding heat to anything increases the entropy by giving the molecules more energy to explore many more different ways of arranging themselves (and become 'more disordered').

3.2 Equilibrium and free energy

Chemical stability and thermodynamic equilibrium represent a balance between two opposing tendencies: first the natural trend for systems to move to lower energies (decrease H), and secondly the equally natural tendency at the molecular level for molecules to explore the multiplicity of states available (higher S) under the influence of disruptive thermal motions. This is represented by the Gibbs Free Energy change (ΔG) expression:

$$\Delta G = \Delta H - T.\Delta S \qquad [5]$$

which tells us how much work must be done to bring about the desired change. (Changes can occur spontaneously if ΔG is negative, but require the input of energy if positive. Systems are in equilibrium if $\Delta G = 0$.)

Free energies and other thermodynamic parameters are relative quantities that depend on an arbitrary choice of reference or standard state. It turns out that the equilibrium constant (K) for any process is related to the 'standard' Gibbs free energy change:

$$\Delta G° = -RT.\ln(K) = \Delta H° - T.\Delta S° \qquad [6]$$

(R = gas constant, 8.314 J K^{-1} mol^{-1} or 1.987 cal K^{-1} mol^{-1}) representing the free energy change $\Delta G°$, together with the constituent enthalpy $\Delta H°$ and entropy $\Delta S°$ changes, that would take place in the (hypothetical) standard state in which reactants and products (initial and final states) were all present at 1 M concentration (or activity). (This convention adopting 1 M concentration for standard states in solution, clearly unrealistic for biomolecular systems, is a consequence of an historical choice of standard units for measuring concentration, but remains an appropriate way of comparing interaction free energies and other parameters on the same scale.) A convenient way to view $\Delta G°$ is simply as the equilibrium constant, K, expressed on a logarithmic energy scale. ΔH and $\Delta H°$ are practically identical under most conditions, but ΔS and $\Delta S°$ will normally differ significantly due to large entropy of mixing effects at different concentrations.

3.3 Temperature dependence of thermodynamic quantities

Changes in enthalpy and entropy (ΔH and ΔS) as a system changes from one state to another (A → B) at constant temperature follow directly from the integral definitions:

$$\Delta H = H_B - H_A = \int_0^T \Delta C_p .dT + \Delta H(0) \qquad [7]$$

$$\Delta S = S_B - S_A = \int_0^T (\Delta C_p/T).dT \qquad [8]$$

where $\Delta C_p = C_{p,B} - C_{p,A}$ is the heat capacity *difference* between states A and B at a given temperature. $\Delta H(0)$ is the ground state enthalpy difference between A and B at absolute zero. Most systems are assumed to have the same (zero) entropy at 0 K (3rd Law of Thermodynamics).

It is both conventional and convenient to relate these quantities to some standard reference temperature T_{ref} (e.g. T_{ref} = 25 °C or 298 K, rather than absolute 0 K), in which case:

$$\Delta H(T) = \Delta H(T_{ref}) + \int_{T_{ref}}^{T} \Delta C_p \, .dT \qquad [9]$$

and

$$\Delta S(T) = \Delta S(T_{ref}) + \int_{T_{ref}}^{T} (\Delta C_p / T) .dT \qquad [10]$$

This illustrates how, if there is a finite ΔC_p between two states (as is the norm, for example, in protein unfolding or other processes involving multiple, weak, non-covalent interactions), then ΔH and ΔS are both temperature dependent.

If ΔC_p is constant, and does not vary with temperature (not altogether true for protein transitions, but usually a reasonable approximation over a limited temperature range), then we can integrate the above to give *approximate* expressions for the temperature dependence of ΔH and ΔS with respect to some arbitrary reference temperature (T_{ref}):

$$\Delta H(T) \cong \Delta H(T_{ref}) + \Delta C_p \cdot (T - T_{ref}) \qquad [11]$$

$$\Delta S(T) \cong \Delta S(T_{ref}) + \Delta C_p \cdot \ln(T/T_{ref}) \qquad [12]$$

Interestingly (16) these temperature effects will largely cancel in the standard free energy expression to give a ΔG that is relatively much less affected by temperature change. This is an example of 'entropy–enthalpy compensation' or 'linear free energy' effects that are often, if not universally, found in systems involving multiple weak interactions (8, 9).

For protein unfolding it is sometimes convenient to take the mid-point (T_m) of the transition as a reference temperature. Here, T_m is defined as the temperature at which ΔG for the transition is zero, so that $\Delta H(T_m) = T_m \cdot \Delta S(T_m)$ and consequently the temperature dependence of the unfolding free energy may be written:

$$\Delta G_{unf}(T) = \Delta H_{unf}(T_m).\{1 - T/T_m\} + \Delta C_p.\{T - T_m - T.\ln(T/T_m)\} \qquad [13]$$

assuming ΔC_p does not itself change with temperature. Typical data showing this temperature variation for thermodynamic parameters of unfolding of a globular protein are shown in *Figure 6*. This illustrates how quite large changes in ΔH and ΔS tend to compensate to give relatively much smaller changes in ΔG. This also illustrates how the T_m for the transition, given by the point of intersection of the ΔH and $T.\Delta S$ lines, can be quite sensitive to relatively small changes in either ΔH or $T.\Delta S$, even though they may make relatively little change to ΔG. It is partly for this reason that prediction of the effects brought about by environmental changes or mutagenesis on protein stability is so difficult.

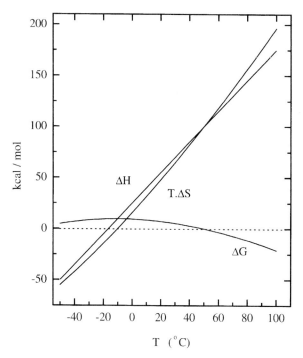

Figure 6 Temperature variation of the thermodynamic parameters for unfolding of a typical globular protein, calculated for a 25 kDa protein undergoing a two-state transition with T_m of 50 °C, $\Delta H_{cal} = \Delta H_{VH} = 100$ kcal mol^{-1}, with a temperature-independent $\Delta C_p = 1.5$ kcal mol^{-1} K^{-1}. The mid-point of the transition (T_m) is given by the point of intersection of the ΔH and $T.\Delta S$ lines, where ΔG crosses the zero axis.

3.4 The van't Hoff enthalpy

Although enthalpy changes (ΔH) can be measured directly by calorimetric techniques, it is frequently convenient to compare these with indirect estimates using the classical van't Hoff equation, arising from the temperature dependence of the equilibrium constant (K) as one might get, for example, from following the thermal unfolding of a protein by fluorescence, circular dichroism, or other indirect technique. Such information can also be obtained from DSC experiments (see below).

Given that:
$$-RT.\ln K = \Delta H° - T.\Delta S° \qquad [14]$$
it follows that:
$$\ln K = -\Delta H°/RT + \Delta S°/R \qquad [15]$$
and, in the general case where $\Delta H°$ and $\Delta S°$ may also vary with temperature:
$$d(\ln K)/d(1/T) = -\Delta H°/R - (1/RT)[d(\Delta H°)/d(1/T)] + (1/R)[d(\Delta S°)/d(1/T)] = -\Delta H°/R \qquad [16]$$
since:
$$d(\Delta H°)/d(1/T) = -T^2.d(\Delta H°)/dT = -T^2.\Delta C_p \qquad [17]$$

and:

$$d(\Delta S°)/d(1/T) = -T^2 \cdot d(\Delta S°)/dT = -T^2 \cdot \Delta C_p/T \quad [18]$$

leads to cancellation of the latter two terms in the above equation.

Thus, a plot of experimental data of $\ln K$ vs. $1/T$ ('van't Hoff plot') gives a line whose slope at any point is the van't Hoff enthalpy ($\Delta H°$ or ΔH_{VH}) divided by R. In simple cases, usually over a limited temperature range, this plot is linear (or is assumed to be so, within experimental error), but in general the temperature dependence of ΔH (due to ΔC_p) will give curvature of the van't Hoff plot that needs more careful analysis (10).

One must be clear about what is meant by the 'van't Hoff enthalpy' and to what it refers. Any van't Hoff analysis is based on a model or hypothesis of the process involved, needed in order to define K. Typically, for protein unfolding transitions, this model will be a 'two-state' picture in which the equilibrium constant K is a dimensionless ratio determined, usually indirectly, from spectroscopic, calorimetric, or other measurements. In such a model the molar van't Hoff enthalpy change, ΔH_{VH}, is the enthalpy change per mole of cooperative unit as defined by the model (3, 11). Comparison of this with the directly measured calorimetric value can frequently be informative. There are two simple ways in which the van't Hoff enthalpy might differ from the calorimetric enthalpy in protein unfolding. First, the model may simply be wrong. For example, if the unfolding transition is not two-state, but involves one or more intermediate steps, then the transition will appear broader than anticipated, and the ΔH_{VH} will be less than ΔH_{cal}. Alternatively, if the protein unfolds cooperatively as a dimer, or higher oligomer, then the transition will be sharper than anticipated for the two-state transition of a monomer, and the ΔH_{VH} will be correspondingly greater than ΔH_{cal}. One must be wary, however, of placing too much reliance on such comparisons in the absence of supporting evidence from other methods, since other factors can affect the shape of the transition. For example, the irreversible and usually exothermic aggregation of thermally unfolded protein can distort and sharpen the DSC transition, leading to incorrect estimates of both ΔH_{cal} and ΔH_{VH}. Moreover, as indicated in an earlier section, any errors arising out of impurities or incorrect concentration estimates will be reflected in ΔH_{cal} (though not necessarily in ΔH_{VH}) and will give rise to erroneous ΔH_{cal}: ΔH_{VH} ratios.

4 Effects of ligand binding

Application of Le Chatelier's principle, or simple equilibrium considerations, shows that if any ligand (small molecule or other protein or macromolecule) binds preferentially to the folded or native form of the protein, then this will stabilize the folded state, and unfolding of the protein will become progressively less favourable as ligand concentration increases. Conversely, ligands that bind preferentially to the unfolded protein will destabilize the fold and will encourage unfolding. Examples of both are commonly seen (12–15).

4.1 Ligand binding and folding equilibrium

4.1.1 Ligand binds to the folded protein

General theories for binding of multiple ligands and multiple protein may be found in refs 12, 13, 16. For the simplest case in which a ligand molecule (L) binds specifically to a single site on the native folded protein (N), the following equilibria apply.

Ligand binding:
$$N + L \rightleftharpoons NL$$
$$K_{L,N} = [N][L]/[NL] \qquad [19]$$

Unfolding:
$$N \rightleftharpoons U$$
$$K_0 = [U]/[N] \qquad [20]$$

where $K_{L,N}$ defines the dissociation constant for ligand binding to the native protein and K_0 is the unfolding equilibrium constant for the protein in the absence of ligand.

In general the effective unfolding equilibrium constant (K_{unf}) is given by the ratio of the total concentrations of unfolded to folded species:

$$K_{unf} = [U]/([N] + [NL]) = K_0/(1 + [L]/K_{L,N}) \approx K_0 K_{L,N}/[L] \qquad [21]$$

where the final approximate form applies only at high free ligand concentrations ($[L] > K_{L,N}$). This confirms the expectation that K_{unf} decreases and the folded form becomes more stable with increasing ligand concentration.

This can be expressed alternatively in free energy terms:

$$\Delta G_{unf} = -RT.\ln(K_{unf}) = \Delta G_{unf,0} + RT.\ln(1 + [L]/K_{L,N}) \qquad [22]$$
$$\approx \Delta G_{unf,0} + \Delta G°_{diss,N} + RT.\ln[L] \qquad ([L] \gg K_{L,N}) \qquad [23]$$

where $\Delta G_{unf,0}$ is the unfolding free energy of the unliganded protein, and $\Delta G°_{diss,N} = -RT.\ln(K_{L,N})$ is the standard Gibbs free energy for dissociation of the ligand from its binding site on the native protein. The approximate form again applies only at high ligand concentrations.

This illustrates how the stabilizing effect of bound ligand can be visualised as arising from the additional free energy required to remove the ligand prior to unfolding, together with an additional contribution ($RT.\ln[L]$) from the entropy of mixing of the ligand when released into the bulk solvent.

In the approximate form at high ligand concentrations the free energy can be broken down into the separate enthalpy and entropy contributions:

$$\Delta H_{unf} \approx \Delta H_{unf,0} + \Delta H°_{diss,N} \qquad [24]$$

and

$$\Delta S_{unf} \approx \Delta S_{unf,0} + \Delta S°_{diss,N} - R.\ln[L] \qquad [25]$$

In many cases the heat of ligand dissociation ($\Delta H°_{diss,N}$) might be quite small compared to the heat of unfolding of the protein, especially in the case of small ligands, and can be hard to distinguish in calorimetric unfolding experiments. This can be further complicated when ΔH_{unf} is also varying with temperature due to ΔC_p effects. Entropy effects, particularly those arising from the ligand mixing term (R.ln[L]), will dominate in such cases. [Similar considerations apply at lower concentrations of ligand, though the algebraic expressions are a little more complicated. In such cases the thermodynamic parameters are intermediate between unliganded and fully-liganded values found in the high concentration limit.]

A typical example of the effects of small ligand binding to a globular protein is described in *Protocol 2* for the case of 2′-CMP binding to ribonuclease A with typical DSC data shown in *Figure 7*.

Protocol 2

Ligand binding to folded protein—DSC of RNase with 2′-CMP

Equipment and reagents

- DSC and related equipment, as in *Protocol 1*
- Protein: ribonuclease A (RNase; Sigma R-5500; M ~ 13 700, $\varepsilon_{280} = 10\,550\ M^{-1}\ cm^{-1}$
- Buffer: 0.1 M Na acetate pH 4.5
- Ligand: cytidine 2′-monophosphate (2′-CMP; Sigma C-7137)

Method

1 Prepare a stock solution of RNase in buffer (sufficient for several experiments) and determine the concentration by UV absorbance. Take portions of this stock solution and add sufficient 2′-CMP (by weight) to give a series of samples containing 0–1.5 mM ligand. Note:

   (a) It is not usually necessary to dialyse the protein solution in this case, since commercial samples of RNase are usually sufficiently pure for these purposes.

   (b) Since 2′-CMP also absorbs in the 260–280 nm region, it is necessary to determine the RNase concentration prior to adding the ligand.

   (c) For detailed work it will also be necessary to prepare identical concentrations of CMP in buffer alone, to use as reference.

2 For each of the RNase/2′-CMP mixtures separately, degas and run DSC experiments as described in *Protocol 1*.

3 Analyse the data using a simple two-state model to determine T_m, ΔH_{cal}, and ΔH_{VH}.

4.1.2 Ligand binds to unfolded protein

The same approach can be applied to situations where ligand binds only to the *unfolded* protein (14, 15):

$$U + L \rightleftharpoons UL$$

$$K_{L,U} = [U][L]/[UL] \quad [26]$$

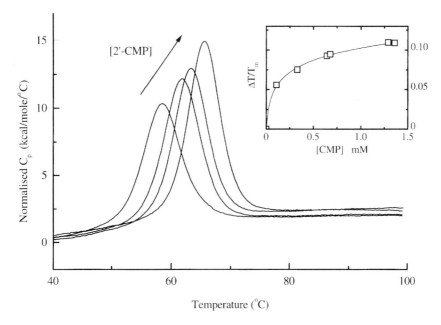

Figure 7 DSC thermograms of ribonuclease A unfolding in the presence of various concentrations of 2'-CMP, illustrating the increase in T_m brought about by binding of ligand to the native protein. The inset shows the relative T_m shift as a function of total ligand concentration, with the non-linear regression fit to Equation 33 (Experimental conditions: 0.1 M Na acetate buffer pH 4.5, 80 µM ribonuclease, 0–1.5 mM CMP).

for which:

$$K_{unf} = ([U] + [UL])/[N] = K_0 \cdot (1 + [L]/K_{L,U}) \approx K_0 \cdot [L]/K_{L,U} \quad [27]$$

and:

$$\Delta G_{unf} = -RT \cdot \ln(K_{unf}) = \Delta G_{unf,0} - RT \cdot \ln(1 + [L]/K_{L,U}) \quad [28]$$

$$\approx \Delta G_{unf,0} - \Delta G°_{diss,U} - RT \cdot \ln[L] \text{ (for high [L])} \quad [29]$$

This illustrates the destabilizing effect of a reduction in unfolding free energy as ligand binds to the unfolded polypeptide. Equivalent expressions for the enthalpy and entropy contributions may be written as above, with appropriate sign changes.

An example of this kind of effect is illustrated in *Figure 8* for the unfolding of globular proteins in the presence of cyclodextrins, with practical details given in *Protocol 3*. The cyclodextrins are a family of toroidal oligosaccharide molecules that form inclusion complexes with small non-polar molecules and therefore bind to exposed aromatic groups on the unfolded protein (14). This gives rise to a decrease in thermal stability of the protein with increasing cyclodextrin concentration that can be analysed in terms of the simple models described here.

Note: the apparent variation in ΔH_{cal} is predominantly due to the inherent variation of unfolding enthalpy with temperature (ΔC_p effect) rather than the result of ligand binding *per se*.

Protocol 3

Ligand binding to unfolded protein—DSC of lysozyme with cyclodextrin

Equipment and reagents

- DSC and related equipment, as in Protocol 1
- Buffer: 40 mM glycine/HCl pH 3.0
- Protein: hen egg white lysozyme (e.g. Sigma L-6876), typically 2 ml at a concentration of 1 mg/ml (for MCS) or 1 ml at 0.1 mg/ml (for VP-DSC)
- Ligand: α-cyclodextrin (Sigma-Aldrich supply a range of cyclodextrins, including some of the more soluble methyl- and hydroxypropyl- derivatives of β-cyclodextrin which may also be used for this experiment.) Cyclodextrins are quite hygroscopic. They should be vacuum dried before use and stored desiccated.

Method

1. Prepare 5–15% (w/v) solutions of cyclodextrin in buffer and use these to make up appropriate solutions of lysozyme, retaining sufficient cyclodextrin buffer solution for use as reference.

2. Proceed with DSC experiments, as in Protocols 1 and 2. Cyclodextrins do not normally absorb in the near UV region, so protein concentrations may be determined directly from A280, as before.

3. Analyse the data using a simple two-state model to determine T_m, ΔH_{cal}, and ΔH_{VH}.

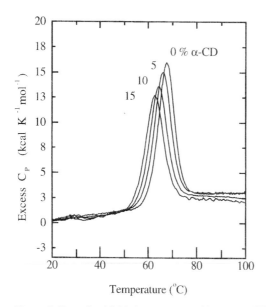

Figure 8 Normalized DSC thermograms of lysozyme (40 mM glycine/HCl pH 3.0) in the presence of 0, 5, 10, 15% (w/v) α-cyclodextrin.

4.2 Change in T_m

In DSC experiments the most obvious manifestation of ligand binding effects is a change in the apparent T_m of the protein under investigation. This can be calculated from the above expressions in the following way. The Gibbs free energy of unfolding at any temperature, T, is given by:

$$\Delta G_{unf} = -RT.\ln(K_{unf}) = \Delta G_{unf,0} \pm RT.\ln(1 + [L]/K_{L,N\,or\,U}) \quad [30]$$

and this will be zero at the mid-point (T_m) of the transition. Consequently, at T_m:

$$\Delta G_{unf,0} \pm RT_m.\ln(1 + [L]/K_{L,N\,or\,U}) = 0 \quad [31]$$

Ignoring any ΔC_p effects (i.e. assuming, for simplicity, that the enthalpy of unfolding is relatively constant over the temperature range considered here), the free energy of unfolding in the absence of ligand binding may be expressed in terms of the enthalpy of the transition ($\Delta H_{unf,0}$) at the mid-point temperature (T_{m0}):

$$\Delta G_{unf,0} = \Delta H_{unf,0} (1 - T/T_{m0}) \quad [32]$$

So that, after substitution in the previous equation and rearrangement:

$$\Delta T_m/T_m = \pm (RT_{m0}/\Delta H_{unf,0}).\ln(1 + [L]/K_L) \quad [33]$$

in which $\Delta T_m = T_m - T_{m0}$ is the change in unfolding transition temperature, and the $\pm$ sign relates to whether ligand stabilizes the folded ($K_L = K_{L,N}$) or unfolded form ($K_L = K_{L,U}$), respectively. (Remember: ligand binding to the native fold, N, will increase T_m, whereas binding to the unfolded polypeptide, U, will decrease T_m.)

Important note: In the above derivations, we have defined T_m as the temperature at the mid-point of the transition, where ΔG_{unf} is zero and we have equal populations of folded and unfolded species. In DSC experiments involving simple single transitions, this usually corresponds to the peak of the thermogram (or is very close to this). However, in some situations with strongly binding ligands at less than stoichiometric concentrations (see below) one may see two peaks in the DSC trace, corresponding to separate unfolding of the free protein and protein–ligand complex. In such cases the 'T_m', as defined here for the entire mixture, lies at some point midway between these two peaks.

For the slightly more general case of multiple weakly binding ligands (15) these equations can be extended to give:

$$\Delta T_m/T_m = \pm (nRT_{m0}/\Delta H_{unf,0}).\ln(1 + [L]/K_L) \quad [34]$$

where n is the number of ligand binding sites on the protein (assumed identical). This is the behaviour encountered in the effects of cyclodextrins on folding stability, illustrated in *Figure 8*, where data fit best to a model assuming multiple binding sites on the unfolded polypeptide. These sites may be identified as the aromatic amino acid side chains exposed during unfolding and to which the cyclodextrin molecules may attach (14, 15).

At low concentrations, with weakly binding ligands ($[L]/K_L \ll 1$) the expression describing the variation in T_m becomes approximately linear in ligand concentration:

$$\Delta T_m / T_m \approx \pm nRT_{m0} [L] / (K_L \cdot \Delta H_{unf,0}) \qquad [35]$$

It is important to note that the T_m shift continues with increasing ligand concentration even beyond levels where the protein is fully ligand-bound. This is a manifestation of the dominant entropy of mixing contribution described above. A common misconception here is that it is the 'bonds' formed between the ligand and the protein that somehow hold the protein in a more stable conformation. But if this were the case, then no further stabilization would occur once all protein sites were saturated, and this is clearly contrary to observation. The thermodynamic rationalization is that enhanced stability arises from the additional free energy required to remove the ligand from the protein prior to its unfolding (or vice versa), and this free energy has an important component arising from the entropy of mixing of dissociated ligand. This has nothing to do with protein conformation, but depends on the relative concentration of free ligand in solution.

4.3 Effects when ligand binds to both N and U

Occasionally cases arise where, contrary to the above argument, the T_m shift does reach a plateau at higher ligand concentrations. This usually signifies binding of L to *both* N and U, albeit with different affinities. For example, a particular ligand might bind strongly to the native protein but less well to the unfolded chain. In such cases the T_m would shift upwards with increasing [L] until the concentration is such that both N and U are fully liganded. An example of this is found with α-lactalbumin, a specific calcium binding protein where increasing [Ca^{2+}] progressively stabilizes the native protein up to a limit where weak, non-specific calcium ion binding to the unfolded chain sets in (*Figure 9*; Robertson, Cooper, and Creighton—unpublished observations). Following the same procedures as before, one can show that the T_m shift in such cases is given by:

$$\Delta T_m / T_m = (RT_{m0} / \Delta H_{unf,0}) \cdot \ln\{(1 + [L]/K_{L,N}) / (1 + [L]/K_{L,U})\} \qquad [36]$$

In the high concentration limit ([L] $\gg K_{L,N}$ and $K_{L,U}$) this becomes constant, independent of ligand concentration:

$$\Delta T_m / T_m = (RT_{m0} / \Delta H_{unf,0}) \cdot \ln\{K_{L,U}/K_{L,N}\} \qquad [37]$$

with the plateau value depending on the ratio of ligand binding affinities to the folded and unfolded protein. This also shows that no T_m shift occurs if the ligand binds equally well to both states, and simply the ability to bind to the folded protein is no guarantee that a ligand will stabilize the fold. As a rough rule of thumb, each factor of 10 in the ratio of ligand binding affinities, $K_{L,U}/K_{L,N}$, will give a ΔT_m of about 5 °C.

4.4 One peak or two?

Analysis of more complex situations involving multiple ligand binding or more tightly binding ligands is generally less straightforward than outlined above,

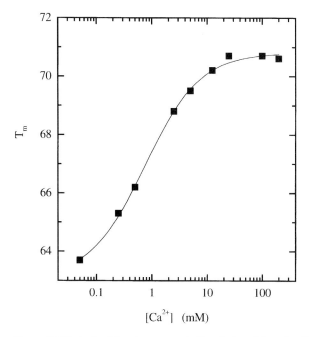

Figure 9 Effect of calcium ion concentration (log scale) on the T_m of bovine α-lactalbumin, determined from DSC experiments in 0.1 M Tris–HCl buffer pH 8.0, containing 0.1 M NaCl. The plateau in T_m at high [Ca^{2+}] indicates binding to the unfolded state, as well as the stabilization brought about by binding to the native fold.

although the same basic principles apply. See refs 12, 16 for details. One interesting problem is related to how many peaks one might expect to see in a DSC endotherm for a mixture of protein plus ligand. In the examples looked at so far, the ligand binding has been relatively weak, and usually the ligand concentration is significantly greater than the protein. In such cases (assuming a simple two-state process) one sees not two peaks corresponding to bound and free protein, but rather a single endotherm, gradually shifting up or down in T_m as the ligand concentration is varied. This is because at any one time the system is in rapid dynamic equilibrium between ligand-bound and ligand-free states, and what we see is a thermodynamic average of the two—assuming, as is usually the case, that on/off exchange of ligand is fast on the DSC time scale.

There are occasions, however, when one might observe separate discrete peaks in a DSC experiment arising from the separate unfolding of apo- and ligand-bound protein, particularly in the case of very tightly binding ligands at sub-stoichiometric concentrations.

One such example of this kind of behaviour has been seen in DSC experiments (*Figure 10*) of the thermal unfolding of a repressor protein (the methionine repressor, MetJ) in the presence of increasing concentrations of a specific DNA fragment corresponding to its consensus target (17, 18). Here, when the protein concentration is in excess over the DNA, we see two peaks in the DSC trace corresponding to separate unfolding of free protein (lower T_m) and protein–DNA complex (higher T_m), respectively. The proportion of the endotherm corresponding to the higher temperature transition increases with increase in DNA fragment concentration until, when the DNA is equimolar with repressor protein, only a single transition corresponding to unfolding of protein–DNA complex is seen. (The melting of the DNA duplex used in these experiments does not occur until even higher temperatures under the experimental conditions used.)

Protocol 4
DSC of protein–DNA complex

Equipment and reagents

- DSC instrument (Microcal MC2, VP-DSC, or equivalent)
- Dialysis tubing or cassettes (e.g. Pierce Slide-A-Lyser®)
- Buffer
- Degassing equipment (vacuum desiccator, magnetic stirrer)
- Protein + DNA: typically 2 ml at a (protein) concentration of 1 mg/ml

A. Sample preparation

1 Prepare the protein:DNA mixture and dialyse several changes of appropriate buffer. Each DSC run will typically require 1–2 ml of protein solution at a concentration of around 1 mg/ml, together with similar stoichiometric amounts of DNA. Depending on conditions, it may be preferable to dialyse the protein and DNA in separate dialysis bags (though in the same pot) for subsequent mixing.

2 Retain the final dialysis buffer for DSC reference, equilibration, and dilutions.

3 Determine the protein and DNA concentrations by 280/260 nm absorbance or other appropriate method.

4 Immediately prior to the experiment, degas portions of the sample mixture and buffer for 2–3 min under gentle vacuum with gentle stirring. Be careful to avoid excessive degassing or frothing of the mixture at this stage.

B. DSC procedure

1 Load DSC sample and reference cells with degassed buffer and collect baseline scan(s) using appropriate temperature range and scan rate (typically 20–100 °C, 60 °C/h).

2 Allow the DSC cells to cool and refill the sample cell with protein:DNA mixture.

3 Repeat the DSC scan(s) using the same parameters as in step 1. (Depending on circumstances, it may be useful to do repeat scans with the same sample to establish reversibility and reproducibility.)

4 After final cooling, remove the sample an examine for turbidity, aggregation, or other visible changes.

5 Process data using instrumental software.

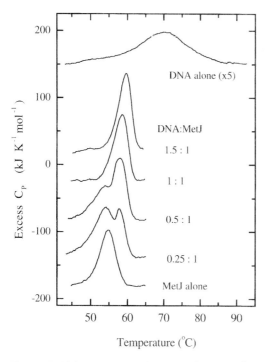

Figure 10 DSC thermograms (concentration normalized and corrected for buffer baselines) for MetJ protein, repressor DNA 16 bp consensus fragment, and protein:DNA mixtures at different stoichiometric ratios; 25 mM phosphate, 0.1 M KCl, 1 mM DTT pH 7; DSC scan rate 60 °C/h. The data for melting of the DNA alone (upper curve) have been expanded for clarity. (Adapted from ref. 6.)

In cases where ligand binding is tight, and where separate unfolding peak might be observed at low ligand:protein ratios, the effects can be calculated as follows.

Assume, for simplicity, a standard two-state unfolding equilibrium in which the ligand binds only to the native state. (Equivalent calculations can be done for other scenarios.) As usual we start with the equilibrium expressions describing ligand dissociation and protein unfolding, respectively:

$$NL \rightleftharpoons N + L$$
$$K_{L,N} = [N][L]/[NL] \qquad [38]$$
$$N \rightleftharpoons U$$
$$K_0 = [U]/[N] \qquad [39]$$

We wish to calculate the concentrations of the different species ([N], [U], [NL], [L]) present at equilibrium under any conditions. The total concentrations of protein and ligand, respectively, are written:

$$C_{tot} = [N] + [NL] + [U] \qquad [40]$$
$$C_L = [NL] + [L] \qquad [41]$$

here we cannot (as we did previously) assume that the free ligand concentration, [L], is the same as the total, C_L, and we must rearrange the above equations and solve for each of the species in turn.

Using the equilibrium constant expressions we get:

$$C_{tot} = [NL] + [N](1 + K_0) = [NL]\{1 + K_{L,N}(1 + K_0)/[L]\} \qquad [42]$$

which, using the expression for C_L, gives a quadratic in [NL]:-

$$[NL]^2 - [NL]\{C_{tot} + C_L + K_{L,N}(1 + K_0)\} + C_L C_{tot} = 0 \qquad [43]$$

with the solutions:

$$[NL] = \tfrac{1}{2}[\{C_{tot} + C_L + K_{L,N}(1 + K_0)\} \pm (\{C_{tot} + C_L + K_{L,N}(1 + K_0)\}^2 - 4 C_L C_{tot})^{1/2}] \qquad [44]$$

By inspection, the negative sign in the $\pm$ is the physically realistic choice, so that the concentration of protein–ligand complex under any conditions is given by:

$$[NL] = \tfrac{1}{2}[\{C_{tot} + C_L + K_{L,N}(1 + K_0)\} - (\{C_{tot} + C_L + K_{L,N}(1 + K_0)\}^2 - 4 C_L C_{tot})^{1/2}] \qquad [45]$$

and the concentrations of other species are obtained by application of the above expressions to give:

$$[N] = (C_{tot} - [NL])/(1 + K_0) \qquad [46]$$

$$[U] = K_0[N] \qquad [47]$$

$$[L] = C_L - [NL] \qquad [48]$$

A procedure for calculating or modelling such behaviour, suitable for simple spreadsheet application, is give in *Protocol 5*, with typical results shown in *Figure 11*. Note how in this particular example, simulating a tight-binding ligand present at half the protein concentration, the unfolding occurs in two discrete steps, corresponding to unfolding of N and NL, respectively. Under these conditions, essentially all the ligand remains bound until released by unfolding of the complex.

Protocol 5

Spreadsheet simulation/calculation

This protocol is provided for those who wish step-by-step instructions to construct a spreadsheet or modelling package to simulate the effects of ligand binding on protein stability. Any proprietary spreadsheet or scientific graphics package should be adequate (e.g. Microsoft Excel™, Microcal ORIGIN™). Here we assume that ligand binds only to the native protein, but the procedure is easily modified for other models.

Constants

1 The Gas constant: $R = 1.987$ (cal K^{-1} mol^{-1}) or 8.314 (J K^{-1} mol^{-1}).
2 $T_0 = 273.15$ (to correct °C to absolute).

Protocol 5 continued

User variables [typical value]

1. T_{m0} — Mid-point T_m without ligand (in °K) — [323 K ≡ 50 °C]
2. $\Delta H_{unf,0}$ — Unfolding enthalpy at T_{m0} — [100 kcal mol^{-1}]
3. ΔC_p — Heat capacity increment — [1.3 kcal mol^{-1} K^{-1}]
4. $K_L(298)$ — Ligand dissociation constant at 25 °C — [10^{-10} M]
5. ΔH_L — Ligand association enthalpy — [−10 kcal mol^{-1}]
6. C_{tot} — Total protein concentration — [10^{-5} M]
7. C_L — Total ligand concentration — [5 × 10^{-6} M]

Spreadsheet columns

A: t — Range of temperatures (e.g. 10–100 °C, in 1° steps).
B: T — $t + T_0$ (convert to absolute temperature).
C: $\Delta G_{unf} = \Delta H_{unf}(T_{m0}).\{1 - T/T_{m0}\} + \Delta C_p.\{T - T_{m0} - T.\ln(T/T_{m0})\}$.
D: K_0 — $\exp\{-\Delta G_{unf}/RT\}$ (unfolding equilibrium at T).
E: K_L — $K_L(298).\exp\{-(\Delta H_L/RT)(1 - T/298)\}$ (ligand equilibrium at T).
F: X — $C_{tot} + C_L + K_L(1 + K_0)$.
G: [NL] ½$\{X - (X^2 - 4C_L C_{tot})^{½}\}$ (concentration of protein–ligand complex at T).
H: [L] $C_L - [NL]$ (concentration of free ligand at T).
I: [N] $K_L[NL]/[L]$ (concentration of native, unbound protein at T).
J: [U] $K[N]$ (concentration of unfolded protein at T).
K: C_p $d[U]/dT$ (unfolding thermogram—arbitrary scale).

Graphical presentation of columns G–K (plotted versus t) will give data such as illustrated in *Figure 11*.

4.5 Hydrogen ions as ligands: the effect of pH on protein stability

The effect observed by varying pH in DSC experiments on proteins is just a special case of the ligand-binding consequences described above for situations where ligand binds to both folded and unfolded states. Here the ligands are aqueous hydrogen ions (H$^+$) that will bind to specific amino acid side chains or terminal groups (acidic or basic) in both folded and unfolded states. Only if the proton binding affinities are different in the two states will pH have any effect on stability.

For proton binding to a single group on the polypeptide, the acid–base equilibrium for folded and unfolded states may be described separately in terms

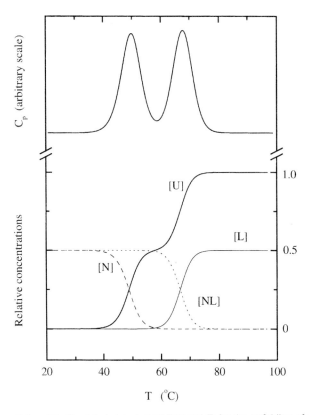

Figure 11 Simulated data (using *Protocol 5*) for the unfolding of a protein in the presence of a tightly binding ligand, using the following parameters: $C_{tot} = 10$ μM, $C_L = 5$ μM, $T_{m0} = 50\,°C$, $\Delta H_{unf,0} = 100$ kcal mol^{-1}, $\Delta C_p = 1.3$ kcal K^{-1} mol^{-1}, K_L (at 25 °C) $= 10^{-10}$ M, $\Delta H_L = -10$ kcal mol^{-1}. The relative concentration of native protein ([N], dashed line) falls with increasing temperature, with a mid-point corresponding to the T_m of the protein in the absence of ligand. The protein–ligand complex ([NL], dotted line) unfolds at a higher temperature, with concomitant release of ligand. The peaks in the C_p thermogram correspond to maxima in the rate of change of [U].

of the usual equilibrium expression involving the acid dissociation constants $K_{A,N}$ and $K_{U,N}$ for the N and U states, respectively:

$$N + H^+ \rightleftharpoons NH^+$$
$$K_{A,N} = [N][H^+]/[NH^+] \qquad [49]$$

$$U + H^+ \rightleftharpoons UH^+$$
$$K_{A,U} = [U][H^+]/[UH^+] \qquad [50]$$

so that apparent or effective equilibrium constant for protein unfolding is given by:

$$K_{unf} = ([U] + [UH^+])/([N] + [NH^+]) = K_0 \cdot (1 + [H^+]/K_{A,U})/(1 + [H^+]/K_{A,N}) \qquad [51]$$

where $K_0 = [U]/[N]$ is the equilibrium constant for unfolding of the unprotonated protein.

Consequently it follows that the stability of the folded protein (with respect to unfolded) can only be affected by changes in pH if $K_{A,N}$ is different from $K_{A,U}$, as might arise for example if unfolding gives rise to a change in environment or electrostatic interactions of the ionizable group.

In more realistic situations with multiple ionizable groups the pH-dependence is somewhat more complex, but the same general principles still apply, and changes in pH can only affect folding stability if the ionizable group(s) have different pK_A values in the folded and unfolded states.

The converse of this argument shows that, where the stability of the folded protein is sensitive to pH, the folding/unfolding transition must be accompanied by an uptake or release of hydrogen ions. This is one example of the general theory of linked thermodynamic functions (19, 20), which can be viewed as the rigorous expression of Le Chatelier's principle. Here it can be used to derive the change in number of H^+ ions bound (δn_{H+}) when the protein unfolds in terms of the pH dependence of the unfolding equilibrium:

$$\delta n_{H+} = -\partial \log K_{unf}/\partial pH \qquad [52]$$

Shifts in pK (designated δpK) also correspond to changes in standard free energy of proton ionization of the group ($\delta \Delta G°_{ion}$) which are numerically related by:

$$\delta \Delta G°_{ion} = -2.303RT . \delta pK \qquad [53]$$

where R is the gas constant (8.314 J K^{-1} mol^{-1}) and T the absolute temperature. This corresponds to a free energy change of almost 6 kJ mol^{-1} for each unit shift in pK at room temperature.

Figure 12 shows typical DSC data for the effect of pH changes on the thermal unfolding of a simple globular protein (lysozyme). The major change in T_m in the low-pH region occurs over the pH 2–3 range, consistent with anticipated protonation of carboxylate side chains, and the variation corresponds to a cumulative uptake (δn_{H+}) of about three hydrogen ions per protein molecule in the middle of this pH range where the effect is most pronounced. The slight decrease in T_m above pH 4.5 may reflect titration of a histidine group, but may also be an artefact arising from the tendency of unfolded lysozyme to aggregate in the neutral pH region.

4.6 Non-specific metal ion effects

In addition to the specific binding effects mentioned above, there are a variety of ways in which non-specific interactions of metal ions in solution may affect the stability of a protein and, therefore, show up in DSC experiments. Some of these are illustrated in *Figure 13*, taking lysozyme as an example, and we mention them here in order to illustrate the caution sometimes needed in interpreting results.

First, there are simple ionic strength effects. In the case of lysozyme, most metal ions (salts) in solution tend to reduce the T_m in ways that might, at first sight, suggest direct ionic interaction with the unfolded polypeptide. Divalent

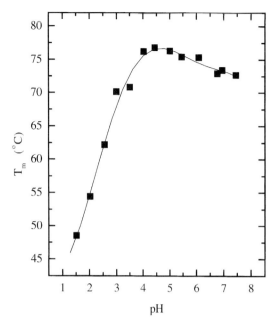

Figure 12 Variation in T_m for unfolding of hen egg white lysozyme with pH, reflecting preferential binding of protons to residues in the unfolded chain. Determined by DSC (*Protocol 1*) in the pH 1.5–7.5 range, using the following buffers: KCl/HCl (pH 1.5–2.0), glycine/HCl (pH 2.5–3.0), sodium citrate/citric acid (pH 3.5), sodium acetate/acetic acid (pH 4.0–5.5), MES (pH 6.1–6.8), and MOPS (pH 7.0–7.5).

metal ions seem to be more effective than simple 1:1 electrolytes in this respect. However, when plotted using the more appropriate ionic strength (I) scale, rather than concentration, we find that the T_m reduction effects for most simple salts fall on a common line, suggesting that this is just a non-specific electrostatic screening effect on protein stability. (Note: ionic strength, $I = \frac{1}{2}\Sigma z_i^2 c_i$, where c_i is the concentration of each ionic species and z_i is its valency (integral charge), is the appropriate quantity replacing concentration in ionic solutions to take account, at least at low concentrations, of ionic screening effects.) This might reflect the weakening of charge–charge interactions between groups in the protein, that might both weaken the native fold and allow greater flexibility of the unfolded chain.

Deviations from this general ionic strength effect are, however seen in some instances. For example (*Figure 13*), addition of lanthanum ions to the solution produces an initial rise in T_m for lysozyme, followed by a more gradual fall in T_m that parallels the ionic strength effect. This is consistent with the known binding of lanthanide ions to native lysozyme, confirmed by separate ITC and fluorescence methods, which will initially stabilize the folded state, superimposed on the more general ionic destabilizing effect common to all salts. Other metal cations—Cu^{2+} and Al^{3+}, for example—show the opposite effect, giving destabilization of the protein in excess of that expected from simple

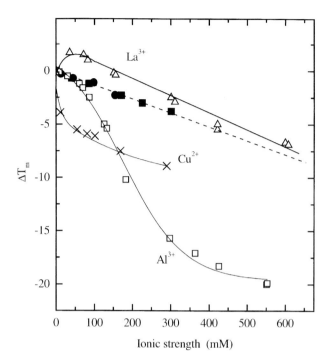

Figure 13 Effects of various metal ions/ionic strength on the thermal stability of lysozyme. ΔT_m is the change in T_m determined from the peak of the DSC thermogram, following experimental procedures described in *Protocol 1* with the addition of appropriate salts (chloride or sulfate) to the buffer. The solid symbols (dashed line) indicate the simple ionic strength effect for simple mono- or divalent salts such as KCl and other alkali metal salts, $MgCl_2$, and $ZnCl_2$. Other data are for $LaCl_3$ (open triangles), $AlCl_3$ (open squares), and $CuSO_4$ (crosses). Data are plotted on a common ionic strength (*I*) scale.

ionic strength effects. This suggests additional binding or other interactions of these particular metal ions with the unfolded polypeptide. In the case of copper ions the effect is the same even in the presence of 100 mM KCl, suggesting that the interaction of these ions with exposed protein or peptide groups is not simply electrostatic. This is consistent with the known complexation properties of Cu^{2+} with peptide groups, known classically as the 'biuret reaction'. For Al^{3+}, the effect is accompanied by irreversible precipitation of the unfolded protein, which probably exaggerates the T_m reduction seen here. Non-equilibrium kinetic effects such as these are difficult to interpret using DSC alone.

Acknowledgements

Much of the work reported here has been done under the auspices of the UK Biological Microcalorimetry Facility in Glasgow, funded jointly by BBSRC and EPSRC

References

1. Privalov, P. L. and Potekhin, S. A. (1986). In *Methods in enzymology*, Vol. 131, pp. 4–51. Academic Press, London.
2. Plotnikov, V. V., Brandts, J. M., Lin, L.-N., and Brandts, J. F. (1997). *Anal. Biochem.*, **250**, 237.
3. Jackson, W. M. and Brandts, J. F. (1970). *Biochemistry*, **9**, 2294.
4. Gill, S. C. and von Hippel, P. H. (1989). *Anal. Biochem.*, **182**, 319.
5. Sanchez-Ruiz, J. M., Lopez-Lacomba, J. L., Cortijo, M., and Mateo, P. L. (1988). *Biochemistry*, **27**, 1648.
6. Galisteo, M. L., Mateo, P. L., and Sanchez-Ruiz, J. M. (1991). *Biochemistry*, **30**, 2061.
7. Haynie, D. T. and Freire, E. (1994). *Anal. Biochem.*, **216**, 33.
8. McPhail, D. and Cooper, A. (1997). *J. Chem. Soc. Faraday Trans.*, **93**, 2283.
9. Dunitz, J. D. (1995). *Chem. Biol.*, **2**, 709.
10. Naghibi, H., Tamura, A., and Sturtevant, J. M. (1995). *Proc. Natl. Acad. Sci. USA*, **92**, 5597.
11. Sturtevant, J. M. (1974). *Annu. Rev. Biophys. Bioeng.*, **3**, 35.
12. Sturtevant, J. M. (1987). *Annu. Rev. Phys. Chem.*, **38**, 463.
13. Fukada, H., Sturtevant, J. M., and Quiocho, F. A. (1983). *J. Biol. Chem.*, **258**, 13193.
14. Cooper, A. (1992). *J. Am. Chem. Soc.*, **114**, 9208.
15. Cooper, A. and McAuley-Hecht, K. E. (1993). *Phil. Trans. R. Soc. Lond. A*, **345**, 23.
16. Brandts, J. F. and Lin, L.-N. (1990). *Biochemistry*, **29**, 6927.
17. Cooper, A., McAlpine, A., and Stockley, P. G. (1994). *FEBS Lett.*, **348**, 41.
18. Cooper, A. (1998). In *Protein–DNA interactions: a practical approach* (ed. A. Travers and M. Buckle). Oxford University Press.
19. Wyman, J. (1964). *Adv. Protein Chem.*, **19**, 223.
20. Wyman, J. and Gill, S. J. (1990). *Binding and linkage: functional chemistry of biological macromolecules*. University Science Books, Mid Valley, CA.

List of suppliers

Amika Corp., 8980 F Route 108, Columbia, MD 21045, USA.

Analytica of Branford, Inc. (Electrospray ionization source), Branford, CT, USA.

Anderman and Co. Ltd., 145 London Road, Kingston-upon-Thames, Surrey KT2 6NH, UK.
Tel: 0181 541 0035   Fax: 0181 541 0623

APL Limited (Circular dichroism suppliers).

Applied Biosystems, 7 Kingsland Grange, Woolston, Warrington WA1 4SR, UK.
Applied Biosystems, 850 Lincoln Center Drive, Foster City, CA 94404, USA.

Applied Photophysics (Stopped flow equipment), 203/205 Kingston Road, Leatherhead KT22 7PB, UK.

Attwater and Sons Ltd., PO Box 39, Hopwood St Mills, Preston PR1 1TA, UK.

AVIV Instruments Inc. (Stopped flow and circular dichroism suppliers), 750 Vassar Avenue, Lakewood, NJ 08701-6907, USA.

Beckman Coulter (UK) Ltd., Oakley Court, Kingsmead Business Park, London Road, High Wycombe, Buckinghamshire HP11 1JU, UK.
Tel: 01494 441181
Fax: 01494 447558
URL: http://www.beckman.com

Beckman Coulter Inc., 4300 N Harbor Boulevard, PO Box 3100, Fullerton, CA 92834-3100, USA.
Tel: 001 714 871 4848
Fax: 001 714 773 8283
URL: http://www.beckman.com

Beckman Instruments (Analytical ultracentrifuge equipment), Progress Road, Sands Industrial Estate, High Wycombe, Buckinghamshire HP12 4JL, UK.
Beckman Instruments Inc., PO Box 3100, 2500 Harbor Boulevard, Fullerton, CA 92634, USA.
Beckman Instruments, Frankfurter Ring 115, Postfach 1416, D-80807 München, FR Germany.

Becton Dickinson and Co., 21 Between Towns Road, Cowley, Oxford OX4 3LY, UK.
Tel: 01865 748844
Fax: 01865 781627
URL: http://www.bd.com
Becton Dickinson and Co., 1 Becton Drive, Franklin Lakes, NJ 07417-1883, USA.
Tel: 001 201 847 6800
URL: http://www.bd.com

BIAcore (Surface plasmon resonance equipment), 2 Meadway Court, Meadway Technology Park, Stevenage, Hertfordshire SG1 2EF, UK.
Biacore Inc., 200 Centennial Avenue, Suite 100, Piscataway, NJ 08854, USA.

LIST OF SUPPLIERS

Biacore AB, Rapsgatan 7, SE-754 50 Uppsala, Sweden (Head Office).

Bio 101 Inc., c/o Anachem Ltd., Anachem House, 20 Charles Street, Luton, Bedfordshire LU2 0EB, UK.
Tel: 01582 456666   Fax: 01582 391768
URL: http://www.anachem.co.uk
Bio 101 Inc., PO Box 2284, La Jolla, CA 92038-2284, USA.
Tel: 001 760 598 7299
Fax: 001 760 598 0116
URL: http://www.bio101.com

Bio-Rad Laboratories Ltd. (Confocal laser scanning microscope, FTIR), Bio-Rad House, Maylands Avenue, Hemel Hempstead, Hertfordshire HP2 7TD, UK.
Tel: 0181 328 2000   Fax: 0181 328 2550
URL: http://www.bio-rad.com
Bio-Rad Laboratories Ltd., Division Headquarters, 1000 Alfred Noble Drive, Hercules, CA 94547, USA.
Tel: 001 510 724 7000
Fax: 001 510 741 5817
URL: http://www.bio-rad.com
Bio-Rad Spectroscopy Group, 237 Putnam Avenue, Cambridge, Massachusetts 02139, USA.

BOMEM, Inc. (FTIR equipment), 450 St-Jean Baptiste Avenue, Quebec PQ G2E 5S5, Canada.   E-mail: metal@bomem.co
URL: http://www.bomem.com

British Drug Houses (BDH) Ltd., Poole, Dorset, UK.

Bruker UK Limited (NMR, diffraction, and FTIR equipment), Banner Lane, Coventry CV4 9GH, UK.
URL: http://www.bruker-axs.com/
Bruker Instruments, Inc., 44 Manning Road, Manning Park, Billerica, MA 01821-3991, USA.
E-mail: optics@bruker.com

Bruker Daltonics (Mass spectrometry equipment), Manning Park, Billerica, MA 01821, USA.

Calorimetry Sciences Corp., 155 West 2050 South, Spanish Fork, Utah 84660, USA.

Clairet Scientific Ltd. (Raman equipment), 17 Scirocco Close, Moulton Park Industrial Estate, Northampton NN3 6AP, UK.
Fax: 0 1604 494499
E-mail: clairet@compuserve.com
URL: http://www.clairet.co.uk

Composite Metal Services Ltd. (sources for capillaries), The Chase, Hallow, Worcester WR2 6LD, UK.

CP Instrument Co. Ltd., PO Box 22, Bishop Stortford, Hertfordshire CM23 3DX, UK.
Tel: 01279 757711   Fax: 01279 755785
URL: http://www.cpinstrument.co.uk

DG Electronics (Benchtop and high resolution spectrofluorimeters), 16/20 Camp Road, Farnborough GU14 6EW, UK.

Dianorm GmbH, Stöcklstrasse 5a, D-81247 München 65, FR Germany.

Dupont (UK) Ltd., Industrial Products Division, Wedgwood Way, Stevenage, Hertfordshire SG1 4QN, UK.
Tel: 01438 734000   Fax: 01438 734382
URL: http://www.dupont.com
Dupont Co. (Biotechnology Systems Division), PO Box 80024, Wilmington, DE 19880-002, USA.
Tel: 001 302 774 1000
Fax: 001 302 774 7321
URL: http://www.dupont.com

Eastman Chemical Co., 100 North Eastman Road, PO Box 511, Kingsport, TN 37662-5075, USA.
Tel: 001 423 229 2000
URL: http://www.eastman.com

LIST OF SUPPLIERS

Fisher Scientific UK Ltd., Bishop Meadow Road, Loughborough, Leicestershire LE11 5RG, UK.
Tel: 01509 231166   Fax: 01509 231893
URL: http://www.fisher.co.uk
Fisher Scientific, Fisher Research, 2761 Walnut Avenue, Tustin, CA 92780, USA.
Tel: 001 714 669 4600
Fax: 001 714 669 1613
URL: http://www.fishersci.com

Fisons/VG (Mass spectrometers), Manchester, UK.

Flow Laboratories, Woodcock Hill, Harefield Road, Rickmansworth, Hertfordshire WD3 1PQ, UK.

Fluka, PO Box 2060, Milwaukee, WI 53201, USA.
Tel: 001 414 273 5013
Fax: 001 414 2734979
URL: http://www.sigma-aldrich.com
Fluka Chemical Co. Ltd., PO Box 260, CH-9471, Buchs, Switzerland.
Tel: 0041 81 745 2828
Fax: 0041 81 756 5449
URL: http://www.sigma-aldrich.com

Hampton Research (Crystallization materials), 25431 Cabot Road, Suite 205, Laguna Hills, CA 92653-5527, USA.
URL: http://www.hamptonresearch.com
xtalrox@aol.com

Hewlett Packard, Cain Road, Bracknell, Berkshire RG12 1HN, UK.
Hewlett Packard, 3495 Deer Creek Road, Palo Alto, CA 94304, USA.
Hewlett Packard, 150 Route du Nant-d'Avril, CH-1217 Meyrin-Geneva 2, Switzerland.

Hi-Tech Scientific Ltd. (Stopped flow equipment), Hi-Tech House, Brunel Road, Church Fields, Salisbury, Wiltshire SP2 7PU, UK.
Tedk@Hi-tech.demon.co.uk

Hybaid Ltd., Action Court, Ashford Road, Ashford, Middlesex TW15 1XB, UK.
Tel: 01784 425000
Fax: 01784 248085
URL: http://www.hybaid.com
Hybaid US, 8 East Forge Parkway, Franklin, MA 02038, USA.
Tel: 001 508 541 6918
Fax: 001 508 541 3041
URL: http://www.hybaid.com

HyClone Laboratories, 1725 South HyClone Road, Logan, UT 84321, USA.
Tel: 001 435 753 4584
Fax: 001 435 753 4589
URL: http://www.hyclone.com

Invitrogen Corp., 1600 Faraday Avenue, Carlsbad, CA 92008, USA.
Tel: 001 760 603 7200
Fax: 001 760 603 7201
URL: http://www.invitrogen.com
Invitrogen BV, PO Box 2312, 9704 CH Groningen, The Netherlands.
Tel: 00800 5345 5345
Fax: 00800 7890 7890
URL: http://www.invitrogen.com

JASCO Limited (Circular dichroism suppliers).

Jobin Yvon / Dilor Raman (Raman suppliers), 2-4 Wigton Gardens, Stanmore, Middlesex HA7 1BG, UK.
Fax: 020 8204 6142
E-mail: jy@jyhoriba.co.uk
Jobin Yvon SA (Head Office), 16-18 Rue du Canal, BP 118, 1165 Longjumeau, Cedex, France.
Fax: (33) 1.69.09.93.19
www.jobinyvon.com

Johnson and Johnson, Clinical Diagnostics, Chalfont St Giles, Buckinghamshire, UK.

LIST OF SUPPLIERS

Kaiser Optical Systems, Inc. (Raman equipment), PO Box 983, 371 Parkland Plaza, Ann Arbor, MI 48106, USA.
Fax: 734-665-8199
E-mail: sales@kosi.com
URL: http://www.kosi.com

KinTek Corporation (Stopped flow equipment), 7604 Sandia Loop, Suite C, Austin TX 78735, USA.

Life Technologies Ltd., PO Box 35, Free Fountain Drive, Incsinnan Business Park, Paisley PA4 9RF, UK.
Tel: 0800 269210   Fax: 0800 838380
URL: http://www.lifetech.com
Life Technologies Inc., 9800 Medical Center Drive, Rockville, MD 20850, USA.
Tel: 001 301 610 8000
URL: http://www.lifetech.com

Lipex BioMembranes (Pressurized extruder apparatus), 3550 West 11th Avenue, Vancouver, British Columbia V6R 2K2, Canada.

Merck Sharp & Dohme, Research Laboratories, Neuroscience Research Centre, Terlings Park, Harlow, Essex CM20 2QR, UK.
URL: http://www.msd-nrc.co.uk
MSD Sharp and Dohme GmbH, Lindenplatz 1, D-85540, Haar, Germany.
URL: http://www.msd-deutschland.com

Microcal Inc., 22 Industrial Drive East, Northampton, MA 01060-2327, USA.

Micromass UK Ltd. (Mass spectrometry equipment), Floats Road, Wynthenshaw, Manchester M23 9LZ, UK.

Millipore (UK) Ltd., The Boulevard, Blackmoor Lane, Watford, Hertfordshire WD1 8YW, UK.
Tel: 01923 816375
Fax: 01923 818297
URL: http://www.millipore.com/local/UK.htm

Millipore Corp., 80 Ashby Road, Bedford, MA 01730, USA.
Tel: 001 800 645 5476
Fax: 001 800 645 5439
URL: http://www.millipore.com

Molecular Dynamics, 4 Chaucer Business Park, Kemsing, Sevenoaks, Kent TN15 6PL, UK.
Molecular Dynamics, 880 East Arques Avenue, Sunnyvale, CA 94086, USA.
Molecular Dynamics, Elisabethstrasse 103-105, D-47799 Krefeld, FR Germany.

Molecular Kinetics/Bio-Logic (Stopped flow equipment), PO Box 2475 C.S, Pullman, WA 99165, USA.

Molecular Probes, PO Box 22010, 4849 Pitchford Avenue, Eugene, Oregon 97402-9144, USA.
Molecular Probes, Poort Gebouw, Rijnsburgerweg 10, NL-2333 AA Leiden, The Netherlands.

MSC (Diffraction equipment)
USA: http@///www/msc.com/
Japan: http@//www/msc.com/

New England Biolabs, 32 Tozer Road, Beverley, MA 01915-5510, USA.
Tel: 001 978 927 5054

Nicolet Instrument Ltd. (FTIR and Raman equipment), Nicolet House, Budbrooke Road, Warwick CV34 5XH, UK.
Nicolet Instruments Corporation, 5225 Verona Road, Madison, WI 53711, USA.
E-mail: nicinfo@nicolet.com

Nikon Inc., 1300 Walt Whitman Road, Melville, NY 11747-3064, USA.
Tel: 001 516 547 4200
Fax: 001 516 547 0299
URL: http://www.nikonusa.com

LIST OF SUPPLIERS

Nikon Corp., Fuji Building, 2-3, 3-chome, Marunouchi, Chiyoda-ku, Tokyo 100, Japan.
Tel: 00813 3214 5311
Fax: 00813 3201 5856
URL: http://www.nikon.co.jp/main/index_e.htm

Nikon Diaphot (Inverted fluorescence microscope), Nikon Inc., Melville, New York, USA.

Nonius Ltd. (Diffraction equipment), Delft, The Netherlands.
URL: http://www.nonius.com/

Nycomed Amersham plc, Amersham Place, Little Chalfont, Buckinghamshire HP7 9NA, UK.
Tel: 01494 544000   Fax: 01494 542266
URL: http://www.amersham.co.uk
Nycomed Amersham, 101 Carnegie Center, Princeton, NJ 08540, USA.
Tel: 001 609 514 6000
URL: http://www.amersham.co.uk

Oxford Cryosystems (Crosystems for crystallography), Oxford, UK.
URL: http://www.oxfordcryosystems.co.uk

Perkin Elmer Ltd., Post Office Lane, Beaconsfield, Buckinghamshire HP9 1QA, UK.   Tel: 01494 676161
URL: http://www.perkin-elmer.com
Perkin Elmer Cetus (The Perkin-Elmer Corporation), 761 Main Avenue, Norwalk, CT 06859, USA.
E-mail: info@pe-corp.com
Perkin Elmer Ltd., Postfach 101164, Askaniaweg 1, D-88647, Überlingen, FR Germany.

Pharmacia Biotech (Biochrom) Ltd., Unit 22, Cambridge Science Park, Milton Road, Cambridge CB4 0FJ, UK.
Tel: 01223 423723   Fax: 01223 420164
URL: http://www.biochrom.co.uk

Pharmacia and Upjohn Ltd., Davy Avenue, Knowlhill, Milton Keynes, Buckinghamshire MK5 8PH, UK.
Tel: 01908 661101   Fax: 01908 690091
URL: http://www.eu.pnu.com
Pharmacia LKB Biotechnology AB, Bjorngatan 30, S-75182 Uppsala, Sweden.

Pierce & Warriner (UK) Ltd., 44 Upper Northgate Street, Chester CH1 4EF, UK.
Pierce Chemical Company, 3747 Meridian Road, PO Box 117, Rockford, IL 61105, USA.

Polymicro Technologies, Inc. (sources for capillaries), 18019 N 25th Avenue, Phoenix, AZ, USA.

Promega UK Ltd., Delta House, Chilworth Research Centre, Southampton SO16 7NS, UK.
Tel: 0800 378994
Fax: 0800 181037
URL: http://www.promega.com
Promega Corp., 2800 Woods Hollow Road, Madison, WI 53711-5399, USA.
Tel: 001 608 274 4330
Fax: 001 608 277 2516
URL: http://www.promega.com

Protein Solutions, Unit 5, The Hillside Centre, Upper Green Street, High Wycombe, Buckinghamshire, UK.
Protein Solutions, 2300 Commonwealth Drive, Suite 102, Charlottesville, VA 22901, USA.

Qiagen UK Ltd., Boundary Court, Gatwick Road, Crawley, West Sussex RH10 2AX, UK.
Tel: 01293 422911
Fax: 01293 422922
URL: http://www.qiagen.com
Qiagen Inc., 28159 Avenue Stanford, Valencia, CA 91355, USA.
Tel: 001 800 426 8157
Fax: 001 800 718 2056
URL: http://www.qiagen.com

LIST OF SUPPLIERS

Renishaw (Raman equipment), New Mills, Wotton-under-Edge, Gloucestershire GL12 8JR, UK.
Fax: +44 1453 524901
E-mail: genenq@renishaw.co.uk
Renishaw Inc., 623 Cooper Court, Schaumburg, Illinois 60173, USA.
Fax: +1 847 843 1744
E-mail: usa@renishaw.com

Rheodyne, PO Box 996, Cotatim, CA 94928, USA.

Roche Diagnostics Ltd., Bell Lane, Lewes, East Sussex BN7 1LG, UK.
Tel: 01273 484644
Fax: 01273 480266
URL: http://www.roche.com
Roche Diagnostics Corp., 9115 Hague Road, PO Box 50457, Indianapolis, IN 46256, USA.
Tel: 001 317 845 2358
Fax: 001 317 576 2126
URL: http://www.roche.com
Roche Diagnostics GmbH, Sandhoferstrasse 116, 68305 Mannheim, Germany.
Tel: 0049 621 759 4747
Fax: 0049 621 759 4002
URL: http://www.roche.com

Savant Instruments, 110-103 Bi-County Boulevard, Farmingdale, NY 11735, USA.

Schleicher and Schuell Inc., Keene, NH 03431A, USA.
Tel: 001 603 357 2398

Separations Group, 1734 Mojave Street, PO Box 867, Hesperia, CA 92345, USA.

SETARAM, 7 rue de l'Oratoire, F-69300, Caluire, France.

Shandon Scientific Ltd., 93-96 Chadwick Road, Astmoor, Runcorn, Cheshire WA7 1PR, UK.
Tel: 01928 566611
URL: http://www.shandon.com

Sigma-Aldrich Co. Ltd., The Old Brickyard, New Road, Gillingham, Dorset XP8 4XT, UK.
Tel: 01747 822211
Fax: 01747 823779
URL: http://www.sigma-aldrich.com
Sigma-Aldrich Co. Ltd., Fancy Road, Poole, Dorset BH12 4QH, UK.
Tel: 01202 722114
Fax: 01202 715460
URL: http://www.sigma-aldrich.com
Sigma Chemical Co., PO Box 14508, St Louis, MO 63178, USA.
Tel: 001 314 771 5765
Fax: 001 314 771 5757
URL: http://www.sigma-aldrich.com

SLT Labinstruments, PO Box 13953, Research Triangle Park, NC 27709, USA.
SLT Labinstruments, Unterbergstrasse 1A, A-5082, Grodig, Austria.

Spectrum Medical Industries, Inc., 1100 Rankin Road, Houston, Texas 77073-4716, USA.
Spectrum Europe BV, European Headquaters, PO Box 3262, 4800 DG Breda, The Netherlands.

Stratagene Inc., 11011 North Torrey Pines Road, La Jolla, CA 92037, USA.
Tel: 001 858 535 5400
URL: http://www.stratagene.com
Stratagene Europe, Gebouw California, Hogehilweg 15, 1101 CB Amsterdam Zuidoost, The Netherlands.
Tel: 00800 9100 9100
URL: http://www.stratagene.com

Supelco, Inc. (sources for capillaries), Supelco Park, Bellefonte, PA 16823, USA.

ThermoMetric Ltd., 10 Dalby Court, Gadbrook Business Park, Northwich, Cheshire CW9 7TN, UK.
Thermometric AB, Spjutvägen 5A, S-175 61 Järfälla, Sweden.

LIST OF SUPPLIERS

Thermomicroscopes Corporation (Atomic force microscopy equipment), Bicester, Oxford, UK.

United States Biochemical, PO Box 22400, Cleveland, OH 44122, USA.
Tel: 001 216 464 9277

Varian Instruments (FTIR equipment), 2700 Mitchell Drive, Walnut Creek, CA 94598, USA.
E-mail: osiusa@varianinc.com

Velonex Ltd. (suppliers of pulse generators), Santa Clara, CA, USA.

Waters Chromatography, 324 Chester Road, Hartford, Northwich, Cheshire CW8 2AH, UK.
Waters Chromatography, 34 Maple Street, Milford, MA 01757, USA.
Waters Chromatography, 6 rue JP Timbaud, F-78180 Montigny le Breton-neux, France.

Wright Instruments (Cooled CCD camera), Enfield, UK.

Index

Page numbers in bold indicate major entries. Computer software algorithms given in italics

acceptor **92-94, 95-98, 116-132, 173-175**
acetyl salicylic acid (aspirin) 30
activation energy 166
activity, thermodynamic **110**, 130
affinities **130-132, 156-168**
affinity chromatography 14
  advancing and trailing elution profiles **48-50**
  frontal **48-60, 66-68**
  frontal vs zonal 48
  Hummel and Dreyer procedure **61, 62**, 182
  limitation of Hummel and Dreyer **62-63**
  non-enantiography **48-50**
  zonal **61-63, 68-70**
alpha-amylase 132
alpha-chymotrypsin 112, 131
AmiC (a cytoplasmic protein) 257
analyte see acceptor
analytical ultracentrifugation see sedimentation equilibrium or sedimentation velocity
antibiotics 30
antibody see also immunoglobulins
  mediated indirect coupling 151
  monoclonals (mAbs), 139, 140, 151, 178
antibody-toxin complex 96
anti-inflammatory drugs 30
Arrhenius equation 166
association constant
  molar **3, 4, 88, 93, 115-126, 156, 267-273, 279, 282, 283**
  weight **4, 115, 127-132**
  note on "units" 3
association rate constant **138-140**, 156, 160, 164, 186
association, self see self-association
ASSOC4 114
atomic force spectroscopy 9, 14
azide 4

B12 cofactor 85
bacterial growth problem 30
bead, bead-shell models **90**, 255, 256
beamtime (neutron, x-ray scattering) 238
BIACore 137, 138, 141, 142, 162, 168
Beer-Lambert law 22, 23, 79, 84
beta-trypsin see trypsin
BIDS see world wide web
binding constant see also association constant
  intrinsic 70, **130-132**
  measurement by chromatography 51-72
binding energy dependence 6
BioSCAN 15
Biotin 187
biphenylacetate 22
BLITZ 15
BSA (bovine serum albumin) 85, 91, 92, 94
Bradford reagent 31, 293
buffer background 228, 242
calorimetry see isothermal titration calorimetry or differential scanning calorimetry
capillary electrophoresis (CE) 7, 14, **171-193**
  buffers 177
  comparison with conventional gel electrophoresis 173
  comparison with other methods 174
  main approaches 180
  pitfalls 192
  wall interaction 179
CD2 266
CD4 150
CD42 95
CD48 95, 266
centrepieces, sedimentation 77
chelating agents 27
charge density 10
chromatography
  affinity see affinity chromatography
  attraction over ultracentrifuge methods 47
  elution volume 50-51
  general experimental aspects 48
  HPLC see HPLC
circular dichroism 8, 12
Claverie method 97, 98
cofactor 85
COLETTE 241
computational methods 14
concentration determination, importance of **275, 276**
concanavalin A 71

327

INDEX

conformation **89-91**, **96-101**, **197-220**, **223-260**
contrast variation see neutron scattering
cyclodextrin 304, 306
cytochrome c 49, 50, 53, 123-125

DCDT, DCDT+ 81
DCDR 25
decanoate 33
decay time constant (electro-optic) see electro-optics
denaturants 4
densimetry 81, 109
density, charge see charge density
density, solvent 77
DETEC 241
detergents 4
dextran 59
  matrix, in SPR 139
Dianorm apparatus 24, 25
difference sedimentation velocity see sedimentation velocity
differential scanning calorimetry (DSC) 13, **287-317**
  baseline problem, 294-297
  calorimeter cell 288
  instrumentation 289
  protein unfolding 289
diffusion coefficient, translational 10, 79, 81, 90, 98, 105, 256
diffusion, rotational 10, **197-220**
dips (SPR) 141, 142
diptheria toxin see toxin
DISCRETE 214
dissociation constant
  molar **3, 4, 88, 95, 139, 156, 157, 159, 164, 165, 174, 187-191, 267, 281, 299, 301-303, 305, 311, 312, 313, 314**
  note on "units"
  weight **4**
dissociation rate constant **139, 156, 160, 164, 181, 186, 190**
DNA (deoxyribonucleic acid) 187, 213, 241, 250, 253, 259, 260, 282, 310, 311
  Holliday junctions 259, 260

Donnan effects 20, 29, 34, 42
drug
  concentration 23
DTT (dithiothreitol) 4

electric birefringence see electro-optics
electric dichroism see electro-optics
electro-optics 8, **197-220**
  birefringence **216-218**
  dialysis requirements **208, 209**
  dichroism amplitude 200
  dichroism decay 201
  dichroism decay time constant **198, 202**
  instrumentation **203-207**
  orientation function 200
  requirements 199
  stationary dichroism 211
electron microscopy 226, 256
electron paramagnetic spin resonance 9, 12
ellipsoid models (revolution, triaxial) 90
ELLIPS1 90
ELLIPS2 **90, 216**
ELLIPS4 **216**
ELLIPSE 90
ELISA (enzyme-linked immunosorbent assays) 283
enthalpy 139, 165, 166, 263, **267-272**, 293, 295, 296, **298-308**, 313
  calorimetric 292, 293, 295, 296, 305, 306
  van't Hoff 292, 293, 295, 296, 306
entropy 139, 165, 166, 264, 272, **298-303**
equilibrium constant
  association see association constant
  dissociation see dissociation constant
  note on "units" 3
  unfolding (molar) 303, 305, 311, 312, 313, 314
equilibrium dialysis 7, 13, **19-46**
  macromomethod 44
  micromethod 44-45

measurement of free ligand 21
  principle 21
  running an experiment **23-25**
  radiolabelling 32-45
ESI (electro-spray ionization) see mass spectrometry
evanescent wave 167
EXCEL 256, 312

factor VIIa (FVIIa) 259
flavoprotein 123
flexibility, problem 10
FPLC (fast protein liquid chromatography) 63
FTIR (Fourier transform infrared spectroscopy) 9
fluorescence spectroscopy 8, 12, 218, 219, 316
flurbiprofen 26
free energy (Gibbs) 156, 165, 264, 267, 272, 293, **299-301**, 303, 305, 307, 313, 315
frictional coefficient (translational) 79, 89, 90, 105
frictional ratio (translational) 89, 90
fringe numbers, Rayleigh 109, 111

GdnHCl 4
gene 5 protein 87
Gibbs free energy see free energy
glycerol 4
Gralén parameter 81, 88-90
GroEL 87
Guinier plots 249

heat capacity C_p 165, 268, 272, 282, **290-294, 296-302, 304-307, 311, 313, 314**
heparin 182, 190
  binding peptide 190
hexanoate 39
hexokinase 101
high resolution methods
  brief comparison of 6-11
Holliday junctions see DNA

INDEX

HPLC (high performance/pressure liquid chromatography) 25, 28, 31
HSA (human serum albumin) 26, 28, 33, 39
human serum amyloid P component (SAP) 224, 258
Hummel and Dreyer procedure – *see* affinity chromatography
Hydration, δ 89, 228

Iasys instrument 72
immobilization, for SPR 149
Immunoglobulins
  Fc domain 150
  IgG, monoclonal 150, 178
  IgG, mouse monoclonal 150
  IgG$_1$, human 150, 151
  IgE, rat 87
Interactions *see also* ligand, acceptor
  binary system 4
  concentration requirements 4
  experimental variables 4
  factors to be taken into account 5
  frontiers of viability 1
  parameters to be defined 3
intrinsic viscosity [η] 90
ion concentration 4
isothermal titration microcalorimetry 8, 13, **263-286**, 287, 316
  advantages 283, 284
  disadvantages 284

Johnston-Ogston effect **85**

K_a, K_b see association constant, molar
k, dialysis rate constant 34-36, 41
k_{off} see dissociation rate constant
k_{on} see association rate constant
kinetics 160-166

lactate dehydrogenase, rabbit muscle 67, 68
Lambert-Beer law *see* Beer-Lambert law

Lamm equation **79**, 96, 97
Langmuir binding isotherm 158, 159
Le Chatelier's principle 105
ligand
  acceptor (or analyte) interactions **92-94, 95-98, 116-132, 152-169, 173-192, 258-260, 303-317**
  immobilization for SPR 143-153
  macromolecular 92
  meaning 2
  mediated conformation changes **89-91**, 96-101, 197-220, 223-260
  preferential binding **130-132**
  small 91-92
light scattering 219
liposomes 145
low-density-lipoprotein receptor 63
Lowry reagent 31
lysozyme 122, 126, 128, 131, 289, 290, 291, 315, 316, 317

MALDI (matrix-assisted laser desorption/ionization) 9, 114
Mark-Houwink-Kuhn-Sakurada relation **88, 89**
mass spectrometry 8
maximum packing fraction 88
MEDLINE see world-wide web
membranes
  for equilibrium dialysis 25
meniscus concentration problem 113
methyl orange 85, 91, 92, 94
methylmalonyl mutase 85
microcalorimetry 165
microchambers **40-43**
modelling *see* molecular modelling
molecular electro-optics *see* electro-optics
molecular modelling 8
molecular weight (molecular mass, molar mass) **77, 78**, 88, 105, 106, **109-116**

monoclonal antibodies (mAbs) *see* antibodies
MSTARA, MSTARI **113**, 114
mutant proteins 140

N-acetylglucosamine
N-acetyltryptophan
NADH 68
neutron scattering (solution) 8, 14, **223-231, 234-260**
  contrast variation & matchpoints 250-252
  distribution function analysis 251-252
  data collection 243-248
  instrumentation 234-237
NMR (nuclear magnetic resonance) 6, **9-11**, 12, 226, 281
  advantages over X-ray crystallography 6, 10
  disadvantages over X-ray crystallography 11
NNPREDICT 15
NOEs 11
non-ideality 81, 128

OMEGA 114
optical path length *see* path length
ORIGIN 157, 279, 312
osmolytes 4
ovalbumin 49, 71, 111, 122, 123-125

partial specific volume **79**, 81, 89, 105, 106, 118
path length, optical 77
PDB (protein data bank) 15
peak appearance times (CE) 173, 174
Perrin parameter, P 89, 90
PHD 15
Phenylalanine 100
phenylbutazone 28
pH 4
PI-3 kinase 280
PRO-FIT 85
PSI 114
PubMed *see* world wide web
pyruvate kinase 100, 101

329

INDEX

radiolabelling 32-45
Raman spectroscopy 9, 13
RASMOL 15
radius, equivalent hydrodynamic 77
radius of gyration 223, 226, 244, 245, **248-250**, 252, **254-256, 258-260**
rate constant *see* association rate constant, dissociation rate constant
rate dialysis **19-46**
Rayleigh interference optics 83
refractive index effect (in SPR) 154, 167
refractive index gradient 83
R-factor 254
RGUIM 241
ribonuclease 305
ribosomes 80
RNA 276
RNILS 241
RPLOT 241
rotational diffusion *see* diffusion
RuvA 253, 259, 260

SAP *see* human serum amyloid P component
SA-PLOT 88, 94, 95
Scatchard plot 26, 39, 52, 65
  non-linear **158**
scattering densities 225
scattering lengths 225
scattering vector 223, 224, 233, 236, 237, 239, 240, **244-249, 251-257**,
Schlieren optics 83, 87, 92, 96
sedimentation coefficient 76-101, 256
  definition 76
sedimentation equilibrium 7, 14, 75, **105-133**
  experimental aspects 106-121
  ligand binding 121-132

meniscus concentration problem 113
omega and psi analyses 118-125
sedimentation velocity 7, 14, **75-103**, 226
  co-sedimentation 85
  difference 99, 101
  fingerprinting 89
  ligand binding 92-94, 95-98
self-association 94, **127-130**
sensors, for SPR 141, 145
sensorgram 138
SIGMAPLOT 157
size exclusion chromatography 7
SOLPRO 90, 216, 256
soybean trypsin inhibitor 49, 53
specific heat capacity *see* heat capacity
SPOLLY 241
stopped flow 9, 12
stoichiometry
  from frontal gel chromatography 56-57
structure 90, 215, 216, **252-256**, 276, 281, 282
  changes **5**
Stuhrmann contrast variation *see* neutron scattering
sulfonamides 22
surface plasmon resonance 7, **137-169**, 183, 283
  angle 167
surfactants 4
Svedberg, unit 77, 80
SVEDBERG 81
SWISS-MODEL 15
SWIS-PROT 15
swollen specific volume 88
synchrotron sources **229-231**

temperature 4
  midpoint transition, T_m 291, 301, 304, **307-310**, 313, 316, 317

toxin, diptheria 96
transferrin 258
translational diffusion *see* diffusion
triaxial shape 90
trimethylamine dehydrogenase 123
trypsin, beta 64, 65

Ultracentrifuge
  high temperature 80
  XL series 80, 82
UNIX 239, 257
ultra-violet optical system 82, 87
unfolding equilibrium constant – *see* equilibrium constant
urea 4

water
  at binding interface 5
weak interactions 94, 137
world wide web
  BIDS 15
  information sources 15
  MEDLINE 15
  protein servers 15
  PubMed 15
  Van't Hoff analysis 139, 263
  virus, elucidation by x-ray crystallography 10
XOTOKO 241
water *see also* hydration
  internal 10
  surfacd 10

X-ray crystallography 6, **8, 10-11**, 12, 226, 281
  advantages over NMR 6, 10
X-ray scattering (solution) 8, 14, **223-227, 229-234, 238-260**
  data collection 243-248
  instrumentation 232-234